Warfare in the Russian Arctic:
The Military History of Chukotka
From the Early 1st Millennium to the 19th Century

Alexander K. Nefedkin

Translated
by
Richard L. Bland

Warfare in the Russian Arctic:
The Military History of Chukotka
From the Early 1st Millennium to the 19th Century

Alexander K. Nefedkin

Translated
by
Richard L. Bland

Academica Press
Washington - London

Library of Congress Cataloging-in-Publication Data

Names: Nefedkin, Alexander K., author. | Bland, Richard L., translator.
Title: Warfare in the Russian arctic : the military history of Chukotka from the early 1st millennium to the 19th century / Alexander K. Nefedkin | Richard L. Bland
Description: Washington : Academica Press, 2020. | Includes references and index.
Identifiers: LCCN 2020933053 | ISBN 9781680531435

CONTENTS

TRANSLATOR'S INTRODUCTION

This book was originally published as *Очерки военно-политической истории Чукотки (начало I тыс. н. э.—XIX в.)* by Alexander K. Nefëdkin (St. Petersburg: St. Petersburg Centre for Oriental Studies Publishers, 2016).

Every translation, and particularly from Russian to English, has the problem of finding a suitable form of transliteration. None of the three systems available (U.S. Board of Geographic Names [BGN], Library of Congress [LOC], or "Linguistic" system [Ling]) was felt to be entirely adequate. I have therefore "created" my own system. In this I use some of the BGN system with a slightly modified version of the LOC. For example, the "ye" of BGN is written as "e" following LOC. The Russian "ë" is also written as "e" (not "yo"), following Ling. (In the ë/e problem, an exception is made for the name of the author of this book). The Russian "э" is written as "e," following BGN. Both the Russian "и" and "й" are transliterated as "i," unlike in any of the three systems. The Russian "ю" and "я" are written as "yu" and "ya," following the BGN. The Russian soft sign, which is often dropped in transliterations or replaced with an "i," is retained here as an apostrophe, following BGN. I have also settled on one ending for words, as the English language forces us to do, rather than providing the appropriate ending (masculine, feminine, neuter, plural/nominative, genitive, dative, accusative, instrumental, prepositional) that can occur Russian. And having 24 possible grammatical endings is not the end of it. In the masculine nominative, for a name ending in "-sky" there are at least five possible endings that can be found in English ("-sky," "-skiy," "-skij," "-skii," "-ski"). In addition, there are aberrant spellings that have been accepted in the literature; for example, Wrangel instead of the Russian Vrangel' (here the soft sign has been replaced by a second "l") has already been adopted in English. Some names are "semi" formalized in English. For names that do not have an accepted English form I have tried consistently to use my system for transliterating. On the other hand, I generally give a (Russian) "i" or "y" plural for plural words that are not translated. This is with the exception of ethnic names, which are given no ending in the plural (following one accepted form found in *Webster's 3rd International Dictionary*, 1965). All this in no way exhausts the problems the translator faces, but rather provides a notion of the difficulties attendant upon any translation project. Nonetheless, I hope the explanation of my method will aid the reader, especially if he or she should want to go back from English to Cyrillic, and I apologize to all whose names I have unintentionally "corrupted."

Some of the Native names resulted from forms that can only be appreciated by dealing with the Cyrillic and the phonetic alphabets. This becomes far too cumbersome without a great deal of explanation. For those interested in name and word origins, it is recommended that the original text be used.

In an attempt to retain some of the flavor of the original reports, many names of individuals and ethnic groups are transliterated directly into Roman characters. An attempt has also been made to keep the phrasing as much like the original as possible, changing only that necessary to make the text intelligible in English.

There are a number of native and Russian terms that are too laborious to translate with each use. Therefore, the native or Russian term has been used, with direction to a glossary at the end of the book that elaborates the term.

I owe a debt of gratitude to Anna Gokhman who proofread the translation, Julia Knowles did an excellent job editing, and I am most indebted to Nan Coppock for much-needed editorial assistance. Without them this work could not have been done. I want to thank Academica Press for their work on the manuscript, and in particular Soumyadev Bose for the layout. Ultimately, we all are very deeply indebted to Dr. Alexander K. Nefëdkin for permitting his valuable book to be published into English.

AUTHOR'S PREFACE

I received several positive reviews for a book I wrote on Chukchi warfare.[1] A letter of 31 March 2006 from the President of the Association of Indigenous Peoples of the North Chukotka Autonomous Region, N. N. Ettyne, noted that the book's publicaiton brought back memories to "those who heard from the mouths of their great-grandfathers and grandfathers about the distant past" and "provided knowledge to the younger generation about the glorious and great past of their people." An appeal from the Regional Ethnocultural Social Union "Chychetkin vetgav" ("Native Word") of 30 March 2006 declared: "We are proud of the fact that thanks to you the Russian people are able to learn about the heroic past of the ethnic Chukchi people. We consider this a significant event." It became clear that the work, dedicated to a military theme, was interesting not only to enthusiasts of military history, but also to the broadest stratum of the population of Chukotka. This was also confirmed by my encounters with readers in Anadyr' at the end of September to early October 2013, when for the first time I succeeded in visiting Chukotka. Since I was asked to continue development of the theme, I was inclined to study the military history of Chukotka that began to be developed from new post-Soviet positions, representing not only a peaceful aspect of the annexation of Siberia, but also its military component: the complex relations of the local peoples among themselves and with Russian authority. Indeed, the military history serves as a distinctive background and necessary addition to a book on warfare—it becomes clear how and when the warriors of Chukotka operated in historical expanse.

In conclusion, I would like to thank those who helped me and contributed to the emergence of this book: Tatyana Ivanovna

[1] Chukchi warfare was originally published as *Voennoe delo chukchei (seredina XVII–nachalo XX v.)*. St. Petersburg: Peterburgskoe Vostokovedenie, 2003. It was translated into English as *Warfare of the Chukchi (Mid-17th to Early 20th Century)*. Anchorage, Alaska: Shared Beringian Heritage Program, National Park Service, 2014.

Bardashevich—ideological inspirer and assistant in this publication; the French scholar on the Chukchi, Charles Weinstein, who provided his materials for translations; the President of the Association of Native Ethnic Peoples of Chukotka (now a member of the Council of Federation from Chukotka) Anna Ivanovna Otke and Executive Director of the nonprofit organization "Fund for Social Development 'Kupol'" Lyudmila Leonidovna Danilova for the possibility of visiting Anadyr' in September–October 2013 within the framework of the project "Ymyl`o cheekei" ("All Together") and in March 2017, which made it possible to gather material there; the Anadyr' radio journalist Margarita Ivanovna Belichenko for translation into Russian of the necessary folklore texts; the journalist and writer Ivan Vasil'evich Omruv'e for help in collecting material; philologist Ideya Vladimirovna Kulikova for consultation on the Chukchi language; the Chukchi writer Samira Yakunina (nee Asadova) and Latvian writer Marina Grigor'evna Kostenetskaya for permission to use their material; Doctors of Philological Sciences from St. Petersburg Aleksei Alekseevich Burykin and Nikolai Borisovich Vakhtin for valuable consultations on philological-ethnological questions; Doctor of Historical Sciences from Novosibirsk Aleksandr Khristianovich Elert for valuable translations from the diaries of G. F. Miller (Müller in the original German); Director of GUK "Museum Center 'Heritage of Chukotka'" Ol'ga Borisovna Rastorgueva for the possibility of working in museum repository; Moscow archaeologists Kirill Aleksandrovich Dneprovskii, Mikhail Mokovich Bronshtein, and Elena Stanislavovna Sukhorukova for the possibility to become acquainted with the Chukchi archaeological collection at the State Museum of Art of Peoples of the East and for valuable comments on Chukchi archaeology.

Minor additions and clarifications have been made to the English translation of the book.

INTRODUCTION

Until recently little was known to non-specialists about the military-political history of Chukotka, in particular on Russian-Chukchi relations. In fact, in Soviet survey courses on Russian history, the small colonial warfare in Northeastern Siberia was overshadowed by the large political events of the second and third quarters of the 18th century—the "War of the Polish Succession" (1733–1738), the Russian-Austrian clash with Turkey (1735–1739), the Russian-Swedish conflict (1741–1743), participation of Russia in the Seven-Years War (1757–1762), the Russian-Turkish War (1768–1774), and finally, the rebellion of E. I. Pugachev (1773–1775). Of course, military activities in Northeastern Siberia are described in essays in Siberian history, though not in detail. The general state of development of the theme in the Soviet period is shown, for example, in the book published in 1991 by the local historian M. I. Kolesov *Istoriya Kolymskogo kraya* [History of the Kolyma Region], which contains in the account of these events several chronological and factual errors, or *Istoriya Chukotki* [History of Chukotka] (1989), where the whole half-century Chukchi war, laid out on two and a half pages, is in essence reduced to the description of the Shestakov-Pavlutskii expedition.[2]

Such inattention to the theme was not dictated by a few gaps in the sources, but rather by ideological attitudes, according to which the annexation of Siberia occurred primarily in a peaceful way. We will characterize the primary sources on military-political history of Chukotka. They can be divided into two large groups—written sources (including recorded folklore) and archaeological. The theme is based, of course, on written sources—documents, dating from the second half of the 17th–18th

---

[2] Kolesov 1991: 22 (Anadyrsk was abandoned in 1746), 43 (D. I. Pavlutskii set off on a campaign in 1720), and so on; Dikov 1989: 81–83 (author of the paragraph—V. V. Leont'ev).

centuries. In the 17th century these are "stories" (statements) and petitions, documents from tax collection, edicts of local authorities, orders sent to expeditions, reports of the Yakutsk governors, and the like. Documents of this century are generally characterized by brevity of content and lack of details of incidents. Documents of the 18th century (reports, official papers, orders, instructions, letters) became more detailed, which was sometimes dictated by regulations of higher authority and the new European tradition of office clerical work. These documents (frequently simply transferred content of diaries of military actions) depict in detail what occurred during a campaign and are the basis for reconstruction of events. In addition, a new kind of source appeared—ethnographic works dated chiefly to the second half of the 18th century, in which administrators, at the request of higher officials, described the Siberian people under their control. It should be noticed that a large part of the documents of the 17th century based on the present theme and a substantial part of the materials of the 18th century have already been published.[3]

Folklore is an auxiliary source for the study of the present theme. In Chukchi folklore a cycle of heroic legends has been preserved about the antihero Yakunin, about the South Chukotkan hero Kunlelyu, about the epic heroes Lyavtyleval, Elennut, and others; also interesting are historical traditions of the Asiatic Eskimos regarding wars both among themselves ("How the Unazik Fought with the Sivukak" and so on) and with neighbors ("The Leader Viyutku," "The Battle of the Naukan People with Foreigners," and the like).[4] However, the complexity of folklore as a source consists of the fact that in it there is no clear chronological connection—different temporal layers shift about and are shuffled; in addition, in local tradition there was no common account of time. Although historical events in folklore could be reinterpreted, some details that point to a cause and effect relationship for events have nevertheless been preserved as a story about important local occurrences, such as wars

[3] Documents of the 17th century: DAI; KPMGYa; MIYa; Orlova 1951; Belov 1952; 18th century: PSI; KPTs; Berkh 1825; Pokrovskii 1941; Divin 1979; Fedorova 1984; Okhotina-Lind and Mëller 2013.

[4] Bogoras 1900; Voskoboinikov and Menovshchikov 1951, 1959; Kozlov 1956; Baboshina 1958; Menovshchikov 1974, 1988; Belikov 1982; Nutekeu 2005; Asadova 2010, 2012. About historical folklore, see Menovshchikov 1964a; Belikov 1956, 1965; Vdovin 1970.

and resettlements. As Baron G. Maydell (1835–1894) noted, the Chukchi, in distinction from the Chuvan and Yukaghir, did not know anything about battles that occurred in Anadyr'.[5] In fact, the folklore of the Anadyr' Chuvan people and the Lower Kolyma Russians is more informative from the historical point of view: many details have been preserved in historical traditions, telling about the activities of the Russian side.

Some historical information is also preserved in the notes of travelers, missionaries, and local historians of the end of the 18th–19th century (K. Merck, F. P. Wrangell, A. I. Argentov, G. Maydell, A. E. D'yachkov). In these works, as a rule, are superficial ethnographic descriptions, but sometimes they provide information on the history of Chukotka, primarily based on data from the same folklore.

Archaeology of extreme Northeast Asia is still comparatively young and has many different problems, among which can be isolated questions of dating (because of the features of bedding of the archaeological layers in the permafrost zone) and ethnic attribution of finds. Archaeology permits tracing the origin of different kinds of weapons, in first order, points of arrows and spears and the development of fortifications in the preliterate period.

One of the prospective directions of study of sources on the military history is the study of the toponymy of Chukotka connected with military actions. Curator Natal'ya Pavlovna Otke and folklorist Margarita Konstantinovna Takakava (1941–1993) identified about 220 toponyms in the territory of Chukotka connected with military actions, chiefly in the former Beringian and Anadyr' regions. Unfortunately, the results of this work have not been published. Of course, toponymic folklore is not always real history, though sometimes it presents some, probably late, interpretation of events. However, data from toponymics can be overlaid on known historical events, which in this case can be newly illuminated, showing the causal connection of events.

A no less prospective direction of research is the study of family histories of the residents of Chukotka, based on stories of people of middle and old age who still remember their family traditions. These two directions still await their researchers.

---

[5] Maidel' 1894: 517.

The historiography that concerns questions of the military-political history of Chukotka now amounts to more than two and a half centuries. We will dwell only on special general works on international relations in the region or the wars—works dedicated to specific events are mentioned in the corresponding parts of this book. The first historiographer of Siberia, Academician Gerhard Friedrich Miller [Müller] (1705–1783), turned to an examination of the subject. During the Second Kamchatka Expedition (1733–1743), as a scholar in his own right or by his own initiative, he copied the most important documents of the 17th–first third of the 18th century from the Siberian Archives. These documents are amassed in the St. Petersburg branch of the Archive of the Russian Academy of Sciences, in 34 volumes, four of which include documents from the Yakutsk Chancellery, which governed Eastern Siberia.[6] On the basis of these documents G. F. Müller wrote a corresponding historical works, first of all his classic work on the history of Siberia, which did not chronologically cover the conquest of its northeast. In the work *Izvestie o severnom morskom khode* [News of the Northern Sea Route] (published in 1742), which was then reworked into the composition *Opisanie morskikh puteshestvii po Ledovitomu i po Vostochnomu moryu, s rossiiskoi storony uchinennykh* [A Description of Sea Voyages along the Arctic and Eastern Sea, Committed by the Russian Side] (1758), G. F. Müller analyzes the history of the opening up of the northern sea route and along the way tells about the interrelations of the service people with the local population in the second third of the 17th–first half of the 18th century, including about Chukchi-Russian relations.[7]

In 1766 G. F. Müller was assigned as the head of the Archive of the Board of Foreign Affairs in Moscow. And in July 1770, based on the recommendation of Catherine II, Captain Timofei Ivanovich Shmalev (1736–1789) arrived to see the academician after a report to the Admiralty Collegium in St. Petersburg. In the course of a long conversation, G. F. Müller urged the captain to gather material on the history and ethnography of Eastern Siberia. Even before the meeting with the academician, the

---

[6] AAN. F. 21. Op. 4. No. 1–34; Yakutskie dokumenty [Yakutsk Documents]: Ibid. No. 31–34.

[7] The first composition, see: Okhotina-Lind and Mëller 2013: 69–103; the second: Miller 1996: 19–126; also see: Elert 1998.

captain was interested in local history, and after it, over the course of ten years (to 1780), he collected and sent to G. F. Müller various materials on the theme—both his own writings and copies of documents. Indeed, T. I. Shmalev occupied command posts in Okhotsk, Gizhiga, and Kamchatka and could copy current and archival documentation (even secret ones). In 1772 the captain, using his official position, created a detailed questionnaire, including 49 questions about the history of the Anadyr' Stockade and the Gizhiga Fortress, and sent it to the commander of Gizhiga Captain Ya. M. Peresypkin.[8] The answers of the latter, written on the basis of extracts from documents and newly taken oral testimony from participants in events, were the basis for the unpublished essay "Notes, Made by Captain Timofei Shmalev, from which Time the Chukchi People Began to be Known to the Russians Both from the Chukchi and the Russian Side, which Notes of this Honorable Continued up to the Incident in the Gizhiga Fortress in the Year [1]775, Committed to the Chukchi by Captain Peresypkin."[9] In this essay T. I. Shmalev examined in detail the history of Russian-Chukchi relations from 1646 to 1770. The work itself was created in chronological order of compilation of copies of documents on the theme, to which the author added his commentary (primarily in the notes) and proposals for improvement of the situation contemporary to him in the region. The second fundamental essay, also unpublished, on the theme dedicated to the Chukchi-Koryak relations is the monograph *Notes, Made by Captain Timofei Shmalev, on the Reason for which Ancient Malice Emerged among the Koryak with the Chukchi and Campaigns, Murders, and Robbery on both Sides, and Especially from the Occurrence in the Year* [1]*757 of Deception by the Koryak of the Chukchi and Murder of the Best of Their People Tatymkin and Arelpuga and their Families; Based on Which, the Chukchi are named by the Russians, Irreconcilable* (1777).[10] This essay was based not only on documents but in significant degree on oral statements of Russians and Koryak. In the "Notes" the Chukchi-Koryak confrontation of the third quarter of the 18th century is described in detail, with a list provided of Chukchi raids on the Koryak in

---

[8] The questionnaire: RGADA. F. 199. No. 528. Ch. 2. D. 3. L. 3–7 ob.; answers: Ibid. L. 8–43 ob.; also see: Zuev 2003a.
[9] RGADA. F. 199. No. 528. Ch. 1. D. 17. L. 1–30; No. 539. Ch. 2. D. 6. L. 63–93.
[10] RGADA. F. 199. No. 528. Ch. 2. D. 9. L. 45 –56.

1759–1774. This essay bears the somewhat more independent character of the author than the first.

The next essay, which should be mentioned, is one of the most important works of the naval historian Aleksandr Stepanovich Sgibnev (1826–1881) "A Historical Essay of the Main Events in Kamchatka, 1650–1856" (1869).[11] However, the content of the local history essay is somewhat broader than the title since events not only in Kamchatka proper are examined, but also in the bordering territories of Anadyrsk and Gizhiga. This work was composed on the basis of materials from the Siberian archives (in first order, the Irkutsk)—part of these documents have at present been lost. Therefore, this essay largely serves as a primary source. The account by the historian is more detailed if the preserved source on which the author bases his narrative includes more details, the source often being a simple retelling of the documents. The information of A. S. Sgibnev is sometimes exceptional, though it is not free of numerous errors and distortion in the retelling.

The founder of modern Chukchi studies, Waldemar Germanovich Bogoras (1865–1936)—after his work as part of the Yakutsk Expedition, outfitted by the means of I. M. Sibiryakov (1895–1898)—and as part of the Russian-American Jesup North Pacific Expedition organized by the American anthropologist Franz Boas (1900–1901)—wrote his basic four-volume research on the Chukchi. In the first volume, dedicated to social organization, there are parts about relationships of the Chukchi and Russians and briefly about the wars of the Chukchi with the Koryak, Even, and Eskimos.[12] The primary source for description of Russian-Chukchi relations were documents, both published and not, while inner-tribal relations are described by the author on the basis of folkloric materials. The present monograph, the first specially based on a Chukchi theme, is, on the one hand, ethnographic description, and on the other—scholarly research.

In Soviet times military-political history was viewed as part of a general-historical process and was not specifically singled out. One of the great Russian historians Semën Bentsianovich Okun' (1908–1972), having studied the archival materials of the senate, wrote on their basis the

---

[11] Sgibnev 2005–2008.

[12] Bogoraz-Tan 1934: 32–78, 162–178.

small monograph *Essays on the History of Colonial Politics in the Kamchatka Region* (1935). One of the chapters of the that book, "War between the Russians and Chukchi," includes a brief description of the subject that interests us from the Marxist anti-monarchist position.[13]

Innocent Stepanovich Vdovin (1907–1996), student of W. G. Bogoras, continued the business of his teacher's investigation of Chukotka. He not only studied the folkloric material in place, but also actively used archival material in his works of 1940–1960s, which, strictly speaking, were the basis of his unpublished candidate dissertation—"Russian-Chukchi Relations of the 17th–19th Centuries," defended in fall 1941, and his doctoral dissertation defended with the monograph *Essays on the History and Ethnography of the Chukchi* (1965). I. S. Vdovin, as a scrupulous historian, introduced into scholarly circulation a large part of the material available in various archives and on this basis in sufficient detail presented the history of the relationship of the Chukchi with neighboring local peoples and the Russian state.[14] These works of I. S. Vdovin largely lay at the base of modern historical research on the theme.

In 1947, a small article was published by the Siberia scholar Valentina Vasil'evna Antropova (1909–1976) that was specially dedicated to inter-tribal relations in Northeastern Siberia. In the beginning of the work the author examines the wars between the Chukchi and the Koryak and Chuvan peoples, allotting special attention to fortifications. After ten years, the author used collected material in her large article (in fact, a monograph) on warfare of the peoples of Northeastern Siberia.[15]

The ethnographer Il'ya Samuilovich Gurvich (1919–1992), in his fundamental research on the ethnic history of the peoples of Northeastern Siberia in the 17th–20th centuries, was not able to put aside the history of the Paleo-Asiatics. In essays based on published and archival sources, he briefly revealed the events of the ethnic history of the Chukchi and Eskimos, including military conflicts.[16]

The historian of the Far East, Boris Petrovich Polevoi (1918–2002), in his summarizing monograph dedicated to new data on the

---

[13] Okun' 1935: 66–86.
[14] Vdovin 1965: 54–78, 102–152.
[15] Antropova 1947: 64–65; 1957: 174–186.
[16] Gurvich 1966: 47–54, 112–117.

opening up of the region, data that he revealed in the process of study of the archival documents, found several additional facts concerning relations with the Chukchi in the last third of the 17th century and interpreted them.[17]

After the collapse of the USSR and abolishment of Late-Soviet ideological attitude in the peaceful annexation of Siberia to Russia, military subjects began to be researched more intensively, and separate works began to be dedicated to them. From the theoretical articles touching upon our problem, it is necessary to distinguish a report by the Novosibirsk historian V. I. Kuz'minykh, who, examining the history of Russian-Chukchi relations as colonization, divided military activities in Chukotka into seven stages beginning from 1649 and ending with adoption of Russian citizenship by the Chukchi in 1778.[18]

One of the most important works based on the theme for today is the monograph of the Novosibirsk historian Andrei Sergeevich Zuev, dedicated to relationships of Russia and the peoples of Northeastern Siberia in the 17th–first quarter of the 18th century. That book was created primarily on previously published materials. Having characterized the socioeconomic structure of the peoples of the region and Russian movement into the northeast, the author moves to a general characterization of the Russian-Chukchi conflict. Especially useful is the extensive appendix called the "Chronicle," in which the author briefly describes the events year after year (1634–1726).[19]

A subsequent monograph by A. S. Zuev was written based on his doctoral dissertation, defended in 2005. This extensive research was dedicated to the annexation of Chukotka to Russia in the 17th–18th centuries. In the work, based on a large mass of archival documents, the policies of the Russian Empire are examined with regard to its relations with the population of Chukotka up to the moment of formal annexation of the latter, including the history of military activities (1720–1780s). In the first part of the book the author briefly describes the events of 17th century, essentially duplicating his previous monograph. The special value of this book is provided by the refinement in the dating of some military-

[17] Polevoi 1976; 2000: 135–139.
[18] Kuz'minykh 1994.
[19] Zuev 2002.

political events and active use of unpublished archival materials. The work presents a fully completed picture of the annexation, including the military aspect of the problem. The monograph of A. S. Zuev is complemented by his articles, a large part of which were incorporated in the text of the book, but in the book the author does not analyze some subjects in detail, in particular those concerning the campaign of D. I. Pavlutskii in Chukotka in 1731.[20]

Foreign historiography is not as extensive as Russian. Of course, the annexation of Chukotka is discussed in essays on the general history of Siberia. For example, in the classic book of essays on the history of the annexation of Eastern Siberia to Russia, the American historian-Russianist Frank A. Golder (1877–1929) examines in particular the expedition of A. F. Shestakov-D. I. Pavlutskii. A theme interesting Western (predominantly American) historians and ethnologists is the interrelations of the Eskimos in Bering Strait—both trade and military, about which more detail will be given in Chapter 5.[21] It is of value to highlight the article by the Danish researcher of the history of Russia in the 18th century, Hans Bagger, specially dedicated to the time of annexation of Chukotka to the Russian Empire, where a separate part is dedicated to a description of the Chukchi war with accent on the campaigns of D. I. Pavlutskii. The article is interesting by the fact that the process of annexation of Chukotka to the Empire is presented by the author in broad geopolitical aspect. The shortcoming of the article is the lack of Russian historiography on the theme of the 1990s–beginning of the 2000s and being supported mainly by literature of the Soviet period.[22]

In distinction from the most closely related works of A. S. Zuev, in the present book the accent is placed not on political, but on the military side of events and, correspondingly, battle actions are examined in more detail, and not negotiations between sides and Russian projects. The primary feature of the present book consists of the fact that in it are examined in a broader chronological segment of time not only Russian-

---

[20] Abstract: Zuev 2005; monograph: Zuev 2009; additional articles: Zuev 1998, 2001, 2001a, 2003, 2004a.

[21] Golder 1914: 153–164; about relations in Beringia: Smith 1970; Malaurie 1974; Hickey 1979; Burch 1988; Sheppard 2009.

[22] Bagger 2005: 39–64.

Chukchi relations but also relations of the Chukchi with neighboring local peoples, beginning from the Old Bering Sea culture, with which the Eskimos are connected, and ending in the 19th century, when clashes still occurred between representatives of different ethnic groups. Also, in distinction from works of A. S. Zuev, folklore as an additional source is extensively brought into the book. With this, it is necessary to note that, as I. S. Vdovin correctly noted, "in all cases of bringing in folklore as a historical source, direct or indirect evidence of the accurateness of the data being reported in the folklore should be sought" in archaeology, images, or etymology.[23] It can be added that in the process of preparing the book two articles were published on the theme, which in somewhat reworked form were included in this book.[24]

* * *

A few technical comments. In the work, I adhered to the principle of initial citation of the source,[25] after which I gave my interpretation of events. This was done in order that the reader could judge about the events based on the primary source and not just on my retelling. If the ethnonym "Chukchi" in the account is placed in quotation marks, this means that the source is unclear whether reference is to the Chukchi or to the Eskimos.

As it turns out, in the two most complete collections of documents of the 17th century, compiled by N. S. Orlova and M. I. Belov, because of Stalinist ideological attitudes in the peaceful annexation of Siberia, information about the wars between the Cossacks and the local residents was dropped and omitted.[26] Therefore, at times it was difficult to restore the course of a military expedition—it was necessary to double-check data in other sources, in documents in archives, for which there was not always the possibility.

All dates before 14 February 1918 in the sources and in the transcription of events are in Old Style.[27] In the process of work with the documents of the 17th century I corrected the dates of those events cited

---

[23] Vdovin 1970: 18.

[24] Nefёdkin 2013, 2015.

[25] With citation of sources their orthography and punctuation were preserved.

[26] Omissions in necessary documents, see: Orlova 1951: 134–135; Belov 1952: 57, 217, 223.

[27] Old Style (O.S.) refers to the Julian calendar, New Style (N.S.) to the Gregorian calendar. See https://en.wikipedia.org/wiki/Julian_calendar.—*Trans.*

by I. S. Vdovin, I. S. Gurvich, and A. S. Zuev,[28] that turned out to be inaccurate, to more reliable ones, sometimes without special stipulation in the text.

For the people who did everything to bring this book to light

[28] Vdovin 1944, 1965; Gurvich 1966; Zuev 2002.

CHAPTER 1

PREHISTORY (BEGINNING IN THE 1ST MILLENNIUM A.D. to the 17TH CENTURY)

1. Archaeological Evidence

In the archaeological history of Chukotka, several cultures have been distinguished that are identified based on forms of the most common hunting tools—harpoons and the artistic style of decorating artifacts. These cultures are connected with the settled population of the region, especially with the Eskimos and their ancestors, who, in the opinion supported by the eminent archaeologist N. N. Dikov (1925–1996), separated from the Aleuts 4,000 years ago. A separate archaeological culture that could belong only to the Chukchi has not been clearly distinguished, though the Ust'-Belskaya culture of central Chukotka and the Pegtymel' sites in the north is connected with them. All these cultures are Neolithic, though the use of iron tools is known from the beginning of the 1st millennium A.D., but metal itself was an extremely rare import; before the 17th–18th centuries weapons and tools continued to be made of stone, bone, and antler.[29]

The remains of an early Paleo-Eskimo culture, discovered at the site on a rocky cape at Chertov ovrag [Devils Gorge] on Wrangel Island, are dated to the middle of the 2nd millennium B.C. (3360 ± 155 years ago). This culture belonged to hunters of sea mammals, who used weapons and tools of siliceous slate and black chert worked by retouch—leaf-shaped spear points with and without stems, stemmed leaf-shaped stone arrowheads, and blunted points (bunts). Investigation was seriously hindered by the fact that no graves were discovered. This culture was

[29] Separation of the Eskimos: Dikov 1979: 180; Mikhailova 1996: 165; the appearance of iron in the Arctic: Okladnikov and Beregovaya 1971: 26; Arutyunov, Glinskii, and Sergeev 1977: 101–102.

typologically close to Dorset in Alaska, and N. N. Dikov viewed its creators as arrivals from America.[30]

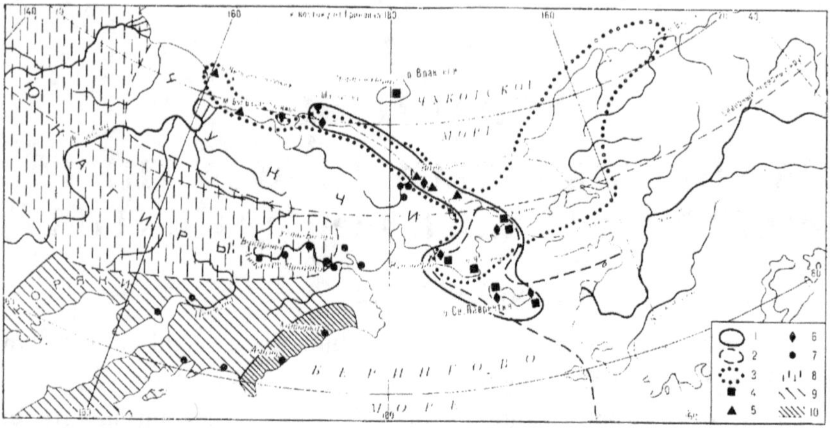

Figure 1. Chukotka in the early Iron Age and Remnant Neolithic: 1—area of distribution of the Old Bering Sea culture; 2—area of distribution of the Okvik culture; 3—area of distribution of the Birnirk culture; 4—Okvik sites; 5—Birnirk sites; 6—Old Bering Sea sites; 7—other sites; 8—Yukaghir sites; 9—Koryak; 10—Kerek. (Reproduced based on Dikov 1989: inset between pages 96 and 97).

At the end of the 2nd millennium B.C. the climate began to change, becoming cooler. The shift of the vegetation to the south of the forest zone and the spread of the tundra to the south set before the remaining local population the necessity of adapting to the prevailing natural conditions: heavy pull-over fur clothing became widespread, pit houses subsequently became the type of dwelling characteristic for the local residents, baidary became the mode of travel, and toggling harpoons appeared as a means of hunting.[31]

The Okvik culture is dated to the first half of the 1st millennium A.D. This chiefly island culture embraced St. Lawrence Island, Southwest Alaska, and Kodiak Island, and archaeological sites belonging to Okvik are sometimes encountered on the shore of Eastern Chukotka from Uelen to Nunligran. Since the residents wore labrets this culture is viewed as Eskimo. The residents hunted animals both in the sea and on land. Among the weapons stone points of arrows and spears stand out. Arrows had

[30] Dikov 1976: 22; 1977: 210–212; 1977a: 129, Fig. 25: 1–2; 1979: 165–168; Tein 1983: 31–33, 52, 65–66, 70–71, 75, 81, Fig.; 1991: 7–8.
[31] Dikov 1979: 164; 1989: 45.

foreshafts that transitioned from the point to the shaft, and in a slot on them were set stone points; points of arrows and darts of stone were retouched, while spear points of slate were ground and made in leaf-shaped or triangular form; slate leaf-shaped knives 22 cm long, on which a handle was fastened through a hole, were also used.[32]

Western historiography usually views the Okvik culture as the first period of the Old Bering Sea culture (OBS-I). In general, both these cultures coexisted during the first half of the 1st millennium A.D., embracing, respectively, the islands and the mainland part of Chukotka. The Old Bering Sea culture proper was spread in Chukotka from Shalaurova Island in the north to Nunligran in the south, including St. Lawrence Island, Ratmanova [Big Diomede] Island, and Western Alaska. The relatively small population lived in pit houses in small coastal settlements located far from each other and situated near walrus haulouts; residents were occupied predominantly with hunting on the sea in baidary for seals, walruses, and whales, and to a lesser degree, hunting on land for reindeer and birds. Clear remains of sledges have not been found, but hand-drawn sleds are known. The custom of the residents wearing labrets has not been found, which might attest not just to a change in custom, but also to the arrival of a new population. Weapons are represented by percussion flaked or retouched stone (chert or slate) points of spears and darts of leaf-shaped or leaf-shaped/rhomboid form, with flat stem, and an overall length of 5.2 to 13.2 cm. Points were inserted into a slot in the shaft and wrapped with baleen; stone knives were also used.[33]

Finds made in the Ekven cemetery, located approximately 15 km from Uelen, reveal the armament of this period, in particular a reinforced bow from Burial No. 302. The total length of its riser amounts to 1.28 m, maximal width of the horns—5 cm, thickness of the grip—1.55 cm; on the ends of the limbs there are long (21.5 cm) pieces of walrus tusk, in slots of which the bowstring was attached. The wrist was protected from the blow of the string by a wrist guard; an example of one, made of reindeer

[32] Dikov 1979: 169–182.

[33] Rudenko 1947: 80–81; Dikov 1979: 183–214; 1989: 47–52; Tein 1991: 9–14; Bronshtein 2007: 31; type of settlement: Krupnik 1981: 109; 1989: 174; Western historiography about Okvik: Fitzhugh 2009: 88–90; spear points: MAZ: 69. Nos. 162, 164, 166, 168, 172, 175.

antler, was found in a dwelling on Cape Billingsa. Various types of arrowheads—the most popular weapon—are represented in the cemetery at Ekven: composite, in which a triangular or pentangular stone point was set in a slot in a foreshaft of reindeer antler (13 to 16.1 cm), the foreshaft ended in a shank, most often with two barbs, the number of which could reach six; long (13.7 to 20.1 cm) rod-like points without barbs or with one to four barbs at the bottom and with a pointed shank; and small (4.9 to 8.1 cm) conical bone points with forked shank. If the first two types could have been used for both hunting and war, the third was clearly only for hunting. The most widespread were composite arrowheads—a type that (with the stone point sometimes being replaced by an iron one) was still being used in the 19th century.[34]

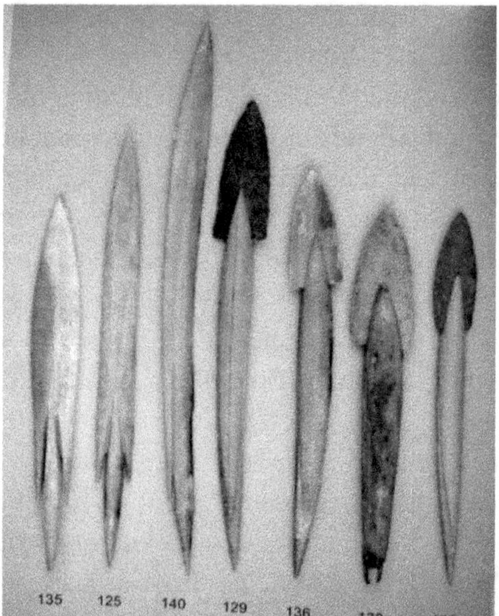

Figure 2. Arrowheads of reindeer antler and walrus tusk; composite with foreshaft of reindeer antler and points of stone or walrus tusk (OBS-I and OBS-II/III). Ekven cemetery, Burials 217, 251, 285, 302. (Reproduced based on MAZ: Nos. 125, 127, 129, 135, 136, 138, 140).

[34] The bow from Ekven: MAZ: 67, 119. No. 119; compare Glinskii 1986: 7; 1987: 125; arrowheads: Collins 1937: 323, Pl. 29: 6–7; 34: 1–6; Arutyunov and Sergeev 1969: 128–130, Fig. 62–65; Dikov 1977: Pl. 137: 1–3; 177: 1–2; arrows from Ekven: Arutyunov and Sergeev 1983: 215, Fig. 10; MAZ: 67–68. Nos. 125, 127, 129, 131, 135–140; arrows of walrus tusk, wrist guard: Tein 1991: 11–12; historical composite points: Nefëdkin 2003: 82–83.

The complex of equipment of a warrior-hunter can be traced in Ekven Burials No. 204, 238, and 250. In the first, very rich burial, 365 items had been placed together with an older man. Two (or even three) more buried men were found in neighboring graves. Along with other tools, each of them had by the left femur—10 compound (with bone foreshaft) arrowheads, by the knee—6 dart points, on the chest—a spear point, and by the right femur—12 more arrows. In the second burial were 10 arrowheads (9 of them with a bone foreshaft and stone point), one stone spear point, and one more dart point. In the third burial were 15 points (3 of them whole, of walrus tusk) and 5 stone insets, 9 spear points, 3 stone dart points, and 2 ice picks, leister insets, bird spears, and runners from hand sleds.[35] According to museum inventories of the GMINV, in Ekven Burial No. 251 were 10 arrowheads, of them 4 were composite, the rest of walrus tusk; in Burial No. 255—13 arrows, only one of them not being composite (OBS-I); in later Burial No. 302—13 points, of them 4 composite, 5 single-unit, and 2 blunt hunting points; in Burial No. 310A—14 points, of them 4 composite, 1 rod-like, and 1 a hunting point; of these points 11 are stemmed, 3 forked; in Burial No. 310B were 2 spear points of walrus tusk and 2 of stone, as well as sets of 10 rod-like and 11 composite arrowhead (OBS-II/III).

In Pegtymel' Cave in a context dated to the period of the Old Bering Sea culture were two types of plates of reindeer antler, worked with iron (5th–6th centuries). The first type of plates have a width of 3.5 cm, the second—2.0 cm, respectively, with two and three pairs of holes for attachment.[36] Such perforation is less complex than on later plates, which have additional holes in the center for attachment of the plates: the upper and lower pairs of holes served for fastening the rows, and central ones for joining plates in one row. Though modification of the blades with an iron tool can attest to a later time for the artifacts, iron was also known in the OBS period and could well be used to work blades for valuable armor.

---

[35] Arutyunov and Sergeev 1983: 206–210, 213, 223–226; Bronshtein 2007: 29; MAZ: 67. Nos. 131–133; 69. Nos. 164, 165–168.

[36] The plates from Pegtymel': Dikov 1971: 43, Fig. 35. A rectangular plate with a slot and three holes from Uelen, which N. N. Dikov (1979: 191, Fig. 72: 2) believed was armor, is hardly such, since it has an unusual perforation and an incision in the middle.

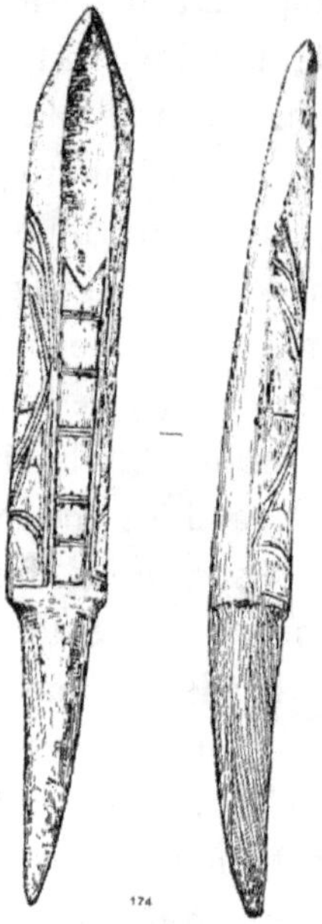

Figure 3. Spear point of walrus tusk 19.4 cm long. Ekven cemetery, Burial No. 309 (OBS-II/III). (Reproduced based on MAZ. No. 174).

The appearance of the reinforced bow, which stipulated a wrist guard that defended the warrior from the blow of the string, attests to the increased role of conflicts and the appearance among warriors of special equipment, primarily among wealthy warriors. Equipment that could have been used not only for hunting, but also for battle, was rather simple: a quiver with 10 to 15 arrows, predominantly with stone points, which were attached to a transitioning foreshaft, and a polearm weapon (a heavier spear and darts). If one imagines that the burials really show a quiver kit, the tactics of shooting can be assumed as slow aimed firing, which points to a rather fleeting battle and alludes to the small number of groups fighting on both sides. The spear was used for finishing off an enemy and probable close combat.

In the 5th–7th centuries, during the next warming period, the new Birnirk culture was established in the north and east of Chukotka, brought from Northwest Alaska by the Eskimos, again wearing labrets. The area of this culture was maximally spread to the west, including Chetyrekhstolbovoi Island (Medvezh'i Islands), Cape Bol'shoi Baranov, and Vankarem. The settlements were now sometimes located in places at lower elevation than formerly, and the basic occupation of the residents became maritime hunting of small pinnipeds (seal and bearded seal), and on land—hunting birds and catching fish with nets. The weapons used were arrows with bone points, spears, and darts with ground stone stemmed points, and knives of stone.[37]

Again, the Ekven cemetery presents specific weapons of this epoch. In Burial No. 297 were preserved a whole spear-ice pick 1.33 m long; a stone leaf-shaped spear point 11.2 cm long with its stem going into a slot in the shaft to a depth of 4.5 cm; in the bottom of the shaft is an ice pick 22.6 cm long of walrus tusk. Spear points from Ekven and Paipel'gak were made of antler or bone and had an awl-like form (21.8 and 24.2 cm). Among the surface materials at Ekven were shafts of arrows 41.5–46.0 cm long and 6.0–8.5 mm in diameter; attached to the shaft with a slot in the butt (6 mm deep) were two feathers 10.5–13.0 cm long; and inserted in the shaft was a socketed point that was fastened with a binding. The points themselves, found at the same Ekven and Paipel'gak, in fact had a pointed sleeve; these points, made from reindeer antler or walrus tusk and equipped with one or two asymmetrical barbs (8.3–12.7 cm), had an elongate-rhomboid (10.0–19.4 cm), blade-like (8.8 cm), or elongate-pentangular (8.7–14.0 cm) form; the last form with wide shank was also made from slate (8.8 cm), whereas the small sub-pentangular tips were made from chalcedony and chert (3.3–5.9 cm) and probably were set, like before, in foreshafts. If barbed points were more probably for hunting, intended for becoming stuck in the body of a game animal, the other forms could fully have been used both for hunting and for war. The wrist was protected from the blow of the bowstring by a rectangular or rectangular with rounded edges wrist guard 9.6–10.5 cm long and 2.7–2.8 cm wide with two or three holes for fastening with a thong; wrist guards were made of walrus tusk or reindeer antler.[38]

---

[37] Dikov 1979: 215–218; 1989: 53–54; Tein 1991: 14; Dneprovskii 2001: 19; compare Raushenbakh 1969: 7–9; Okladnikov 1947: 181; Okladnikov and Beregovaya 1971: 22.

[38] Spears: MAZ: 69, 119. No. 160, 177, 179; arrow shafts: MAZ: 68, 119. No. 147; GMINV, ChAE-97—666/1–3; 98—438/16; arrowheads: MAZ: 67–69. No. 142–146, 148, 150–159; wrist guards: MAZ: 67. No. 122–124; GMINV, ChAE-97—762/1; B 2002—94, 196.

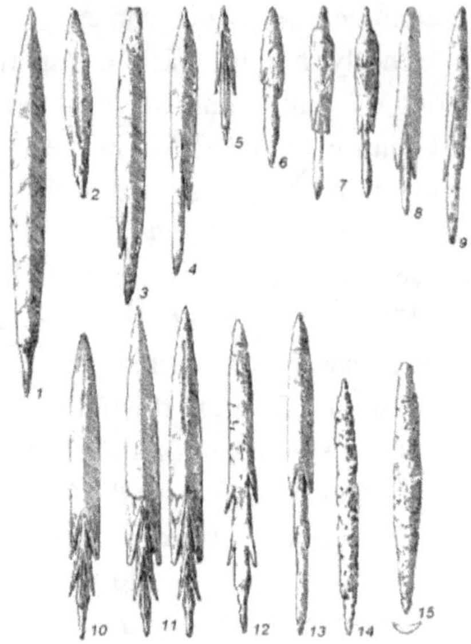

Figure 4. Bone stemmed arrowheads from the Uelen cemetery (OBS): 1—from Burial 16 (67); 2, 4—surface material; 3—from Burial 9 (58); 5—from Burial 17 (57); 6—from Burial 12 (57); 7—from Burial 13 (57); 8—from Burial 6 (59); 9—from Burial 19 (57); 10, 11—from Burial 15 (59); 12—from Burial 12 (57); 13—from Burial 19 (59); 14, 15—from Burial 17 (57). (Reproduced based on Arutyunov and Sergeev 1969: 133, Fig. 62).

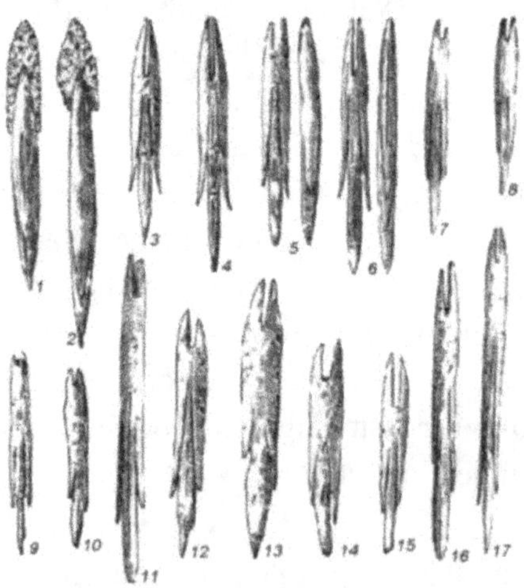

Figure 5. Composite arrowheads, consisting of a foreshaft and stone inset, from the Uelen cemetery (OBS): 1—from Burial 18 (59); 2—from Burial 1 (55); 3, 5, 6—from Burial 15 (59); 4—from Burial 16 (59); 7, 8—from Burial 6 (59); 9, 12–15—surface material; 10—from Burial 2 (60); 11—from Burial 18 (57); 16—from Burial 19 (57); 17—from Burial 7 (59). (Reproduced based on Arutyunov and Sergeev 1969: 134, Fig. 63).

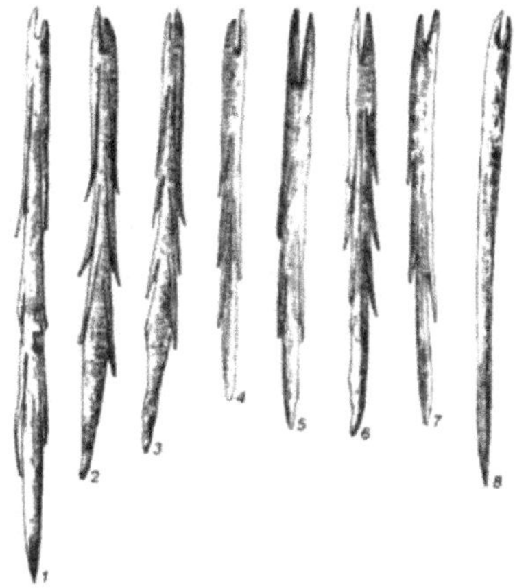

Figure 6. Composite arrowheads, consisting of a foreshaft and stone inset (absent), from the Uelen cemetery (OBS): 1—from Burial 22 (59); 2, 3—from Burial 19 (58); 4, 5—from Burial 19 (57); 6, 8—from Burial 19 (58); 7—from Burial 19 (57). (Reproduced based on Arutyunov and Sergeev 1969: 135, Fig. 64).

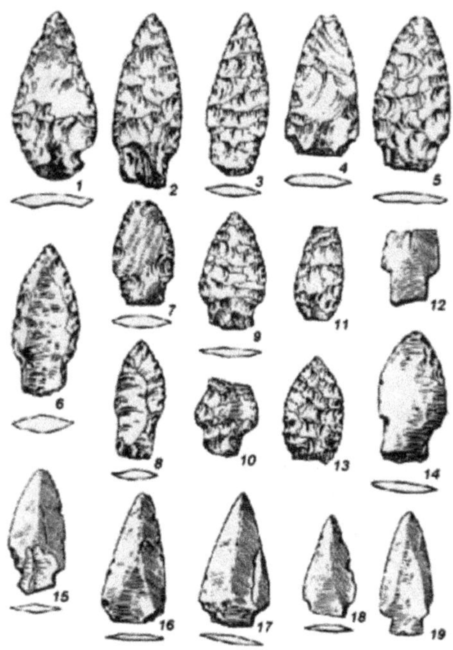

Figure 7. Stone points of spears and darts from the Uelen cemetery (OBS): 1, 4–6, 9, 14—from Burial 7 (57); 2—from Burial 8 (58); 3, 8—from Burial 23 (58); 7, 12, 15, 17—from Burial 22 (58); 10—from Burial 9 (58); 11—from Burial 7 (58); 13—from Burial 7 (59); 16—from Burial 19 (57); 18—from trench backdirt; 19—from Burial 6 (59). (Reproduced based on Arutyunov and Sergeev 1969: 137, Fig. 66).

Assigned to purely military equipment are armor plates. Two sets of eight and nine long plates of reindeer antler were found in the Ekven site (4th century) and cemetery (5th–8th centuries): in House No. 18 was a set of plates 29.5–37.0 cm long and 1.8–3.2 cm wide, with four holes and grooves for attachment at the ends, and in the cemetery among the surface materials were plates measuring 31.5–37.5 by 2.0–2.7 cm 1 cm thick; one armor plate was found in the rich Burial No. 130.[39]

Figure 8. Compound bow with overlay of walrus tusk; total length 128 cm. Ekven cemetery, Burial No. 302 (OBS-II/III). Arrow 59 cm long from Ekven (Birnirk-early Punuk). Spear-ice pick 133.2 cm long. Ekven cemetery, Burial No. 297 (Birnirk-early Punuk). (Reproduced based on MAZ: No. 119, 147, 160).

Figure 9. Arrowheads of walrus bone and reindeer antler, Paipel'gak, House No. 1 (Birnirk-early Punuk). (Reproduced based on MAZ: No. 148, 152, 154, 156–159).

The Punuk culture (6th–beginning of the 16th century) replaced the Old Bering Sea; it coexisted with Birnirk and supplanted it by the 12th

[39] Ekven: Arutyunov and Sergeev 1975: 50–51; armor sets: GMINV, KP 53404/121–128, 379; KP 53185/1728–1735; Dneprovskii 2001: 13; Bronshtein and Dneprovskii 2001: 611, Pl. 9: 6 (dating: Dneprovskii 2007: 19: Birnirk-early Punuk).

century. Geographically the Punuk sites are located from Aion Island and Cape Shmidta in the northwest to the site at Wharf No. 7 at Anadyr' in the southeast. Since a new warming was observed in the first part of this period, the Punuk people, living in large settlements with a men's house (kazhim), returned to hunting large bowhead whales as the chief occupation; large land game (reindeer and mountain sheep) was also hunted. The production of stone weapons did not move forward: there were the same ground slate knives and spear points. At the same time, a rise in the bone industry is observed.[40] Made of walrus tusk, antler, and bone were three-faceted arrowheads of rod-shaped or elongate pyramidal form; points of the latter form sometimes had a barb below; points were set in a shaft with a pointed stem; to the three-faceted stemmed ones belong barbed points with one or two barbs on one side; other types of points were lanceolate with two barbs below and with a broad shank, and elongate-rhombic ones with forked stem. A large part of these types of points could have been used not only for hunting large game but also in battle. That the elongate pyramidal points with a barb were for battle is attested by a burial near Gambell (on St. Lawrence Island), where a buried individual had been wounded with just such arrows. The wrist was protected from blows of the bowstring by trapezoidal, oval, or rectangular wrist guards with rounded top, made most often of walrus tusk, as well as from bone, baleen, or antler; sometimes wrist guards were decorated with simple geometric designs.[41] Late-Punuk rectangular bone armor plates differ by their perforation from earlier ones since they have a more complex system of paired holes that is similar to later historical ones. Later, in the 18th–19th centuries, bone armor was a characteristic element of Eskimo armament.[42]

---

[40] Dikov 1979: 218–226; 1989: 55–57; Tein 1991: 15–16; Dneprovskii 2001: 19; on population: Krupnik 1981: 110; 1989: 174. The American archaeologist W. Fitzhugh (2009: 91) dates the Punuk to 800–1200 years.

[41] Arrows: Collins 1937: 323, Pl. 74: 6–10; Dikov 1979: Fig. 89; Bandi 1995: 170–171, Fig. 6; wrist guards: Machinskii 1941: Fig 12: 3; Rudenko 1947: 82, Pl. 10, Fig. 13 (Naukan); Pl. 13, Fig. 16 (Cape Chaplino); Pl. 21, Figs. 4–7 (Sireniki); Pl. 31, Fig. 11 (Nunligran); Pl. 33, Fig. 25 (Enmylen); Wardwell 1986: 105–107. No. 132–134, 138.

[42] Punuk plates: Rudenko 1947: 96, Pl. 6, Fig. 17 (Uelen); Pl.15, Fig. 19 (Cape Chaplino); Pl. 18, Fig. 22 (Avan'); Antropova 1957: 218, Fig. 32; Dikov 1979: 187, 222, Fig. 90; Nefëdkin 2003: 69–70.

Figure 10. Eskimo wristband of walrus tusk 11.5 cm long. Stylized in the form of walrus tusks; Punuk culture (c. 500–1200). University of Alaska Museum of the North (Fairbanks), 84-150-1. (Reproduced based on Wardwell 1986: 105, Fig. 133).

Figure 11. Eskimo wristband of walrus tusk 6.4 cm long. Stylized in the form of a walrus head. Probably from the middle period of the Punuk culture (c. 900). University of Alaska Museum of the North (Fairbanks), 84-150-1. (Reproduced based on Wardwell 1986: 105, Fig. 134).

Battle actions in the region are attested by the mentioned burial near Gambell, investigated in 1967 by the Swiss archeologist H.-G. Bandi. There, one of the buried individuals of Punuk times—a man 35 to 40 years old—died from 15 arrows that struck him in the body and face, and from a basalt spear point, that struck the right clavicle. Since two points were made of antler (the others of walrus bone) it can be supposed that the enemies who attacked the man were connected with Siberians.[43] These wounds clearly attest to the tactics of an enemy: intensive shooting of an enemy with subsequent transition to hand-to-hand combat. Probably the deceased clashed with superior forces of the enemy and was riddled with arrows, not managing to dodge them. In an attempt to escape he was struck

[43] Bandi 1995: 168–170. Also, numerous points were discovered in one body from Deering (north of the Seward Peninsula in Western Alaska); and a kazhim on Cape Kruzenstern (Western Alaska) was burned (Mason 1998: 296). During the course of excavations in the 1989 or 1990 season in the Ekven cemetery, one of the skeletons found had been struck in the region of the carotid artery by a point of walrus tusk (date: Birnirk-early Punuk) (oral communication with M. M. Bronshtein).

with a spear; or the warrior was initially wounded in battle by an arrow in the leg and was not able to flee from the shooters, and was then struck by numerous arrows. Such intensive shooting was characteristic for Eskimos even in historical times.

Based on the assumption of the American archaeologist H. B. Collins (1899–1987), supported by the Soviet archaeologist S. I. Rudenko (1885–1969), the appearance in the Punuk period of armor, wrist guards, and new types of points identical to later historic specimens attest to the appearance in the region of a new population. However, judging by the archaeological materials, the Punuk culture developed from the Old Bering Sea. Though a large part of the components of armament enumerated above had already appeared in the preceding periods, but their spread attests to the increasing role of war in the life of society. In general, the Old Bering Sea culture and the Birnirk and Punuk are viewed by researchers as Eskimo. That the first Eskimo culture emerged suddenly, without a clear historical background, indicates that it came from outside, and it is believed—from Alaska. The Punuk culture faded in the 14th–16th centuries during the fluctuating cold period, when the coastal population again transitioned to hunting pinnipeds as the primary economic branch and to hunting wild reindeer, which is already well known in historical times.[44]

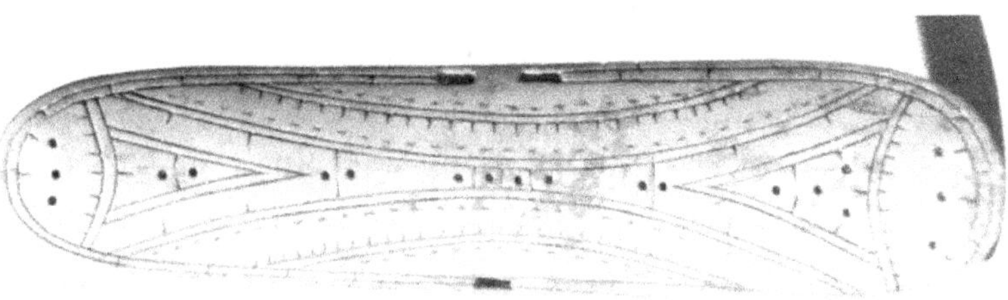

Figure 12. Eskimo wristband of walrus tusk 13.8 cm long from St. Lawrence Island; Punuk culture (c. 500–1200). American Museum of Natural History (New York), 60.1.9017. (Reproduced based on Wardwell 1986: 105, Fig. 138).

[44] Collins 1937: 331; Rudenko 1947: 97; pro: Bandi 1995: 170–171; Mikhailova 1996: 171; Mason 1998: 270; Cooper and Gabriel 2013: 1, 3; increase in the role of war in the Punuk period: Arutyunov and Sergeev 1975: 196; Fitzhugh 2009: 91; Eskimo cultures: Arutyunov and Sergeev 1975: 193; Mikhailova 1996: 167; Dneprovskii 2001: 23; fading of Punuk: Mikhailova 1996: 172, 184; Eskimos from Alaska: Reuse 1994: 296.

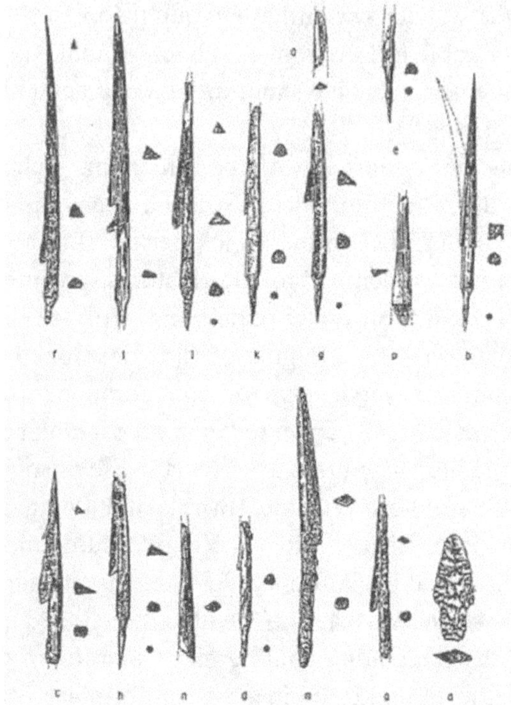

Figure 13. Arrowheads of walrus tusk and reindeer antler and spears of basalt from a burial near Gambell on St. Lawrence Island (Punuk time). (Reproduced based on Bandi 1995: 171, Fig. 6).

At the end of the 18th century Dr. Carl Merck (1761–1799) noted the presence in Chukotka of four Eskimo tribes, which correlate to the Sireniki people, Chaplino people, Naukan people, and Uelen people. According to linguistic data, the earliest Eskimo population of the Chukotkan coast was the Sireniki people, who spoke in their separate dialect. During the Little Ice Age (14th–16th centuries), it is believed, a resettlement of Eskimos from the islands to the mainland occurred. In particular, in the opinion of ethnographer M. A. Chlenov, at that time the Chaplino people arrived in Chukotka from St. Lawrence Island, and the Naukan people arrived from the Diomede Islands. In any case, in 1945 some male Eskimos (evidently Chaplino people) still had tattooing under the lower lip, imitating labrets—a clear remnant of the true decoration that hints at the origin of the population. At least, in 1871, Eskimos (probably Chaplino people) told a member of the Chukotka Expedition, astronomer K. K. Neiman (1830s–1887), that their ancestors arrived from America.

Later, in the 18th century and up to the 1920s, the Chaplino people resettled from Chukotka to St. Lawrence Island, where of 11 clans four had an Asian origin.[45]

2. Development of Reindeer Herding

Still in the first quarter of the 19th century travelers unsuccessfully tried to learn from the Chukchi what they knew about their past. The British captain J. D. Cochrane (1793–1825), who met and conversed at the Anyui fair in March 1821 with the Chaun, Belomorsk, and Nose Chukchi, noted: "They know nothing about their origin, nor about the first settlement in this country." Later, in 1871, K. K. Neiman also wrote: "The Chukchi do not have any idea about their origin and in this matter, refer to their ancestors, who allegedly recently still knew where they came from, but they did not communicate it to them."[46] Travelers probably wished to obtain specific historical information but could not converse with local residents or simply asked people who did not want to tell their traditions to foreigners or simply were incompetent.

Nevertheless, there is historical information in Chukchi folklore, but in altered form. According to a tradition transmitted by this same K. K. Neiman, the Chukchi were at first sea mammal hunters and then god gave them reindeer. Based on Chukchi legend, which explains the origin of reindeer herding, at first man was a hunter and then, in order not to be away from his family and to guard it, he began to tend reindeer. I. S. Vdovin, based on folkloric data, concluded that the ancestors of the Chukchi were nomadic hunters of reindeer and sheep, and changed to reindeer herding when the Yukaghir occupied their hunting grounds.[47] It is quite logical that reindeer herding would emerge from hunting. It is

---

[45] Neiman 1871: 25; Merck 1978: 98–99; Dolgikh 1960: 551; Krupnik and Chlenov 1979: 19, 24; Krauss 1981: 153–154, 173; Krupnik 1981: 98, 111–113; Chlenov 1988: 64–75; Mikhailova 1996: 172; compare: population on St. Lawrence Island arrived from Chukotka: Tan-Bogoraz 1935: 353. Tattooing: Rudenko 1949: 149, 154. Based on legends told in 1924, the Eskimos from Alaska resettled in Naukan one hundred years ago (Galkin 1929: 11, 73).

[46] Cochrane 1824: 203; Neiman 1871: 16; compare: Neiman 1871: 25; Maidel' 1871: 68. Based on the assumption of I. I. Krupnik (1981: 100; 1983: 70), oral tradition was preserved by the coastal residents for only two centuries.

[47] Tradition: Vdovin 1961: 31, 35; 1970: 17; 1972: 110; 1975: 143; other traditions: Neiman 1871: 16; Vdovin 1961: 31–35; Dioneo 1895: 144; compare Leont'ev 1972:

believed that the ancestors of the Chukchi appeared on the Chukchi Peninsula in an environment of people of the Ust'-Belaya culture (proto-Yukaghir) who hunted on foot.

In the 1890s the Reindeer Chukchi translator Ainanvat told W. G. Bogoras this about the past of his people: "In the old days the Chukchi were so poor that they had only two or three or four reindeer each, the rich had 10 each, but they lived beyond Chaun and ate seal and walrus only in winter; if there was no moss close by the sea, those who had reindeer went farther. Now both herds and people have multiplied."[48] Having outlined the ancient region of settlement of the Chukchi to the east of Chaun, the informant indicated that in the old days each Chukchi had a few reindeer, evidently for transportation, whereas the economy of the coastal residents was complex: together with reindeer herding they hunted sea mammals. Indeed, only with the presence of a hundred head in a herd could a small family now live only by reindeer herding.[49]

There is no consensus among researchers concerning the time of transition of hunters of Chukotka to reindeer herding: some believe that this transition dates to the boundary between the 1st and 2nd millennium A.D., others indicated that the transition to reindeer herding as the basis of the economy occurred in the time most favorable for this, the period of cooling from the end of the 16th to the beginning of the 18th century.[50] It is evident that it was a matter of developing large-scale reindeer herding. Indeed, at the first encounter with Cossacks in 1642 the Reindeer Chukchi drove up to the Alazeya cabin to sell reindeer, that is, the Chukchi already had a surplus of reindeer that they could sell. In the words of G. F. Müller (1730s), the richest in Siberia were the Yuraki Nenets and Chukchi, who

86; Chukchi tradition: Bogoras 1900: 159, 164, 175–176; Kozlov 1956: 48–50 (the legend was recorded in 1955 in Enmelen from G. Tegryne by L. I. Solov'eva).

[48] AAN. F. 250. Op. 1. No. 124 / 10. L. 2; about Ainanvat, see: Kolesnitskaya 1971: 141.

[49] Kulikov 2002: 64. According to V. I. Zadorin (2009a: 123), a herd of 2,000 head could feed 15 to 20 people.

[50] The first opinion: Vol'fson 1991: 44; Mikhailova 1996: 166, 175; the second opinion: Vdovin 1950: 86 (end of the 17th–beginning of the 18th century); 1975: 152, 154 (15th–16th centuries); Simchenko 1970: 322 (second half of the 18th century); Syroechkovskii 1986: 62–63 (from the 17th century—sled reindeer herding, from the second half of the 18th century—large-scale herding); Krupnik 1989: 151–155; Dikov 1996: 28; Mikhailova 1996: 175–176; Zadorin 2009: 18 (15th–18th centuries).

possessed herds numbering more than 200 to 300 head, but by the 1750s the rich Chukchi now had 1,000 or more reindeer each, whereas the poor had only 10 to 20 each, and the average was several hundred animals each; in the 1820s the rich had 1,000 head each, while the poor had up to 100 each.[51] In general, during the course of the 18th century the size of the reindeer herds substantially increased. Later, in the first quarter of the 19th century, even the Settled Chukchi, having become richer, acquired reindeer and became nomadic. And in the second half of the 19th century, in the process of impoverishment of marine resources as a consequence of hunting by American whalers, some Settled Chukchi began to transition to small-herd reindeer herding, often giving their animals to the nomads to pasture.[52]

3. Wars of the Chukchi and Eskimos

In 1924, the Kamchatka merchant Petr Kosygin, who lived 12 years in Chukotka, told the Danish traveler Knud Rasmussen (1879–1933) this about the past of the country: "Initially the Chukchi were a mainland people; they arrived on the shore from within the country when there were no Eskimos there. The Chukchi were a reindeer herding people who knew nothing about maritime hunting."[53] This information is probably correct not in the sense that the Chukchi came out to the coast earlier than the Eskimos, but in the fact that the Chukchi appeared in those places where there were no Eskimos.

According to the supposition spread among researchers, Chukchi reindeer herders began to appear on the coast in the middle of the 1st millennium A.D., when also the first contacts between them and the Eskimos occurred. This happened in those places where pastures came into

---

[51] 1642: Belov 1942: 57; opinion: Miller 2009: 224; Kiber 1827: 197; compare: Merck 1978: 113 (the rich Chukchi had 300 female reindeer); Elert 1998: 126; 1999: 96, 121; KPTs. No. 70: 184; Bogoras 1991: 9, note 5. In the third quarter of the 19th century a rich Chukchi had 16,000 head, and a poor Chukchi 200–300 ("Notes on the Chukchi," 1877: 4). In 1756 Toion Leuchi, who lived beyond Kanchalan, had 10,000 head of reindeer (RGADA. F. 199. No. 528. Ch. 2. D. 3. L. 82 ob.; No. 539. Ch. 2. D. 6. L. 81).

[52] Kiber 1824: 96; 1827: 197; De Windt 1898: 216; Tolmachev 1911: 96; Bogoraz-Tan 1933: 243; Reuse 1994: 302.

[53] Rasmussen 1952: 144; compare: story by L. I. Ainana (b. 1934) in Anadyr' in October 2013 (see Nefëdkin 2016: 224); Galkin 1929: 40, 63.

contact with the coast: in Southeastern Chukotka between Mechigmen Inlet and Provideniya Bay, on the north coast of Chukotka, and also on the northwest coast of the Gulf of Anadyr'.[54] The Chukchi began to settle in Eskimo settlements, adopted Eskimo culture, gradually assimilating the Eskimos, and in general forcing the latter into places of difficult access.[55]

Figure 14. Asiatic Eskimos in festive dress before their yaranga. (Reproduced based on Rubtsova 1954: 550, Fig. 50).

In known Russian documents, military conflicts between the Asiatic Eskimos and the Chukchi are not noted, but they are recorded in folkloric sources. W. G. Bogoras supposed that clashes of the Chukchi with the Eskimos occurred in the 12th–13th centuries, while the Chukchi themselves are represented in folklore as wanderers without specific names.[56] According to Chaplino traditions, told by Ludmila I. Ainana (October 2013), the Chukchi, already being reindeer herders, did not once

---

[54] Vdovin 1950: 80 (3rd–5th centuries); 1975: 146–147; Arutyunov and Sergeev 1975: 195; Reuse 1994: 296 (the Chukchi reached Anadyr' in the 4th–5th centuries); Mikhailova 1996: 165, 183.

[55] Vdovin 1961: 38 (assimilation by the Chukchi and Koryak of the Eskimos in the 3rd–4th centuries), 49; Reuse 1994: 296 (settling of the Chukchi in the 16th–17th centuries); Mikhailova 1996: 183; compare: Zolotarev 1938: 81; Vdovin 1961: 27–63.

[56] Tan-Bogoraz 1930: 69–70; Bogoraz-Tan 1934: XXIII; pro: Dioneo 1895: 143; Vdovin 1961: 36; Gurvich 1980: 224; Mikhailova 1996: 172; compare: in the 12th–16th centuries: Arutyunov and Sergeev 1975: 196; Reuse 1994: 296.
[Waldemar Bogoras (his English name) was born Natan Mendelevich Bogoraz and for various reasons (he was Jewish and a dissident) changed his name to Vladimir Germanovich Bogoraz. At times he used the thinly veiled “pen names” of “Tan-Bogoraz,” “Bogoraz-Tan,” and “N. A. Tan.” Thus, the variety of names.—*Trans.*]

come to the coast for the purpose of attacking and seizing the property of the Unazik Eskimos. In 1926 someone indicated to the Yukaghir writer Tekki Odulok (1906–1938) a mountain area in the region of Cape Dezhneva and Naukan where clashes occurred between the Chukchi and the Eskimos. There Odulok himself found the occipital bone of a skull with a bone arrowhead stuck in it. According to Uelen traditions, the Chukchi—maritime hunters—arrived in Uelen from the north. More detail about this is given in a family tradition presented by the Chukchi writer Yu. S. Rytkheu (1930–2008): the ancestor of the author, Ermen, was the first to arrive in Uelen, eleven generations before. The latter, with his fellow tribesmen, attacked the neighboring Naukan people at night and seized their women. In the valley between the mountain chains the Chukchi repelled a retaliatory attack of the Eskimos at Ekven. After about a year pacification of the sides occurred.[57] If 30 years is accepted as the life span for one generation, the time of arrival of the Chukchi Ermen in Uelen is the first third of the 17th century. Even later, at the end of the 18th century, C. Merck named one of the dialects of the Eskimo language "Uelen," as if to emphasize that the main population of the village was still Eskimo.[58]

Figure 15. Eskimo family at their yaranga. Staroe [Old] Chaplino (1929). Photograph of the ethnographer A. Forshtein. From the MAE collection. (Reproduced based on: *http:// nabiraem.ru/adm_imgs/ 2695.jpg*).

[57] Tekki Odulok 1933: 36; traditions: Leont'ev 1972: 86; Rytkheu 2004: 9, 17–23; Nefëdkin 2016: 223–226. As is believed, the Chukchi began to settle in Uelen in the 17th–18th centuries (Alekseeva et al. 1983: 18).

[58] Merck 1978: 99; Bogoras 1934: 105, note 2; Dolgikh 1960: 551; Leont'ev 1972: 87; Chlenov 1988: 66.

In the winter of 1865–1866 the Eskimo Nokum, from a village in Provideniya Bay, told the following story about the Chukchi and Eskimos, in the English language to the American naturalist William H. Dall (1845 1927): "The deer men'' were the original inhabitants, and "bowhead men"... had come, long ago, from the islands (the Diomedes) to the northeast.. . . the reason they came was because there was war between them and the people who wore labrets. The latter proved stronger, and the former were obligated to come to the country of the "deer men". The latter allowed the "bowhead men" to settle on the barren rocky coast and formed an offensive and defensive alliance with them against the invaders from the eastward." In general, based on W. Dall's estimate, the emigration of the Eskimos from America to Asia occurred 300 years ago.[59]

The story clearly contrasts the Asiatic Eskimos and their fellow tribesmen who wore labrets, that is, the Alaskans. Researchers suppose that on the Diomede Islands the Eskimos spoke in the Eskimo Yupik language, which was then supplanted by the American Inupiaq language. It was evidently as the result of this war that the initial Eskimo population was forced to the Asian shore, where the Reindeer Chukchi summered, the latter permitting the Eskimos to live there. Having formed an alliance, the Reindeer Chukchi became allies with the new arrivals.

At the boundary of the 1870s and 1880s the local residents probably told the American ethnologist Edward W. Nelson (1855–1934) about this same event: "In ancient times the Eskimo of Bering Strait were constantly at war with one other; the people of the Diomede islands being leagued with the Eskimo of the Siberian shore against the combined forces of those of King island and the American shore from near the head of Kotzebue sound to Cape Prince of Wales and Port Clarence." If it concerns the same war, some details of the ongoing conflict can be identified: the conflict occurred between the islanders of the Diomede Islands and King Island; the former were supported by Siberians, and the latter by Alaskan Eskimos of the west coast of the Seward Peninsula. Judging by W. Dall, during the course of the war, alliances were evidently finally formed after the expulsion of the Eskimos and their resettlement on the coast of

[59] Dall 1870: 375. As Dall himself explained, the "Whale people" (bowhead men) were called this based on the name of the Greenland (bowhead) whale—the primary hunting prey of the Eskimos (Dall 1870: 375, n.).

Chukotka, where the Reindeer Chukchi also joined the alliance. On the basis of information obtained from Eskimos and Reindeer Chukchi, W. Dall calculated that the beginning of emigration of the Eskimos and the war occurred approximately in the middle of the 16th century. At least, during the voyage of S. I. Dezhnev "Chukchi" who wore labrets lived on two islands of the strait.[60] Subsequently, in the mid-17th century Alaskan Eskimos were already established on the Diomede Islands. Waves of resettlement from the American shore could also have occurred later; indeed, in 1779 there was an elder on Little Diomede who came from America, that is, the latter appeared on the island approximately in the second quarter of the 18th century. At least, on T. Perevalov's 1744 map the Diomede Islands were named Peek, which alludes to the name of the Naukan people (Peeki), with whom these islands were connected at that time.[61]

By 1718 islanders with a different language were in peaceful relations with the Siberians, as the Nose "Chukchi" told of this in Anadyrsk.[62] In the 18th century the Asiatic Eskimos were permanent military allies of the Chukchi, with whom strong economic ties were established. And in the case of great danger, which the campaigns of the Russians in the first half of the 18th century presented for the Chukchi, residents of Chukotka accepted the help of their former enemies—residents of the islands.

4. Fortification Structures

Settlements in Chukotka usually were not fortified; however, the village was constructed in a convenient place, on a hill on a cape projecting into the ocean, which could be easily defended, having discovered the approach of the enemy, and easier to see prey in the sea. For example, the

---

[60] Quote: Nelson 1899: 330; information of S. I. Dezhnev: DAI. 1851. T. IV: 21; Okhotina-Lind and Møller 2013: 87; compare: Dall 1870: 375; Polevoi 1967: 108; Efimov 1971: 86; on the date of resettlement, compare: Chlenov 1988: 68–69, 73–74 (19th–20th centuries); Mikhailova 1996: 172–173 (14th–16th centuries).
[61] On replacement of the population of the Gvozdev [Diomede] Islands: Chlenov 1988: 68–69; on the elder: Efimov 1948: 230; map of T. Perevalov: Orlova 1957: 125, Fig. 3: 126; Peeki: Bogoraz-Tan 1934: 8.
[62] AAN. F. 21. Op. 4. No. 33. L. 134–135; Efimov 1948: 224. ["Nose" refers to the residents of Chukotskii Nos [the Chukchi Peninsula].—*Trans.*]

early Eskimo village of Naukan was situated on a plateau 10–12 m above sea level. For the sake of security, the whole settlement could even be moved to a more favorable place.[63]

With direct threat of attack, the Eskimos constructed or remodeled a stockade, digging pits there for preservation of meat; and they set a watch on the hills on the side of the expected attack. However, on the whole fixed fortifications and even stone forts (Eskimo *umky*) served the residents of the village as a temporary refuge, in which the population took shelter in times of direct threat.[64]

In the coastal regions of Chukotka, a whole series of remains of fortifications are known. On a cliff at Nunligran that steeply drops into the sea were preserved the remains of an earthen fortification. The trapezoidal fortification, measuring about 10 x 7 m, was enclosed on three sides (except the side facing the cliff) by an earthen wall about 70 cm thick, which reached to more than a meter in height in the mid-20th century.[65]

It seems that more characteristic (or simply better preserved) were fortifications made of stone. G. A. Menovshchikov thus described the typical Eskimo fort: in a wall formed of stones with a height a little more than that of a man there were two kinds of loopholes: narrow for shooting from a bow and large for dumping stones. The entrance into the fortification was closed by a fence and a stone slab. On the coastal mountain Sanluk (Chukchi Chenlyun) between Naukan and Uelen, at an elevation of more than 50 m and in a saddle between a steep cliff on the east and another summit on the west, is an area 50–60 by 105–110 paces, the entrance to which is along the southeastern talus slope facing the sea; the fortification is enclosed on the two sides undefended by precipices by walls formed of local stones without mortar 1.5–2.0 m wide and a height preserved in the mid-20th century of about 1.0 m. On this area were seven pit houses and the remains of several round bases of yaranga dwellings. If

---

[63] Baboshina 1958: 166; Merck 1978: 106; Naukan: Chlenov 1988: 71; movement of a settlement: Vakhtin 2000: 124.

[64] Repair of fortresses: Menovshchikov 1985: 321–322; Chaplino traditions told by L. I. Ainana (Anadyr', October 2013; see Nefëdkin 2016: 224–226); watches: Menovshchikov 1985: 307, 322–323; strongholds for refuge: Bogoras 1900: 389, compare: 332–333.

[65] Report by A. N. Kergintval' (b. 1945) from Anadyr' (31 January 2014, St. Petersburg).

my correlation of the Naukan historical tradition named "Invasion" with the events of D. I. Pavlutskii's campaign in Chukotka in 1744 is correct, then the Sanluk fortification was made at that time. On Cape Chin'gan at the edge of Sinrak village a ringed fortification with walls 1.5 m high formed of boulders was preserved. Since the village was abandoned and the residents resettled in Sireniki, in I. I. Krupnik's opinion, at the boundary of the 18th and 19th centuries at the latest, then the fortification also should be dated earlier than this time. We also see a fortification in the form of an open circle of stacked stone in an engraving on tusks, executed by residents of Chukotka.[66]

A fortification was also preserved in the interior of Cape Chaplino, on the high hill of Guigungu three kilometers from the remains of the Eskimo village of Uniyramkyt. It is an oval area 20–50 m wide and 70 m long, lined by a solid wall of unworked stone 1.0–1.5 m high with the entrance on the west side. In the center of the fortification, on a natural rise, there is a low round tower. Another kind of fortification was preserved on the same Cape Chaplino, on the top of the low hill Rygnakhpak. At present, it is not a complete whole, but rather consists of a series of individual angles or cells lined with unworked stone that turns in the southwest into a trench. From this fortification, the vicinity could be easily viewed, and it served precisely for the observations, as suggested.[67] However, in order to track an enemy, it is not necessary to build a fortification. It is more probable that it served as a place of defense. Indeed, even if it was not a solid wall, because of the angles a crossfire could be carried out at the enemy and thereby not permit them into the fortification. The tower served as the last stronghold of the defenders.

The fortifications on Cape Chaplino are located near river valleys and thus served as obstacles for movement from the central part of the peninsula to the coast at Unazik (Staroe Chaplino). Though it is supposed that these fortifications were erected against the Cossacks in the 18th century,[68] the Cossacks used seize shields when material was available,

---

[66] Description of a fortification: Menovshchikov 1987: 171, note 3; Sanluk (Senluk): Dikov 1958a: 41; 1977: 173–175; Naukan tradition: Nefёdkin 2003: 307–310; Sinrak: Krupnik 1983: 70; images on tusks: Antropova 1957, Fig. 34; Shirokov 1968: Fig. 7; Fisinina 1995: 78, Fig. 53.
[67] Arutyunov et al., 1982: 72–74, Fig. 61.
[68] Arutyunov et al., 1982: 74.

which permitted them to quickly take such forts, and the fort could have been built against local opponents. Indeed, according to Unazik traditions, told by the informant and Eskimologist L. I. Ainana in October 2013 in Anadyr', these fortifications were used against the Reindeer Chukchi, who carried out raids on the Eskimos.

On the whole, it can be said that the coastal residents having certain skills in constructing permanent dwellings, also used their skill for erecting permanent fortifications from improvised material: from unmodified stone. In distinction from the Koryak and Itel'men, the residents of Chukotka coast did not practice digging ditches and erecting palisades, which is explained by the lack of structural material, and since the population was not numerous, it was incapable of erecting monumental structures. Forts were located on hills dominating the surrounding locality. Fortifications were erected and remodeled in case of direct danger—the fortification itself was a simple refuge, where the residents tried to wait out the raid, since, as a rule, the latter was brief. Construction of fortifications in Chukotka was possibly a late phenomenon brought about by the Russian penetration into the region. Indeed, local warfare usually did not imply assault and thereby siege of a fortification.

* * *

In conclusion, it must be noted that in this chapter I am deliberately avoiding a detailed discussion of such important problems of Chukchi history as the arrival of the Chukchi onto the peninsula, questions of time and location of the appearance of reindeer herding, and the arrival of the nomads onto the coast—all this would have led away from the military-political history. The development of Chukchi culture, like other early cultures, was determined by changes in climate. Archaeological cultures of the coast of Chukotka were created by the Eskimos and their ancestors, and not by the Chukchi. The latter are represented in Eskimo folklore as reindeer herders, which, incidentally, can be a simple anachronism. Initially, interaction of the Chukchi with the Eskimos bore a confrontational character, but the trade established later between the coastal residents and the nomads contributed to consolidation of the local dualistic model of the economy. The settled Eskimos and nomadic Chukchi became allies in the struggle both with neighboring peoples and with the Russians, against whom the bulk of the preserved fortifications in

Chukotka were probably erected. Indeed, local warfare implied not the siege of fortifications, but rather open battle outside the dwelling, which was under the protection of the patron spirit.

CHAPTER 2

MILITARY ACTIVITY IN THE REGION OF THE KOLYMA RIVER AND CHAUN BAY (17TH TO THE BEGINNING OF THE 19TH CENTURY)

1. Appearance of the Chukchi in the Great Tundra between the Alazeya and Kolyma Rivers

Since the Chukchi appeared in the Great Tundra between the Alazeya and Kolyma before the arrival of the Russians, the written history containing information about this even is lacking. However, some information from Chukchi folklore can be connected with this migration. As early as 1944, I. S. Vdovin proposed that the Chukchi legend about the flight of the elder Krekhai from Chukotka to the west reflects precisely the resettlement of the Chukchi to the interfluve between the Alazeya and Kolyma rivers.[69] We will examine the evidence about the main character of this legend.

In 1791, uncertain information about the resettlements in the region was successfully obtained by Dr. C. Merck (1761–1799), who passed through Chukotka as part of the J. Billings expedition:

> Among the Chukchi there is a legend that in antiquity, when they were much more numerous and conducted internecine wars, one small group consisting of consanguineous Chukchi fled from Cape Shelagskii over the ice with their baggage and reindeer in the direction of the mainland and spent only one night on the ice; the strait in this place was so narrow, that in clear weather it was possible to see the mainland from Cape Shelagskii. . . . In passing, the Chukchi tell that some of their kindred crossed in winter to the mainland in one day; there they

[69] Vdovin 1944: 254–255; pro: Dolgikh 1960: 409.

> found people who spoke their language and possessed reindeer, but since the latter did not inspire trust in them, they did not remain among them to spend the night and secretly returned home.[70]

The information was most probably obtained by C. Merck from Reindeer Chukchi, with whom the expedition of J. Billings traveled through Chukotka. The events of the flight were perceived at the end of the 18th century as occurring in antiquity, in a distant period of internal warfare, and consequently the information bore a folkloric character. According to legend, one clan of the Reindeer Chukchi fled. Those who fled spent one night on the ice, that is, the whole route lasted at a minimum of two days. The direction of the flight, implicit in the legend, was the land opposite Cape Shelagskii, that is, the legendary northern land, by which can be understood either Wrangel Island, or a cluster of ice ridges taken by the Chukchi as land. Merck did not provide any names of the participants; this flight of the reindeer people could be assigned both to the group of Krekhai and to a later emigration of the Shelagi people. Also, the Chukchi told that some of them reached these fugitive fellow tribesmen in winter on sleds. This information is evidently assigned to the recent past, though it could also be assigned to folklore. Since it is unclear where the travelers left from and where they arrived, it could be a matter of going on a winter trip to Alaska or to the islands in Bering Strait, where Eskimos lived, that is, from the point of view of the Europeans, "Settled Chukchi," whose language was understood by the Asiatic Eskimos. We note that the "Chukchi" arrived at the reindeer people, which also can be explained by the legend of the flight of Krekhai or the Shelagi.

Later, in 1823, F. P. Wrangell (1796–1870), through a translator, heard a more detailed account from the Chukchi of the north coast of the Chukchi Peninsula, the content of which he delivered in this way:

> According to legend, 200 years ago the Onkilon people occupied the whole Asian shore from Cape Shelagskii to Bering Strait. . .. Severe hostility between the Onkilon elder Krekhai and the chief of the Reindeer Chukchi Errim turned into civil war. Krekhai was defeated and fled with his people. From that time, the shore has been deserted. The residents of the village of Ir-Kaipii, where, according to legend, Krekhai hid for some

[70] Merck 1978: 99.

> time, told me the following about him: he killed the Chukchi elder Errim and was pursued by his son; for a long time, he wandered and finally hid on Cape Ir-Kaipii, where up to now has been seen a natural wall, behind which he settled. But the young Chukchi Errim, craving vengeance for the death of his father, found a way to break in there and kill the son of Krekhai. Though, according to the understanding of the residents here, the blood revenge was satisfied, Krekhai probably nevertheless had reason to fear pursuit by his implacable enemy. Therefore, at night he descended the cliff on a thong to a boat ready for flight and, thinking to deceive his enemies, sailed first to the east, but then turned to the west and reached Shalaurova Island, where he fortified himself in a pit house, the ruins of which we saw. There little by little all his relatives gathered and soon escaped in 15 baidary to an unknown land that can be seen from Cape Yakan on clear sunny days. In the following winter another Chukchi fled, a relative of Krekhai, with his household and reindeer; they think that he also departed to the same overseas land.[71]

As we see, F. P. Wrangell rather clearly presented the historical tradition of the departure of the Onkilon population from the northwest coast of Chukotka. The robustness of the legend was also provided by the geographic binding of the event. Based on the place of residence on the north coast of Chukotka, F. P. Wrangell named the coastal Asiatic Eskimos as the Onkilon people (from the Chukchi *an'k'al'yn*—"coastal"). Though the traveler does not identify the ethnic association of the Onkilon, he nevertheless speaks of a separate people. Judging by C. Merck's description, by the end of the 18th century the Uelen Eskimos occupied the north coast of Chukota to Cape Shelagskii. If F. P. Wrangell gave the name of the leader of the Onkilon correctly, he turned the social status of the head of the Reindeer Chukchi—*erym*—into the proper name of the second hero and his son—Errim.[72] The name of the leader of the fugitives Krekhai is clearly connected with a Chukchi ethnonym. In 1765, after a trip through Chukotka, N. I. Daurkin created a map, where in the northern land is written: "There are also reindeer people on the north side, and they, the Chukchi, call them Khrakhai. And these people have spears and knives

[71] Wrangell 1948: 311; cf.: Dioneo 1895: 145; Kondratov 1983: 155–157.

[72] Merck 1978: 98–100; Dolgikh 1960: 551; about names: Burykin 2000: 52; 2006: 199.

of copper, and their copper is red; and they speak in Chukotski."[73] Explanation of this description must come from the Chukchi understanding of geography, according to which the northwestern part of Alaska continued to the north from Chukotka. That is why residents had copper weapons, known particularly in Southeast Alaska.[74] The very presence in this northern land of the Khrakhai, who speak Chukchi, can be assigned precisely to the same legend about the flight of the reindeer people to the north. In addition, on this map of N. I. Daurkin is designated the island Nym-nym (translated from Chukchi: "village"), located opposite Chaunskaya Bay. The island is incidentally no longer on a map created by this same author in 1771. As the American linguist M. Krauss proposed, this is Wrangel Island.[75] The same M. Krauss compared the ethnonym Khrakhai with the name of a later leader of the Shelagi, Kopai. However, as linguist A. A. Burykin indicated to me in a personal communication, these two names are not etymologically connected. In addition, the Khrakhai were never matched with the Shelagi, and Kopai is a historical person, in distinction from the semi-legendary Krekhai (Khrakhai). A. A. Burykin connects the name Khrakha itself (Khrokha or Khryakha) with the hypothetical ethnonym *ak'oraka*—"those not having reindeer," which, the scholar believes, coastal residents, Chukchi or Eskimo, were called.[76] However, such explanation contradicts the description of N. I. Daurkin, in which the Khrakhai are called reindeer people.

We return to F. P. Wrangell's description of events: a conflict emerged between the leaders of the Reindeer Chukchi and the sedentary Onkilon, developing into an open confrontation, as a result of which Krakhai and his fellow tribesmen fled, leaving home. The residents of Ryrkaipii, where the final stage of the events unfolded, told F. P. Wrangell the local legend: fearing blood revenge, Krakhai had to flee—first to

---

[73] RGADA. F. 199. No. 528. Ch. 2. D. 3. L. 35; cf.: Argentov 1861a: 30; Gel'mersen 1876: 462; Berg 1946: 109; Grekov 1960: 209, Fig. 39; Efimov 1964: Nos. 128, 131; Vize 1933: 11.

[74] About the use of copper needles by the Kygmit people: KPTs: 185; about copper in Alaska: Wrangell 1835: 612; Berg 1946: 111, note 3.

[75] Burykin 2000: 51; Krauss 2005: 172, 177; Efimov 1964: No. 128, point No. 116; No. 130.

[76] Krauss 2005: 173; Burykin 2006: 193.

Schmidt's Cape (Ryrkaipii), which, joining the mainland by a gravel isthmus, rises above sea level 80–90 m. The eastern slope of the isthmus is precipitous, while the western is gentle; on it were discovered ten pit houses of late Punuk time. There the expedition of A. E. Nordenskjöld conducted exploratory excavations as early as 1878. In 1957 archaeologist N. N. Dikov discovered on the edge of the cape an area of 100 sq. m. lined with a wall of stones to a meter high; lower along the slope three groups of dwellings were arranged; and in 1975 on the top of the Kozhevnikova Cliff, on the north of the cape, a pit house and three cache pits lined with stone were found, in one of which was a harpoon of Punuk times. It is also believed, according to local tradition, that the last pit house was the dwelling of Krekhai. Then Krekhai fled to Shalaurov's Island (1.2 km from the shore between capes Shelagskii and Billings near Nol'de Bay), where the expedition of F. P. Wrangell discovered pit houses. Here the relatives of Krekhai gathered, having filled 15 baidary. Considering the baidary as large campaign ones—20–30 people each on a boat, not counting livestock—their total would have been from 300 to 450 people. Further, the adherents of Krekhai moved to an unknown land, as the Chukchi themselves proposed—to Wrangel Island, which could be seen from Yakan (Billings' Cape); a family of Reindeer Chukchi also went there later in winter over the ice. This legend, reflecting the Chukchi view, is supported by the presence of the remains of a dwelling found in 1937 on Thomas' Cape (southwest of Wrangel Island), where among other artifacts a blue bead was also discovered, which clearly attests to the late residence of people at this place.[77]

Another member of F. P. Wrangell's expedition, Dr. August-Erich Kiber (1794–1855), also left a brief report about this movement: "During our journey in Kulichin-bai we found traces of such dwellings and heard that they belonged to some tribes of sitting Chukchi, who, waging war with the Reindeer Chukchi, sought refuge in them; but finally, in time moved to an unknown country, lying beyond the sea. The Chukchi still clearly

[77] Archaeology: Dikov 1977: 201; Tein 1983: 4, 6; about these remains of dwellings discovered by F. P. Wrangell: Nordenskjöld 1879: 221–222; 1881: 434; Shental'skii 1978: 42–50; remains of dwellings on Wrangel Island: Gromov 1941: 75–77; 1957: 155–156; Akkuratov 1984: 97; Kuvaev 1988: 239; cf.: Tein 1983: 25; Shalaurov's Island: Beregovaya 1954: 312.

remember this occurrence." And in February 1823 at the fair in Ostrovnyi, the Chukchi Yatgirgin told that "to the north lies an uninhabited land, where the Chukchi Karkhai moved with all his belongings and reindeer."[78] A. Kiber confirms F. P. Wrangell's information about the expedition finding empty pit houses on the north coast of Chukotka and about the presence of the established tradition that in them lived sedentary "Chukchi," who fled beyond the sea to an unknown land because of war with the Reindeer people. This story evidently refers to Krekhai, who fled with his family and herds to some uninhabited northern land.

Later, in 1844 and 1854, the Chukotkan missionary Andrei Ivanovich Argentov (1816–1896) heard legends from the Chaun Chukchi about two cousins: "Once in Yakan lived Erruva and Krepkai—two patriarchs. They were of restless disposition, did not get along with their neighbors and, having quarreled with everyone, went beyond the sea. They set off in flight from Yakan directly to the mentioned island: first they followed a sled route, and then, upon reaching the water, traveled in two baidary with all their fellows together with the dogs and reindeer." In another place, A. I. Argentov presents the story differently, basing it on another variation of the legend: two Chukchi, or Shelagi captives maintained by the Chukchi, Erruva and Kreekai, were servants of a rich reindeer herder, who was left by them on an island. Having been there several weeks, he was saved by a baidara crew. At this time, Erruva and Kreekai, declaring that the master had drowned, seized his belongings and rallied a certain group of people around themselves. Fearing revenge, they fled from Yakan, first over the ice on reindeer, and then in 16 baidary.[79]

Since A. I. Argentov learned the Chukchi language, he could hardly have misunderstood the Chukchi. He probably recorded local Chaun variations of the legend in which former enemies were turned into like-minded people who fled from Yakan to Wrangel Island because of a dispute with their neighbors. The fugitives now travel in two baidary, including dogs and reindeer, that is, animals of sedentary and nomadic residents. This trip probably corresponds to the first flight of Krekhai with his not numerous (40 people) followers to Shalaurova Island related by

---

[78] Kiber 1824: 123; 1827: 197.

[79] The first story: Argentov 1857: 82, cf.: 105; the second variant of the story: Argentov 1861: 612; 1861a: 19–20.

F. P. Wrangell. In the second variant of the legend, which examined the social context of the conflict, the slaves or workers Erruva and Krekhai, having seized the belongings of their master, fled with their followers in 16 baidary, that is, there were about 130–320 dissatisfied people. There was possibly some contamination of different legends into one—about the abandoned master and Krekhai's flight beyond the sea that is close to the Ryrkaipii legend, as told to F. P. Wrangell. Both heroes now were either Chukchi or captive Shelagi who grew up among the Chukchi. The latter circumstance is a definite chronological indicator: the Shelagi were still in the region or had just disappeared (that is, events before the second quarter of the 18th century).

In the spring of 1870, the Chukchi of Cape Shelagskii told Baron Gerhard von Maydell (1835–1894) the following:

> Previously Settled Chukchi lived there (at Cape Shelagskii) who spoke another language than the current ones . . . they were called . . . Ankali, since they lived on the shore of the sea. . .. The Chukchi here heard from their ancestors that at the cape and Chaunskaya Bay a battle occurred, but they positively denied that the departed part of the population moved to unknown lands beyond the Arctic Ocean. They knew well that in this last was a land that they all saw from Yakan to the east from Erri, but they assert just as positively that it is impossible to land there in baidary, and what is more, those countrymen who left also took herds with them. . .. The Chukchi also told that their vanished countrymen set off to the west along the shore and settled first on the Medvezh'i Islands, but then they moved again to the mainland and went still farther to the west; where they finally stopped, the Chukchi here do not know.

The residents of Yakan reported to G. Maydell that "they do not know about the fact that in earlier times some horde, as a consequence of a struggle, moved to Wrangel's land [Wrangel Island]; the struggle and battles, of course occurred there, and the hordes departed; in this case, they always set off to the west and first, went across the Kolyma to the Indigirka, and from there passed to the Medvezh'i Islands."[80]

In general, in the legends told by the Chukchi from Capes Shelagskii and Billings, two facts of history are reported; with this, the

[80] The first quotation: Maidel' 1894: 273–274, cf.: 517–518; the second quotation: Maidel' 1871: 62, cf.: 69.

participants are not named, and it is assumed that those who fled were Chukchi: 1) earlier on the northeast coast of Chukotka lived "Chukchi"-Ankalin (=Onkilon) of different language, that is, Eskimos; 2) a group of Reindeer Chukchi who lived at Chaunskaya Bay and Cape Shelagskii fled. Information about old battles in the region could be assigned to any of the three wars with the Russians or with the Shelagi, or with Krekhai. Since Chukchi contemporary with G. Maydell had a better understanding of geography, they began to critically assess also what happened: because of wars the Reindeer (!) Chukchi did not go to Wrangel Island, but drove their herds to the west, going along the shore, passing the Kolyma, reaching the Medvezh'i Islands, and farther again moving to the mainland. The fugitives had to pass the Indigirka later, during their migration to the west. The visit to these islands in fact could be an episode of the flight of the sedentary population; indeed, the remains of structures were discovered on Chetyrekhstolbovoi Island, in which the Eskimo population lived before the boundary of the 1st and 2nd millennium; later, in the 17th–18th centuries, the dwellings were only sporadically visited during hunting, as Kaliba told about this.[81] This route could correspond to the course followed by the Reindeer sympathizers of Krekhai, though, on the other hand, this information might be the mixing of events of the flight of the Shelagi, in favor of which speaks the territory where the conflict occurred—Chaunskaya Bay.

After a quarter of a century, in the time of W. G. Bogoras's (1896) investigations, a variant of the legend narrated by the Kolyma Chukchi about Krekai, told that the latter unleashed the Koryak on the Chukchi, because of which the hero himself was forced to flee on reindeer to the Great Land (Alaska), where his descendants became immortal.[82] As we see, this variant of the legend transformed into a tradition, where the Koryak, eternal enemies of the Chukchi, became the participants, and Krekhai himself fled now not to a northern island, but beyond Bering Strait, to Alaska, which was well known even by the western Reindeer Chukchi at the boundary of the 19th and 20th centuries. And again, the story about this flight could also be the result of contamination of different variants of the legend.

---

[81] Raushenbakh 1969: 39, 45–48; Okladnikov and Beregovaya 1971: 22.
[82] Bogoras 1900: 388–389; cf.: Bogoraz-Tan 1934: 10.

In 1941, on Wrangel Island, the Eskimo Anakul' told Arctic pilot Valentin Ivanovich Akkuratov (1909–1993) that "to the west is Cape Shelagskii, to the east is Schmidt's Cape. Grandparents say that the battle of the Onkilon with the Chukchi occurred here. And here, from Cape Shelagskii, they went in baidary in the ocean."[83] Consequently, Anakul' (that is, Anal'ko of Provideniya Bay) believed that the war of the Onkilon with the Chukchi occurred along the north coast of Chukotka and they sailed off to the north from Cape Shelagskii.

After 35 years, the Naukan Eskimo archaeologist Tasyan S. Tein (1938–2005), conducting excavations on Schmidt's Cape in 1975, provided this folkloric material about the pit house and three cache pits found there: "The Chukchi-Eskimo legend states that once upon a time Krekhai and his family lived there in the upper pit house—not Eskimo and not Chukchi, but a man of unknown nationality, who arrived here from the Anadyr'. He ran over the tundra and mountains, paving a way that was called the 'track of Krekhai.' But he was at odds with the local residents who lived below and, if they tried to attack, he shot people with the bow. Then Krekhai launched the baidara in the sea and together with his family fled in an unknown direction. It was supposed that he went to Wrangel Island."[84]

In general, the legend has undergone further evolution. Now Krekhai becomes a foreigner who arrived from Anadyr', that is, a Koryak, which, it seems, corresponds with the information of W. G. Bogoras about the calling of the Koryak to help, or the Yukaghir (the Yukaghir in historical times occupied the Anadyr' Basin). Krekhai now acts alone, only with his family, nothing about a large migration. The hero was in constant conflict with the local residents and was forced to flee in a baidara in an unknown direction, perhaps to Wrangel Island. To the historicity of the personage attest not only his pit house, but the "track" or road laid among the rocks.

On the whole, it is evidently possible, following I. S. Vdovin, to believe that the flight of Krekhai is the folkloric memory of the migration of the coastal settled and reindeer populations of the Chukchi to the west. These traditions were based in first order on *memoris loci*, on the memory

---

[83] Akkuratov 1984: 95; Burykin 2013: 452.

[84] Tein 1983: 6.

of concrete geographic objects—the remains of dwellings of sedentary coastal residents; naturally, the material evidence of emigration of the nomadic reindeer herders did not remain. Therefore, the folklore preserved some striking evidence about the emigration. If the settled followers of Krekhai could easily flee to Wrangel Island, the reindeer herders could hardly drive their herds over the ice, where there was no food for the reindeer, so movement most probably went along the coast.

Advancing his hypothesis about the origin of the Kolyma-Alazeya Chukchi, I. S. Vdovin did not turn attention to information from traditions about the fact that it was not only and not so much reindeer herders who moved but also settled residents of Chukotka. Traces of the latter should also have been recorded somewhere. In fact, mention of them can be found in early documents. The Yakutsk serviceman M. V. Stadukhin, who visited the Chukoch'ya River, related in his report in 1646 what he himself saw: "And along this Chyukhchi river live their kind, called Chyukhchi, also the Samayed, sedentary Reindeer. . .. And among the Chyukhchi there are no sables, because they live in the tundra by the sea."[85] From the captive Kolyma woman Kaliba, who lived three years among the Chukchi, M. V. Stadukhin obtained additional information: "And the Chukhchi on this side of the Kolyma go on the reindeer from their dwellings at this stream in winter across to the island in one day. And at that island they kill sea mammals walrus and bring to themselves walrus heads with all the teeth. And in their way pray to the walrus heads."[86]

[85] The terms "Chukoch'ya," "Chyukhchi," "Chukhchi," and others refer to the Chukchi.—*Trans*.

[86] Orlova 1951: 221–222; Belov 1952: 59–60; cf.: Miller 1996: 23; Krasovskii 1894: 155; Zuev 2002: 194. A. A. Burykin (1998: 83; 2004: 38–39; 2006: 71–74) believes Kaliba was not a Yukaghir woman, but rather a coastal Chukchi by the name Kelev'i. However, we note that Kaliba only lived among the Chukchi three years, and was taken hostage on the Kolyma from the Yukaghir (Belov 1952: 59).

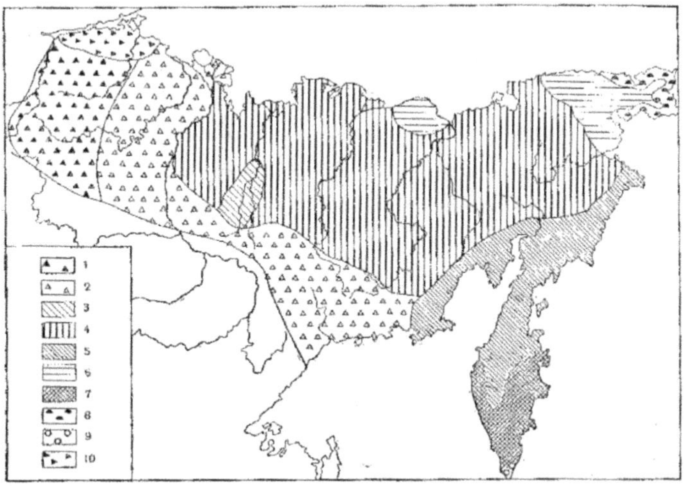

Figure 16. The resettlement of people in northeastern Siberia in the mid-17th century. 1—Tungus (Evenk); 2—Lamut (Even); 3—Yakut; 4—Yukaghir; 5—Koryak; 6—Chukchi; 7—Kamchadal (Itel'men); 8—Eskimos; 9—Kuril people (Ainu); 10—Tavgii people. (Reproduced based on Gurvich 1966: 267).

It turns out that both Reindeer Chukchi, who are compared in the document with the better-known Samoyed at that time, and sedentary residents lived in the tundra by the Chukoch'ya River.[87] Based on the words of Kaliba, in winter the Chukchi went to one of the Medvezh'i Islands to hunt walrus and then celebrate the festival of walrus heads in honor of Keretkun. This was one of the most important hunting festivals of the coastal residents, being celebrated at the start of summer before the beginning of the maritime hunt. During this festival (in Uelen) an experienced hunter went to the shrine, where several walrus skulls with tusks were set out, and the hunter asked one of the heads about the success of the upcoming hunt; in case of a successful hunt the head of a newly killed animal was added to the previous. As W. G. Bogoras noted, still in the 19th century the Chukchi were going in spring over the ice to the Medvezh'i Islands to hunt walruses.[88] Perhaps Sergeant Stepan Andreev,

[87] W. G. Bogoras (1921: 58) noted that the western Chukchi, according to Cossack reports, were Reindeer and Settled. About the Chukchi population of the north coast of Chukotka see: Burykin 2013: 453.

[88] Bogoras 1921: 54; Bogoraz-Tan 1934: 33; 1939: 84, 95–96; Vdovin 1944: 253–254, 262; Gurvich 1966: 48; cf.: Merck 1978: 135 (about seal skulls); Nordenskjöld 1881: 435. About the festival of walrus skulls at Uelen: Leont'ev 1972: 90; also see: Vdovin 1961: 50; 1977: 158; Tein 1992: 17; Tevlyankau 2011: 48–49.

on the way from the Medvezh'i Islands on 22 April 1764, having seen traces of eight reindeer sleds on the ice that belonged to Chukchi hunters, turned his expedition back, fearing encounters with the Chukchi.[89] The sedentary residents in the west probably fell with time into decline because of unfavorable natural conditions—a cooling—whereas the reindeer herders, by contrast, progressed.

Of all the sources that mention Krekhai, only F. P. Wrangell gives the chronology of the events: 200 years ago, that is, the first quarter of the 17th century. The traveler himself did not indicate where he had obtained the date; it can only be supposed that he made calculations based on information he received based on the number of generations that passed from the event.

2. First Contacts of the Western Chukchi with the Alazeya Yukaghir and the Russians

Although the Reindeer Chukchi occupied the tundra territories generally not needed by Yukaghir hunters, it nevertheless seems that from the beginning relations of the natives of the Alazeya and the new arrivals did not develop peacefully. This can be clearly judged based on Yukaghir folklore. Thus, we learn that the Alai hero Edilvei (Idilvei) was born on the Bol'shaya [Big] Chukoch'ya River. His parents were killed by Chukchi who took his father's weapons. The hero, as an adult, set off in the search of the weapons. In total, these Chukchi came three times from the east along the seacoast. They took the women and children captive and forced them to tend their reindeer. Even the wife of Edilvei fell captive and became a servant. Having caught Edilvei, the Chukchi almost sacrificed him to the land. The hero himself periodically fought off the Chukchi, who were tempted by his iron armor. If it is considered that the Chukchi did not yet have iron, it can be imagined what value this armor represented for them! However, the hero was so strong that the Chukchi feared him and did not want to test their strength against him. Only the son of Edilvei, Elnge, was able to establish peaceful relations with the

[89] Sokolov 1852a: 110; Gel'mersen 1876: 471; Zubov and Badigin 1953: 97–98, 119. Incidentally, in 1769 companions of S. Andreev—Yukaghir Efim Konovalov and Lieutenant Fedor Tatarinov—officially stated that they did not see any traces of sleds (Zubov and Badigin 1953: 109–110).

Chukchi. Initially Elnga, fearing the Chukchi, selected a hill for his house. When the Chukchi could not kill him, they, respecting his strength, proposed making peace with him, and then they fled. Therefore, this hill in the Alazeya tundra began to be called "The Place of Flight of the Chukchi"—this explanation for the name of the locality came about at the end of the 19th century. The tundra locality of Elnga was near Irie Mountain to the west of the Alazeya. This locality is evidently connected with the region of the action in the story.[90]

We note that the Chukchi, based on Alazeya Yukaghir traditions, arrived from the east, from beyond the Kolyma, and carried out three campaigns on the Alai. The first generation of Yukaghir was killed by them, the second found itself in a stage of confrontation, and only the third was able to make peace. In the 1640s relations between the local Yukaghir and Chukchi were not unfriendly. If the folklore is followed, then, presuming that only in 1640 the Chukchi and Yukaghir made peace, it can be thought that relocation of the Chukchi occurred at least three generations ago (the parents, Edilvei himself, and his son), that is, with an absolute date of the second half of the 16th century.

In 1642, the first Russian explorers appeared on the Alazeya River. Here for the first time they encountered the Chukchi. In summer two ships with fifteen servicemen under the leadership of the Yenisei Cossack Dmitrii Zyryan (also known as Yarilo) after a two-week voyage under sail arrived from the Indigirka to the Alazeya. In a petition to the tsar, Cossack Ivan Erastov, a member of the campaign, thus described the course of the expedition:

> And having reached, Majesty, we, your servants, the Alazeya River, and we, your servants, encountered at the Alazeya River many Alazeya people of the princes Nevgocha and Mundita. And with them, Sovereign, were Chyukhchi men from the tundra with their families and with kin. And we, your servants, told them about your Tsar's Majesty and your kind words, that they, the Alazeya men, might be obedient and submissive. . . . And taught them your request, Sovereign, for tribute in the 151st year. And they, Sovereign, the foreigners, the Alazeya Yukaghir and Chyukhchi, refused to pay tribute to your

[90] Stories about Edilvei: Okladnikov 1975: 226–230; Kurilov 2005: 126–179; about Elnga: Jochelson 1900: 210–211; Kurilov 2005: 526 (about the locality), 564–565.

> Sovereign and passed along both sides of the Alazeya River, and on both sides, they, the Alazeya, began to shoot at us, your servants. And we fought with them a close combat battle the whole day until evening.

Suffering losses, including one of the princes [*knyazech*] Nevgocha, the Yukaghir fled. The Cossacks had nine wounded.[91]

The story by the Lena Cossack Fedor Chukichev of the same detachment of D. M. Zyryan about this campaign points only to the battle with the Yukaghir; the Chukchi are not mentioned at all: “We ran along this sea by sail to the mouth of the Alazeya River, up it by sail and by foot to the Yukaghir Prince Noochichan for half of the third day. And with this prince with a Kovyma [Kolyma?] shaman, there were three hundred men of their people, and on this the Alazeya River they were also encountered by the servicemen, there was a battle, and in this battle nine servicemen men were wounded, and we killed many of them and took in this battle an *amanat* [hostage] man of the Yukaghir tribe by the name Shaman.”[92]

S. I. Dezhnev also participated in this battle. After one or two days, he arrived by sea at the Alazeya from the Indigirka in composition of the small (13 man) detachment of M. V. Stadukhin and was sent ahead with another Cossack for negotiations with D. M. Zyryan for the purpose of uniting the two detachments. Relating in the report to the tsar in 1662 about his past services, S. I. Dezhnev mentions only the battle with the Yukaghir: “And I, your servant, with him, Dmitrii, found the non-tribute-paying Alazeya Yukaghir men, took as amanaty the son of Prince Mantsit, Toitu, and with these Yukaghir in this place we, your servants, there had a battle, and I, your servant, was wounded in the left shoulder by an arrow.”[93]

As we see, at the lower reaches of the Alazeya the Russians encountered the Alai Yukaghir and people from a Chukchi nomad camp where their families and “captives”—probably all the other inhabitants of the nomad camp—were. Since the Russians arrived unexpectedly, the Chukchi could hardly have quickly appeared specially to help the

---

[91] Ogloblin 1903: 54–55; Orlova 1951: 134–135; cf.: Belov 1952: 57; Ivanov 1999: 102; Zuev 2002: 191–192; Nefëdkin 2003: 22.

[92] Report of F. Chukichev: Belov 1973: 65; cf.: Belov 1952: 57 (with a lacuna in the necessary place in the text).

[93] Report of S. I. Dezhnev: Samoilov 1945: 143.

Yukaghir. They and their families simply happened to be there based on their affairs or were simply on their summer wandering by the shore of the sea, not far from the Alai nomad camp. Both peoples rejected the demand of the Cossacks who had arrived to pay tribute. Evidently the servicemen attempted to take amanaty and captured the son of Prince Mantsit (probably Mundit, mentioned by I. Erastov). In the course of the battle the Yukaghir, using their numerical superiority, began to surround the Russians, carrying out intensive fire from bows. The clash lasted until evening; this clearly indicates the fact that the combat was distant, and not close, in which 17 servicemen (Cossacks of D. M. Zyryan and two newly arrived servicemen) could scarcely long oppose 300 Yukaghir. The Cossacks probably shot back at the attackers, with a large part of the Cossacks being wounded. One of the nine wounded Cossacks was S. I. Dezhnev. That the Chukchi participated in the battle is not mentioned in the documents, evidently as a consequence of the fact that they either actually did not take part in the fight, or were so few in number that they were lost in the general troops of the Alai.

After six days, the Cossacks at the edge of a forest by the river established the Alazeya winter camp for collection of tribute from the Yukaghir. The same F. A. Chukichev tells about the events of the end of this year: "To them, the servicemen, at the winter camp arrived Reindeer Chyukhchie on reindeer to sell reindeer. And for them, the service people, it was impossible to catch them without people."[94] Evidently there was already sufficiently developed reindeer herding among the Chukchi, and they could sell reindeer, wishing to obtain Russian wares. For this the Chukchi arrived on sleds to the southern border of the tundra, to the winter camp itself, offering their wares, probably knowing that the Russians were in need of food. But the Cossacks, since in summer there were conflicts with the Alazeya, allies of the Chukchi, believed the latter were enemies whom it was at least necessary to capture and collect tribute from the relatives of the prisoners. However, since the servicemen were few, they preferred not to leave the winter quarters. In general, the first contacts with the Russian explorers were confrontational. The servicemen did not react

---

[94] Belov 1952: 57.

to attempts by the Chukchi to establish trade relations, possibly because of insufficient wares and/or as a consequence of distrust of the Chukchi.

A. A. Burykin proposes that in the summer of 1642 the Yenisei lieutenant Elisei Buza brought three amanaty to Yakutsk, who were in fact not, as is believed, Yukaghir, but were coastal Chukchi; indeed, the names of all three hostages can be etymologically connected with the Chukchi language. Thus, it turns out that the encounter of the Russians with the Chukchi occurred earlier than the arrival of E. Yu. Buza in Yakutsk. We will see whether data from linguistics correspond in this case with history. In the *nakaznaya pamyat'* [instructions] to E. Yu. Buza it is mentioned that in the year 7150 from the creation of the world (1642) he brought to Yakutsk "3 Yukaghir amanaty, shaman Balganch and Imgonta, and Dalgunta." Before this, the detachment of E. Yu. Buza spent three years (1638–1641) at the mouth of the Chondon River, east of the Yana where these three amanaty-princes were seized, and was going to obtain tribute from their kinsmen. However, this territory, as far as we know, was occupied by Yukaghir, and it is unlikely that the three seized tribal representatives were not from the native population. In addition, in other documents Bylgai is directly called "shaman of the Yukaghir people of the Kromyi clan," and about two others it is said: "One amanat of the Tarabaltsy clan by the name Engota, and another amanat of the Yandagir clan, and by the name Dolgota." Consequently, these were clan princes of the Koromo Yukaghir and Yandii people.[95]

The number of Chukchi in the Bol'shaya Tundra between the Alazeya and the Kolyma can be judged based on statements by F. A. Chukichev from 1646: "And these Chyukhchie live between the Alazeya and Kolyma rivers in the tundra; they say there are four hundred or more people." Thus, with the appearance of the Russians there were more than 400 Chukchi in the tundra. B. O. Dolgikh, without providing any argument, believed that this number was exaggerated, and that in the tundra there were more probably 40–50 families, that is, 150–200 people.

---

[95] Burykin 1998: 81; 2006: 60–64; pro: Zuev 2002: 65, note 2; 2009: 33, note 8; instructions of 1642, see: DAI. 1846. Vol. II: 262; about the activity of E. Yu. Buza, see: Okhotina-Lind and Møller 2013: 70–71; Ogloblin 1903: 49–52; Vize 1948: 11–12; Ivanov 1999: 100; about Yukaghir clans with citations, see: Dolgikh 1960: 382; cf.: Belov 1952: 41, note 2; about amanaty in Siberia in general: Lyutsidarskaya 2006: 121–124.

Researchers usually trust the source and accept the number of Chukchi in the Bol'shaya Tundra as 400 people. It would be possible to think that this is in fact correct, considering that at the end of the 19th century to the west of the Kolyma the Indigirka and Alazeya Chukchi numbered 150 souls in 13 nomad camps, and the Kolyma—400 souls in 35 nomad camps.[96]

However, it pays to turn attention to other facts that are contained in this document. F. A. Chukichev notes that in the clan of the Yukaghir Prince Ivanda on the Indigirka there were more than 100 people, and they took tribute "of two times forty plus thirty sables," and on the Alazeya Prince Noochichan had "about three hundred people" who brought tribute of "seven times forty sables." It turns out that the Yukaghir paid tribute from more than 100 people—110 pelts of sables, and from 300 people—280 pelts. Later the Yukaghir usually paid tribute by the method that the prince brought it for all the clan, and it is impossible to speak reliably about some norm that later among these Yukaghir fluctuated from one to ten sables. But the first payments of tribute, in order not to embitter the new tax payers, could have been established lower, according to the recommendations of the authorities.[97] In general, it seems that the Yukaghir on the Indigirka and Alazeya at the boundary of the 1630s–1640s paid on average one sable per person, that is, per man. It is unlikely that within one document F. A. Chukichev has different ways of computing the number of the population; indeed, at this time usually only men were considered tax-payers, otherwise some reservations would be expected. F. A. Chukichev probably gives only the number of men—potential tribute payers. In such case, the total in the Kolyma-Alazeya

---

[96] Document 1646, see: Belov 1952: 57; Dolgikh 1960: 409, cf. 548 (where the author is inclined to the total number of 400 people, from whom 100 where males). The credibility of the source: Kolesov 1991: 42; Erunov 1993: 108 (400–500 people); Mikhailova 1996: 179; Zuev 2002: 21; 2009: 27; Vakhtin, Golovko, and Shvaittser 2004: 45 (to 500 people). The number of trans-Kolyma Chukchi at the end of the 19th century: Bogoras 1921: 87; Bogoraz-Tan 1934: 12; Gurvich 1966: 190. [The reader should be advised that "trans-" ("beyond") indicates the perspective from Moscow/St. Petersburg, the "home" of the authors. For example, "trans-Kolyma" refers to the area east of the Kolyma River.—*Trans*.].

[97] About tribute from the Yukaghir, see: Dolgikh 1960: 396–408; the Anadyr' Yukaghir paid one sable each: Dolgikh 1960: 425; about initial light tribute, see: DAI. 1848. Vol. III: 351–352; PSI. Bk. 2: 526; Orlova 1951: 239; Ivanov 1999: 116; Slëzkin 2008: 37.

tundra was about 1,600 Chukchi, counting all the adult male population of nomad camps for a quarter of the total number of inhabitants. What is more, in 1656 to Nizhnekolymsk came a militia of more than 200 Chukchi—followers of one Chief Mita, in 1685—more than 100 men, and in 1671 warriors from the clan of the interpreter Afanasii that counted 70 men.[98] The indicated numbers of warriors completely correlate with the total number of the adult male population of four hundred men. But this apparent similarity with the number of the Chukchi population in the tundra in the mid-17th and at the end of the 19th century is not supported by the difference in number of the reindeer livestock that could be fed by one and the same ecological niche: indeed, at the boundary of the 19th and 20th centuries the livestock herd among the owners reached 5,000, which was clearly not the case two and half centuries before. Indeed, at the end of the 1730s F. G. Müller believed that the Yukaghir and Chukchi had the most numerous herds of reindeer in all Siberia: they had herds that numbered 200–300 and more head each, while some Yukaghir had livestock that reached 2,000–3,000 head.[99]

3. Warfare on the Lower Kolyma River during the Second Half of the 17th to the Beginning of the 18th Century

In 1642–1650 the efforts of the Russians in the region were directed at subjugation of the Yukaghir. The Chukchi, who wandered in the tundra, on the whole interested them only little. In 1644, the Stadukhin winter quarters were founded on an island in the Kolyma floodplain—the future Nizhnekolymsk [Lower Kolyma fort] redoubt, which became the base for subjugation of the territory and for movement farther to the south and east.

As believed, in 1644 or 1645 M. V. Stadukhin, going from the Nizhnekolymsk quarters, traveled to the Chukoch'ya River—center of residence of the Kolyma Chukchi—where he received some information

[98] The proportion of adult male population: Vol'fson 1991: 50. 200 warriors: AII. F. 160. No. 383. Sst. 76; Belov 1952: 258; Vdovin 1965: 104. 100 Chukchi: AAN. F. 21. Op. 4. No. 32. L. 130; about the detachment of Afanasii: Vdovin 1944: 257; 1965: 72. B. O. Dolgikh (1960: 408) counted 580 Alazeya Yukaghir.

[99] Data of the boundary of the 19th and 20th centuries: Bogoraz-Tan 1934: 12. Miller 2009: 224; date of G. F. Miller's work: Elert 2009: 18.

about this people. However, the historian of the Far East, B. P. Polevoi, suggests that M. V. Stadukhin traveled to the Chukoch'ya River earlier, in 1643, sailing from the east to the Kolyma.[100] At least, at the time of the first arrival of the united detachments of M. V. Stadukhin and D. M. Zyryan on the Kolyma, there was among the Russians one "Ofon'ka Chyukhcha," who in the preface of a report to the tsar and in other documents is named as the hunter "Ofon'ko Yakimov." If this is not simply a nickname that somehow got attached to a Russian but, as believed, is the real designation of ethnic association, then it is clear that by the beginning of the campaign to the Kolyma, Afanasii had been baptized and received a Christian name. This should have taken some time and a priest had to be present. Possibly Afanasii was in the detachment of D. M. Zyryan and joined it earlier on the Alazeya. If everything is reliable, then this was the first case of a Chukchi going into Russian service. On 25 July 1643, during a battle at the Omok redoubt, he was wounded.[101] Later, Afanasii continued to serve the tsar. In 1650 he, with other traders and hunting people, accompanied tribute from the Indigirka to Yakutsk.[102] B. P. Polevoi compares this Afanasii Yakimov with Afanasii Fedotov, a former *amanat* [hostage], who later became a Cossack and actively worked in the 1660s–beginning of the 1670s.[103] However, it should be noted that the patronymic is nevertheless different, and Yakimov was not at all an amanat, but a hunter.

Figure 17. Nizhnekolymsk. Illustration by Luka Voronin, artist on the J. Billings Expedition (1785–1792). (Reproduced based on: Wrangell 1948: 152–153; Kolesov 1991: 20–21).

[100] 1644 or 1645: Zuev 2002: 194; 1643: Polevoi 2000: 34.

[101] Document, see: Polevoi 1965: 287–288; 2000: 36.

[102] Belov 1952: 186.

[103] Atributation of Afanasii, see: Polevoi 1965: 290; 2000: 41.

With the instability of the surrounding Yukaghir population, open confrontation with the Koyma Chukchi was at first undesirable. The Chukchi could openly join a camp of the Yukaghir, thereby strengthening the forces of the latter. In 1648, Cossack Ivan Pinega proposed to the Nizhnekolymsk leadership to establish relations with the Chukchi. Here is how I. B. Pinega himself described the expediton later (1652) in a report: ". . . [I] filed a petition to the service people Ftor Gavrilov and Mamon Martynov, in order that they release me, your servant, to go to the non-tributary people, to the Chyukchi, to their Princes Aun and Tyka, to call them to your sovereign mercy, in order that they, the Chyukchi, were to You, Sovereign, submissive and obedient." The steward V. Gavrilov and Cossack M. Martynov gave I. B. Pinega a special *nakaznaya pamyat'* (instruction) and on this same day, 26 November 1648, he was sent on his way. I. B. Pinega writes: ". . . and they, the Chukochie princes, I called to your sovereign mercy. And thirty men of the Chyukchi came with me to the Kovyma [Kolyma] River to the tribute winter quarters, and stayed at winter quarters half a verst [~0.5 km] beyond the river." In addition, the serviceman, returning on 23 December, brought five walrus tusks that he had bartered for with his wares.[104]

As we see, the first attempt to establish relations—clearly not demanding tribute or amanaty—ended very successfully. I. B. Pinega and his traveling companions stayed almost a month with a Chukchi elder and evidently told him about the power of the Moscow Sovereign. Quite successful also was mutual trade, during which the Cossack bought five walrus tusks for the treasury. Later, in 1650, this purchase of five tusks with a total weight of 17 *grivenki* (6.9 kg) arrived in Yakutsk. In order to establish relations with the Russians, 30 Chukchi (probably primarily princes) arrived on sleds to Nizhnekolymsk and set up camp half a kilometer from the winter quarters. These princes were probably military leaders or heads of nomadic groups. The best known of them to the Russians was Aun, since in another petition I. B. Pinega mentions only him. But since three campaigns of servicemen had ensued against the Chukchi in the next winter, it is clear that the latter refused either to pay tribute or to provide amanaty.

---

[104] Belov 1952: 217, 219, Note 6; cf.: Gurvich 1966: 48; Ivanov 1999: 122.

In "instructions" of 1647 for the steward Vasilii Vlas'ev, the son of a boyar, newly assigned to the Kolyma, it was mentioned: ". . . and to send him, Vasilii, to the foreigners Chyukhchi and call them to the Royal Sovereign's high hand and, having persuaded, to take from them amanaty of the best people, as many comely people," and to take tribute of walrus tusks weighing more than a *grivenka* (409 gr.), furs, and fat.[105] To fulfill the governor's orders, expeditions for the tribute began to be organized.

To clashes on the Kolyma in 1648 attests a petition of the captured Chukchi Apa, who requested of the tsar that he be assigned as interpreter to the Nizhnekolymsk winter quarters: ". . . in the 156th year at the mouth of the Kolyma River I, Apa, was taken to the Yakutsk redoubt by the son of a boyar, Vasilii Vlas'ev. And from that time, Sovereign, and my Father, and Mother, and Family, the tribe abandoned me, and your Sovereign tribute they do not pay on my behalf. And first, Sovereign, I, your orphan, want to serve you, Richeous Sovereign, and in everything right, and will bring the Chyukhoch, my relatives, lead under your, Tsar, high hand."[106]

Possibly the clash of the Russians with the Chukchi occurred in summer at fishing grounds at the mouth of the Kolyma, where a young man by the name Apa was taken captive as an *amanat.* However, it turned out that the Chukchi, in distinction from other Siberian peoples, did not care about their hostages and did not pay tribute for them. Thus, neither his family, nor clan, nor tribe paid tribute for Apa. If the social attribution of the family and clan are understood, then it is not at all clear what significance is embedded in the idea "tribe," which, as it is believed, did not exist among the Chukchi. It was probably a matter of a group of clans or in general all the Kolyma Chukchi together. Since Apa, because of his relatives abandoning him, was formally left an orphan, then he offered himself to Tsar Aleksei Mikhailovich as a translator able to pursuade the Chukchi to become Russian citizens. After a year as an *amanat,* Apa was able to learn the Russian language, showing himself to be a man with linguistic abilities.

---

105 Orlova 1951: 239; Vdovin 1965: 103. About the weight of the tusks: Ogloblin 1903: 30; Vdovin 1965: 103.

106 Orlova 1951: 254–255; Ivanov 1999: 118; Zuev 2002: 62; Burykin 2004; 42. A. S. Zuev (2002: 172) proposes that Apa later became the interpreter Afanasii.

The campaigns of the beginning of 1649 are described in a petition in the name of the tsar by *desyatnik* [foreman] Vasilii Bugor:

> . . . with Vtorko Gavrilov [we] went on two campaigns to your sovereign's disobedient peoples, the Chukhchi, and those, Sovereign, Chukhchi we could not reach; and a third time, Sovereign, to those same Chukhchi went only twelve people in service in the spring, by the latter route; and those were Chukhchi and we went to the sea on the tundra to a clear place; and they, Sovereign, having spied, fled from the yurts with both the women and children, and the light people on reindeer stayed at the yurts; and at an attack on the yurts, Sovereign, the serviceman Vaska Shchukin was shot, and having arrived, Sovereign, at the winter quarters this serviceman died from those wounds; and we took, Sovereign, two Chukhochi women, and these women we took with us as guides.

In a later petition Vasilii Bugor and Evsevii Pavlov (1655) note: ". . . and your sovereign service we served, in campaigns we went on the Chyukchi, and there among us they killed the serviceman Vaska Grigor'ev Shchukin and the Chukchi took his musket."[107] It was most probably in this campaign that the traveling companion of M. V. Stadukhin also participated, hunter Ivan Fedorov Kazanets, who in his petition to the tsar writes: ". . . and in the campaign in your sovereign service I went to the non-tributary Chukhochi people, and there was a fight with them, and in this fight I, your orphan, was wounded."[108]

These three expeditions were evidently sent out primarily to take tribute from the Chukchi pursuant to the provisions of provincial decrees. Since there were too few Cossacks in Nizhnekolymsk, the people of V. E. Bugor took part in military actions. In the first two campaigns the expedition was led by the steward V. Gavrilov himself. Since these campaigns were the first large expeditions of the Nizhnekolymsk Cossacks against the Chukchi, then the servicemen were not yet clear

---

[107] DAI. 1848. T. III: 212; 1851. T. IV: 8; cf.: Gurvich 1966: 48; Zuev 2002: 201; Nefëdkin 2003: 23.

[108] AAN. F. 21. Op. 4. No. 31. L. 122; Belov 1952: 245–246. I. S. Vdovin (1965: 103) dates the campaign, described by I. F. Kazanets in the petition to the time of the writing of the petition itself, that is, 1655, whereas A. S. Zuev (2002: 209) assigns it to 1656. However, in the document, judging by the chronological sequence of the description of events, it is a matter of events in the winter of 1648/9, before the maritime expedition of M. V. Stadukhin from Nizhnekolymsk to the east in the summer of 1649.

where to search for the enemy and how to act against it, and therefore twice returned to winter quarters without results. They probably searched along the sea, from which the bulk of the Chukchi had already wandered to their winter camps.

The third campaign followed in May 1649. At the head of the expedition now was V. E. Bugor, but the fighters amounted to only 12 men, which again speaks most probably of the tribute character of this campaign. Evidently the Chukchi at this time had now wandered back to the sea, where their camp was discovered by the servicemen. The number of yarangas found is not indicated in the document. Seeing the Russians from afar, part of the men, as well as the women and children, hurried to hiding places. The battle was taken to the men who remained, who had reindeer sleds standing in readiness. The Cossacks attacked. V. G. Shchukin was probably directly in the spearhead of the attack, and he received several wounds from arrows. Since the melee had not yet started, the Chukchi managed to pick up his musket. After the Chukchi took the musket they evidently retired to their sleds, taking the weapon with them. Because of this retreat, the Cossacks did not know about the losses of the Chukchi—those killed, wounded, and captured, at least the document does not mention it. In the remaining yarangas the Russians discovered two women, whom they took captive. The servicemen knew the road poorly and forced the Chukchi women to take them from the tundra to the Nizhnekolymsk winter quarters.

On the whole, the detachment of V. E. Bugor, consisting primarily of Cossacks who rebelled in Yakutsk in 1647, upon finding a small Chukchi nomad camp, did not conduct negotiations about tribute but rather immediately attacked, seeing that the enemy was prepared for battle. No tribute was obtained, but the servicemen acquired some booty. During the battle at least one of the servicemen was wounded—I. F. Kazanets—and V. G. Shchukin received a mortal wound.

About the subsequent campaign, the servicemen V. E. Bugor and E. Pavlov briefly mention in this same petition to the tsar: ". . . in the 158th year we, servants of your sovereign, served your sovereign service together with Mikhail Stadukhin, went on a campaign to the Chukhchi, and there they killed among us the retired serviceman Ganka Mikhailov Komarov." I. B. Pinega also writes in his petition about his participation in this campaign: "And in the 158th year, in fall, I, your servant, in your sovereign service, went along the first route to the sea to the Chyukchi in a campaign of sixty men and found a few people and were not able to take them. . . ."[109]

In this campaign in September–October 1649 a rather large number of servicemen (60 men) went. They went on the first snow in dog sleds and/or skis hoping to find the Chukchi at the sea. At the head of the campaign now was M. V. Stadukhin. The Russians went out to the sea, discovered a small nomad camp of Chukchi, but were not able to capture them, possibly as a consequence of the latter fleeing.

The death of serviceman Gavrila Komarov attests to the occurrence of the clash with the Chukchi. The former was a retiree but nevertheless participated in the expedition, probably because of the general lack of servicemen at Nizhnekolymsk. Since this part of the document does not mention tribute obtained or prisioners, the campaign most probably had little success.

The situation on the Lower Kolyma in 1649 is outlined by the report of the Kolyma steward Vasilii Vlas'ev and the minor official (*tseloval'nik*) Kiril Kotkin to the Yakutsk military governors V.N. Pushkin and K. O. Suponev of 1650:

> In the present 158th year, on the Kolyma River from the Nizhnoe [Lower] tribute winter quarters, the serviceman Mamonko Martynov and comrades wrote to the Verkhnoe [Upper] tribute winter quarters: In the past, in the 157th year, on the 6th day of June, from a captive of the serviceman Ivashka Permyak, from an Omok girl who was on the Duvannyi sand it became known, that on the Kolyma River at the Lower tribute winter quarters the Omok Princes Eyuk and Timka, and Nirpa, and Ermo, and Ail, son of Kigich, and Igol'nik with relatives and all his people betrayed the sovereign and called the non-tributary Omok and the Chyukhchi from the sea and wanted to kill the servicemen at the fishing place and to plunder the sovereign's treasury, and to recapture the amanaty.

Investigation of this affair was conducted by the next steward, Cossack Timofei Buldakov:

---

[109] Campaign of M. V. Stadukhin, see: DAI. 1851. T. IV: 8; cf.: Gurvich 1966: 48; Zuev 2002: 201; Nefëdkin 2003: 23; the petition of I. B. Pinega, see: Belov 1952: 217.

> And in the past in the 159th year, as I, Timoshka, went to the Upper tribute winter quarters, and at the Lower winter quarters foreigners seeing that in the Lower tribute winter quarters of the servicemen and hunting people of the previous years few were left, and in fall they, the foreigners, wanted to kill the servicemen. And they, the servicemen, found out about the murder. And they, the servicemen seized them, the foreigners, the best people, breeders Timka and Panyuika, and Chyuk and put them in jail. . .. And I, Timoshka, beat those foreigners Timka and Panyuika, and Chyuk with canes mercilessly. And before this those Nizhnekolymsk foreigners in all years thought to kill servicemen. . . .[110]

In general, on 6 June 1649, the head of Nizhnekolymsk, Cossack Mamon Martynov, learned from a captive Yukaghir woman that 6 of 12 tributary Nizhnekolymsk Yukaghir clans urged the non-tributary Omok and Chukchi to overthrow Russian dominion on the lower reaches of the river. The Chukchi were thus allies of the Omok in opposition to the Russians. In the fall of 1650 the Yukaghir again wanted to take advantage of the small number of Russians in Nizhnekolymsk (in 1651, according to T. M. Buldakov, eight Cossack and the same number of *promyshlenniki* [hunters/trappers/traders]) and kill them. This idea, as we see, was fostered by the Omok over the course of several years. Three of the alleged ringleaders of the plot were placed under arrest and, after confession, were beaten with canes. We note that of these three only Timka figures in the reply of V. Vlas'ev. Thus, the uprising was nipped in the bud, that is, it did not take place.

In 1652 the Yakutsk governor, steward I. P. Akinfov, sending the *pyatidesyatnik* [commander of fifty.—*Trans.*] Ivan Rebrov to the Kolyma, set for him in the instructions the same task as earlier in the directives of 1647 to V. Vlas'ev: ". . . and him, Ivashka, send to the foreigners, the Chyukhchi, and call them under the sovereignty of the Tsar's high hand, and, having persuaded, to take from them as amanaty the best kind of people, as many comely people, and to instruct them to pay the sovereign tribute with large and medium walrus teeth and walrus furs, and to have fat . . . and to have from them the walrus fish teeth, looking at their affairs

---

[110] Document of 1650: Tokarev 1948: 9–10; cf.: Tokarev 1948: 9–10; petition of T. M. Buldakov of 1652: Orlova 1951: 296–297. About Yukaghir clans that paid tribute in Nizhnekolymsk in 1651, see: Dolgikh 1960: 415.

and seeing the people, how much who can pay, and therefore to write in books by name."[111] In general, it was again planned to obtain large walrus tusks, furs, and fat, as well as hostages—well-known people—from the Kolyma Chukchi.

Because of these campaigns of the Cossacks and the clashes accompanying them hostility emerged, fueled by blood revenge among the latter, a state of permanent conflict emerged. In 1653, on the basis of a report from the boyar's son Ivan Erastov, the situation in the region was thus described in an edict of the tsar to the Yakutsk military governor: ". . . and in the tundra between the Kovyma and Alazeika rivers live the foreigner Chukhotskie men, and there are three princes among them: Aun and Tyko, and Kencheka, and these Chukhotskie princes and with their people do not pay our tribute, and in concert they kill and rob the Russian people at the edge of the sea, and do not provide game, and create great damage for our treasury in the sable trade." To oppose the Chukchi, I. Erastov was ordered to be sent to Kolyma from the Lena redoubt with 20 servicemen.[112]

As we see, together with two former leaders, probably the heads of nomadic groups, Aun and Tyka, there appeared also Kencheka, who had in the eyes of the Russians, judging by the documents, the same social status as the two other princes. The Chukchi not only hindered the hunting business but hunted by piracy, attacking ships stuck in the ice. Such situation became a threat to the economic interests of the Russians in the region. Indeed, from 1645 intensive northern navigation began between the Lena and the Kolyma. For example, in the season of shipping, in June–July of 1647, the Yakutsk customs office issued 404 permits for such passage.[113] In order to correct this situation, the Yakutsk administration had to act.

Still in 1651, recently recruited Cossack Yurii Seliverstov received orders from the Yakutsk governor D. A. Frantsbekov to "go by sovereign edict by sea beyond the Kolyma River and hunt bone of fish teeth and non-tributary foreigners and go to the Chyukhchi in campaigns

---

[111] DAI. 1848. T. III: 352; cf.: Vdovin 1965: 103; Zuev 2002: 204.

[112] AAN. F. 21. Op. 4. No. 31. L. 5; cf.: Vdovin 1944: 254; 1965: 85.

[113] About the nomadic groups among the Alazeya Chukchi: Gurvich 1985: 53; permits from the customs office: Belov 1956: 155.

and to bring them under the Sovereign's high hand." The specific reason for the campaign of 1653 was the murder of Russian promyshlenniki and traders by the Chukchi at the mouth of the Alazeya. At the Nizhnekolymsk fair Yu. Seliverstov recruited 22 promyshlenniki joining them to his 11 newly recruited servicemen. Further, Yu. Seliverstov "went to the mouth of the Kovyma River in the summer season of the present year of 161, July on the 7th day, onto the Sovereign's disobedient people and traitors non-tributary Chukhchi people." He, "on the Kovyma River, bowed to the steward, Pyatidesyatnik Ivan Ivanov Rebrov, asked for a guide that the Yukaghir might take us to the non-tributary people the Chyukhchi. And he did not give me, Yushka, a guide, did not listen to me, unknowingly indulging the foreigners, and I, Yushko Seliverstov, went in the Sovereign's service onto the Chyukhoch without the guide." The day before this, 6 July, Yu. Seliverstov unsuccessfully requested of the tribute collector Semen Shubin the translator, a "Yukaghir woman by the name of Alevaika," as a guide. At the mouth of the Anyui, at the fair, Yu. Seliverstov sent the Cossack Gavrila Alekseev on a ship to I. I. Rebrov whom the pyatidesyatnik detained until evening on his ship. "And I wanted to go to those foreigners non-tributary Chyukhchi people, and he, Rebrov . . . did not give me a guide, unknowingly indulging the foreigners, and caused great damage to the state's service and the treasury, and those foreigners non-tributary people Chyukhchi killed many trading and hunting people at the mouth of the Alazeya River." The reply is completed with the words: "And I, Yushko, because of this interpreter did (not) go to the sea, no one to hunt fish teeth, because that woman was on the sea before and knows different languages."[114]

In general, becoming a Cossack from the promyshlenniki, Yu. Seliverstov gathered in Yakutsk the same newly recruited servicemen, to whom volunteers were added at the Nizhnekolymsk fair, famous at that time. The expedition of Yu. Seliverstov, which had a *koch* [a sail-paddle boat.—*Trans.*], could easily travel by land or sea. Therefore, on 7 July 1653 he went to the mouth of the Kolyma, probably, in the koch. For the maritime part of the expedition he lacked a guide, whom the Kolyma steward and tribute collector refused to send. Evidently the steward and

[114] AAN. F. 21. Op. 4. No. 31. L. 45 ob.; Belov 1952: 223–224 (with lacunae); cf.: Vdovin 1944: 254; 1965: 103.

the tribute collector did not want to be without the Yukaghir translators, which would complicate their communication with the local population. Whether the expedition moved on is not entirely clear, based on this complaint-document. However, judging by the fact that Yu. Seliverstov intended the expedition to the Chukchi to join in the hunt for walruses, because of lack of guides this expedition did not take place.

It is believed that in spring 1653 several hundred Chukchi besieged the Nizhnekolymsk winter quarters. However, upon checking, it turned out that in this case reduplication of the 1656 campaign simply occurred among historians. In a description of this siege I. S. Vdovin refers to a document preserved in the archive of the St. Petersburg Institute of History, Russian Academy of Sciences. However, this document is a draft of a petition of the son of the boyar K. S. Dunai of 1659, where in fact the siege of the winter quarters in 1656 is described. Two hundred Chukchi were changed by I. S. Vdovin to "several hundred," and Konstantin Stepanov Dunai to Kostik Stefanov. A. S. Zuev also distorted the date of the event: instead of spring 7164 (1656) the siege was assigned to 1659—the year of the drafting of the petition.[115] In 1653 there obviously was no siege of Nizhnekolymsk by the Chukchi.

In 1656, an escalation in the Russo-Chukchi conflict occurred. In the report of steward Ivan Erastov to the Yakutsk governor of 1659, the events are described in the following way:

> In the past, in the 164th year, in winter, a week before Christmas, the best lower Kovyma Prince Timkui invited and caught the best Chyukhchi Prince Mita and gave him as an amanat to the son of the boyar Kostyantin Stefanov in the Lower tribute winter quarters, and he, Konstyantin, kept this Mita until spring, and they, the Chyukhchi, relentlessly came with their clans to the tribute winter quarters, and he, Kostentin, did not ask the best lower Kovyma princes, released this Mita, and took boys in his place: they said to him that his own brother

[115] Date of the siege 1653, see: Vdovin 1965: 104; Mikhailova 1996: 189; Zuev 2002: 64, 206; 2009: 40, 294; Nefëdkin 2003: 23; Gorosov 2011: 14. Draft of the petition, see: AII. F. 160. No. 383. Sst. 74–76; publication of the petition: Belov 1952: 256–258. Date 1659: Zuev 2002: 212; cf.: Vdovin 1965: 104, Note 13. Reference of A. S. Zuev (2002: 206, Note 6) to the mention of the campaign of 1653 in DAI (1862. T. VIII: 9) is inaccurate; there it actually tells of events at the boundary of the 1670s and 1680s.

> and his clan brother, and his son, a child, not knowing if they are serfs, there was no one who knew them. And he, Mita, arriving in fall, killed the best tribute man Ermo and his son and nephew, and they, Ermo, were going with the state's tribute to the tribute winter quarters, and they (Mita and his comrades) took 50 of the state's sables and carried them away to their land on the Chukhoch'ya River; and he, Konstentin, collected the sovereign's tribute from those Yukaghir, and did not write to you in the Yakutsk fort anything about it. And presently in the 167th year on the Kolyma River at the Lower tribute winter quarters the non-tributary foreigners Chyuchki, he Mito, they committed wickedness, and killed the service people of the Yakutsk fort at the fishing place: the service man Ivan Savin Yachmenev and service man Fedot Matveev, and wounded other service people Ivan Perfir'ev, Vasil'ii Oshurkov, and the hunting service man Fil'ka Sevast'yanov. And now in the 167th year on the 8th day of July I, Ivashko, assembled on the Kolyma River at the Lower Fair of the Yakutsk fort 19 men of service and hunting people and one hundred and fifty foreign tributary Yukaghir and sent them in the state service on a campaign in gun and bow combat to those non-tributary killers and villainous foreigners the Chyukhoch.[116]

As we see, I. R. Erastov rather clearly, without too much detail, describes the situation around the Nizhnekolymsk in the winter of 1655–summer of 1659, criticizing with this the activities of the previous Kolyma steward Konstantin Dunai. Now the head of the Chukchi is the "best Chukchi Prince" Mita, and not two or three "princes."[117] Possibly as a consequence of the Russian threat, the Kolyma Chukchi elected one military leader, called in the source the "best," and not simply, as earlier, a prince, or, which is less probable, only one leader, who lived near the Russians on the Chukoch'ya River, was involved in the conflict. It all began with the fact that the Lower Kolyma Yukaghir Prince Timkui invited Mita and took him captive on 18 December 1655. Possibly, Timkui wanted to curry favor with the Russian administration or he had a

---

[116] KPMGYa: 64; cf.: Vdovin 1944: 258; 1965: 79, 104; Gurvich 1966: 22, 48–49; Kolesov 1991: 29, 42; Zuev 2002: 63, 209, 212; Nefëdkin 2003: 23.

[117] Though B. O. Dolgikh (1960: 409) supposes that Tyka and Mita are one and the same name, and correspondingly one and the same person, I. R. Erastov knew both princes and hardly would have been mistaken in the names, which, in his view, two different men had.

preliminary agreement with the steward K. S. Dunai. Mita arrived to Timkui, not expecting unfriendly actions by the Yukaghir-Omoki and violation of traditional rules of northern hospitality. The latter unexpectedly captured Mita and handed him over to K. S. Dunai, who placed him in an amanat hut in the Nizhnekolymsk winter quarters.

The Chuchi decided to free their leader. They went to Nizhnekolymsk. I. R. Erastov briefly writes about further military actions, in which he did not participate, pointing to repeated attacks: "And they, the Chyukhchi, by their clans persistently arrived at the tribute winter quarters." However, K. S. Dunai himself, participant of military actions, describes the events in detail in a petition to the Tsar:

> And in this same year of 164, after your, Sovereign, tribute collection, two hundred and more villainous non-tributary foreigners behind shields arrived at the Nizhne [Lower] winter quarters in the spring and proceeded resolutely to the winter quarters. And I, your servant, served you, Sovereign, with your service people, not sparing our heads, and went out in a foray against those foreigners and fought with those foreigners, and in this foray among them, the foreigners, among the Chukhchi, we killed the best man by the name Pimavatka, and wounded many among them. And by the grace of God and your Sovereign fortune we repulsed the foreigners and saved your Sovereign tribute winter quarters. And after this we were under siege from those foreigners for a long time, which, Sovereign, they, the foreigners, did not let us, your servants, few people, to get either firewood or water or with nets to catch fish for ourselves and for amanaty.[118]

The Chukchi rather carefully prepared for war. They planned to take the unfortified winter quarters by storm. Substantial forces were gathered for the storming—more than 200 warriors, that is, approximately half of all the fighting men. Even in spite of a possible increase in the number of the Chukchi, it can be noted that we do not know of such a large militia among the Kolyma Chukchi. In order to sustain the fewest losses from bullets during the siege, the Chukchi made thick siege shields made of wood. Understanding that shooting would not repulse the Chukchi storm, the Cossacks, led by the steward, made a sortie. This was probably directed at Pimavatka, the leader of the assault. His death and the losses

---

[118] Belov 1952: 258; cf.: Vdovin 1965: 104.

born stopped the coordinated storming. The Chukchi, using their numerical superiority, proceeded with the siege. Of course, this was not so to say a "correct" blockade from the military point of view. The Chukchi camp was probably set up not far from the winter quarters, and because of the danger of constant attack the servicemen did not have the possibility to stock up on firewood or to get water and replenish supplies of food by fishing. The last, incidentally, attests to the fact that the threat of attack continued to exist until late spring.

Finally, in the same spring of 1656 an agreement was reached between the Russians and Chukchi about the exchange of hostages: in place of Mita three of his relatives were given, two brothers and a son. I. R. Erastov in fact doubted whether these were relatives of Mita or were only presented as such. In 1658, the servicemen held two Chukchi amanaty, and in January 1661, in a list of five amanaty of Nizhnekolymsk, only the "Chyukchi amanat Chochyuga Mitov" is mentioned, that is, the unbaptized son of Mita.[119] Since the interests of the Yukaghir were affected in the events that occurred, K. S. Dunai was supposed to consult with their princes about this exchange, which he did not do, resolving the question independently.

In the spring of 1656 Mita proceeded to avenge his capture. He attacked the Yukaghir Ermo, who, accompanied by his son and nephew, carried tribute for his family to the winter quarters, and killed all three. Judging by the documents of tribute collection, Ermo did not belong to the clan of Timkui, and paid tribute for his hostage at the winter quarters. According to the documents the tribute amounted in 1651 and in 1659 to 20 sables,[120] and not 50, as the document mentions. Probably the latter value of the tribute is a simple distortion. Nevertheless, K. S. Dunai took the tribute at the standard rate and did not notify the Yakutsk governor about this, probably considering the incident insignificant. Moreover, the robbery was not viewed by the steward as a reason to march against the Chukchi.

The unfriendly actions of Mita toward the Russians continued even when I. R. Erastov became the Kolymsk steward. In the summer of 1659 at the fishing place near Nizhnekolymsk the Chukchi killed two

---

[119] AAN. F. 21. Op. 4. No. 35. L. 400 (Document No. 101); Orlova 1951: 326; Vdovin 1965: 104.

[120] Dolgikh 1960: 415.

servicemen and wounded three other men. The men fishing probably resisted the Chukchi and were thereby saved. The fact that the attack occurred at the winter quarters itself can attest to a deliberate attack and not to a chance encounter of the Chukchi with the Russians on hunting trips, which incidentally also cannot be rejected.

I. R. Erastov acted more forcefully than K. S. Dunai. At the summer fair in Nizhnekolymsk he, based on the existing custom, began to call together volunteers for a campaign, people who were thirsting for adventure and booty. The total, together with the Cossacks, amounted to 19 men armed with flintlock muskets. Simultaneously the steward gave the Kolyma Yukaghir instrusctions to prepare for the campaign a contingent armed with bows. Of such, 150 warriors turned out. Probably not only the Omoki from the Lower Kolyma were summoned, but also other Yukaghir who paid tribute to the Middle and Upper tribute winter quarters and were subject to the Kolyma steward. Indeed, based on the tribute books B. O. Dolgikh calculates the number of Yukaghir who paid tribute at Nizhnekolymsk in 1659 at only 100 men.[121] The total number of the exedition of 169 fighting men attests to the preparation of a rather large campaign, based on East Siberian standards. On 8 July 1659, the army set out. Of course, it is unclear whether on foot or in boats [kochi]. Also, not reported in the document is who the leader was, but it was not the steward himself.

Since I. R. Erastov does not later write in a petition about successes attained in the campaign, but asks the governor to send him powder and lead, it is quite possible to suppose that the campaign was not a success. However, the expedition contributed to an escalation in military activities and primarily split the Chukchi from the Kolyma Yukaghir, and not the Alazeya people who were administratively subject to the Alazeya winter quarters.

In 1662, military actions between the Chukchi and the Russians continued. The *desyatnik* [leader of ten] Mikhail Kolesov thus told about his services in a petition to the Tsar: "... in the 170th year the non-tributary people Chyukhchi besieged the Kovyma lower tribute winter quarters and the tributary Yukaghir and began to kill the servicepeople who were

---

[121] *Ibid.*: 414.

fishing. And the mandative son [i.e., with mandatory service] of the boyar Vtoroi Kataev sent me, your servant, with servicepeople to those villainous non-tributary Chyukhchi, and I, your servant, went to those Chyukhchi on a march and those Chyukhchi we killed and defended the tribute winter quarters, and in that battle, they wounded me, your servant, in the right leg at the knee."[122]

The information about military actions was rather sparsely described by M. Kolesov since this operation is only one of many campaigns of the serviceman. Mita himself or his descendant possibly continued hostile actions. It is probably possible to view the siege there not as a direct blockade of the winter quarters (though such cannot be excluded), but rather as a hindrance to the servicemen and Yukaghir at fishing, fish being the primary food for the population. Upon encountering the fishermen, the Chukchi killed them. The action occurred in summer. The Kolyma steward Vtoroi Kataev sent M. Kolesov on a campaign from Nizhnekolymsk. The desyatnik with a detachment attacked the camp of the Chukchi. A battle occurred in which he was wounded in the leg by an arrow or spear, but the Russians won, and siege of the winter quarters was terminated.

It is possible to trace the fate of one of the amanaty that was given in exchange for chief Mita. Four years (until 1660) he was kept in Nizhnekolymsk, where he mastered the Russian language and became an interpreter. He was baptized and given the name Afanasii Fedorov. Four Cossacks vouched for him and he received the status of a serviceman. In 1666, being in Yakutsk, Afanasii reported that

> his relatives and people of the Chukchi under him, Ofon'ka, wanted to pay tribute to the great sovereign with reindeer, but the steward and servicemen who were on the Kolyma River at the Lower tribute winter quarters in those years did not dare to take from his relatives and people of the Chukhoch reindeer tribute without a decree of the great sovereign, and asked for tribute of sables based on the edict of the great sovereign, but they, the Chukhchi, cannot get sables anywhere, because they live by the sea, not in forested places, and in the tundra, and do not go into the forest to hunt sables, therefore the foreigners of other different clans from these forest places of the land and of

[122] KPMGYa: 69; cf.: Vdovin 1944: 254; 1965: 104; Nefëdkin 2003:23.

> hunting do not let them, the Chukhchi, hunt sables in the forest and their hunting grounds and beat them to death. And for the sable scarcity, his relatives and people the Chukhchi, of the great sovereign departed from tribute payment and eternal slavery and him, Ofonka, abandoned in amanaty.

Tribute could be taken from the Chukchi in furs of foxes, wolves, Arctic foxes, and reindeer. Based on this petition, Afanasii was sent back to the Kolyma.[123]

In general, according to the petition, which justified the rebellion of the Chukchi, the latter some time after giving amanaty in 1656 wanted to pay tribute, but not with sables, which were not found among them in the tundra, but rather with reindeer. When the hunters went south into the forest zone and tried to hunt sables, the Yukaghir killed them, defending their hunting grounds. However, the Kolyma stewards did not dare to take tribute of reindeer without orders from above. Since such orders did not follow, the situation did not change. The Chukchi, as usual, abandoned their amanaty. It is possible, of course, that Afanasii simply made up his explanation in a manner clear for the Russians, and the Chukchi were not even going to pay tribute. Indeed, this same Mita had begun in fall 1656 to attack the Yukaghir, and in 1659—the Russians.

In the summer of 1669, the Chukchi Afanasii sailed from Yakutsk in a koch with the expedition of desyatnik Ivan Ermolin, who set off to investigate Penzhina. In the fall of that year the koch moored at Nizhnekolymsk, and from it they began to take cargo. I. Ermolin tells in a petition about the events that occurred later: ". . . on these same dates the newly baptized interpreter Ofon'ka Chyukhcha, and with him a Kovyma Yukaghir of the Lower tribute winter quarters Kereto, and Chernous, and Nagan . . . and Tuyuk with nephews and with relatives and with all the men betrayed the great sovereign, killed the service people and captured the weapons and all the treasury of the great sovereign, and the miscellaneous property, and took the regulations and written documents." In the Yakutsk "account book" it is mentioned that "Fomka Chyukhcha in the 178th year on the Kovyma . . . betrayed, drove off to his relatives to the Chyukhchi."[124]

---

123 Excerpt: Gurvich 1966: 49; about Afanasii: 2000: 40, 130–131.

124 Polevoi 2000: 131; cf.: KPMGYa: 147 (about the campaigns of I. Ermolin); Dolgikh 1960: 554–555; Zuev 2002: 217.

As we see, Afanasii, together with the Lower Kolyma Yukaghir, rebelled. The uprising was probably organized earlier, since rebellion of the whole clan of Tuyuk is mentioned. They killed the servicemen, seized the weapons, the treasury, grain supplies, and documents. On the whole, the Lower Kolyma Yukaghir revolted, whom Afanasii simply joined, returning to his relatives.

In 1671, the Yukaghir of different clans went to Nizhnekolymsk to deliver the tribute shortage with sable; but they were attacked "on the way, not allowed to the Lower winter quarters . . . by the traitor interpreter Afon'ka Chyukhcha with his relatives of seventy people on reindeer and with them Yukaghir. This Afon'ka engaged in battle and killed two men of the tributary Yukaghir, and wounded many others, and the . . . shortage tribute of sable took back from those tributary Yukaghir and went in the tundra to sea."[125]

Thus, Afanasii's relatives took him back, and he in turn joined them and perhaps also talked them into seizing the tribute convoy of the Yukaghir. Afanasii's group numbered 70 people on sleds, possibly not only blood relatives but also fellow tribesmen—a rather substantial number of warriors for the region. Joined to them were Yukaghir, comrades of Afanasii in the rebellion. According to the report of the Kolyma steward Pyatidesyatnik Petr Aksent'ev to the Yakutsk governor I. P. Boryatinskii of 1 July 1671, a campaign was carried out against Afanasii and his kinsmen. And this in spite of the fact that in 1671 P. Aksent'ev had only 21 servicemen in the three Kolyma winter quarters, while, in his opinion, there were supposed to be 20 Cossacks in Nizhnekolymsk, 7 at Srednekolymsk [Middle Kolyma], and 10 in Verkhnekolymsk [Upper Kolyma]. We note that the order of distribution of the forces clearly attests to the degree of danger threatening the winter quarters. In 1672 alone 11 tributary Yukaghir were killed by the Chukchi.[126]

In 1676, the Chukchi carried out an attack on the Alazeya people. In a petition in the name of the Tsar, 12 tributary Alazeya Yukaghir tell (1679): ". . . in the 184th year the unpeaceful and non-tributary Chyukchi foreigners, arriving in our tributary camps, killed many of our best people

---

[125] Vdovin 1944: 257; 1965: 72.

[126] Campaign of 1671: AII. F. 160. No. 583. Sst. 3; number of servicemen: Belov 1952: 308; 1672: Dolgikh 1960: 415.

and men and took wives and children into captivity and ran off the reindeer. And we, your servants, became without reindeer and without tribute. And now we, servants, go to distant hunts for sable tribute, fearing those Chyukchi and other strange hordes." For defense against the Chukchi the Yukaghir asked the Tsar to order the Yakutsk governor F. I. Bibikov to send servicemen. In November 1678, the Alazeya Yukaghir also explained their delay with tribute by the fact that the Chukchi had run off their hunting reindeer, and that their best people had been killed by the Chukchi.[127]

Thus, in the course of the raids the Chukchi penetrated into the forest-tundra, in the territory of the Yukaghir, along the Alazeya and Rossokha. The Chukchi probably wanted to hunt this same sable in order to trade them through the Yukaghir for necessary Russian iron artifacts. Taking up the hunting grounds, the Chukchi killed the men they encountered, and took captive the women and children, and drove off the domestic sled reindeer. The Russians were not able to actively defend their subjects. Indeed, it should be noted that in 1676 there were ten servicemen with six amanaty at the Alazeya winter quarters, while there were supposed to be 20 Cossacks.[128] It is unknown whether there were retaliatory sanctions by the Tsarist administration against the Chukchi. Probably not.

The conflicts of the Alazeya people with the Chukchi continued even later. Cossack Mikhail Mnogogreshnyi, sent to the Alazeya winter quarters to collect tribute, thus described to the Yakutsk Governor the situation in 1688: ". . . in the past and present 196th year many tributary foreigners died, killed by the unpeaceful non-tributary Chyukchi foreigners. ... And in this Alazeya winter quarters, because of the tributary foreigners and their previous infidelity, robbery, and unsteadiness, and of the unpeaceful non-tributary Chyukchi, it is frightening to collect tribute and to live in the Alazeya winter quarters with the few people of the great sovereign."[129]

---

[127] Events of 1676: KPMGYa: 239; AII. F. 160. No. 722. Sst. 21–22; cf.: Vdovin 1944: 254 (with an unreliable date of 1672). Document 1678: AII. F. 160. No. 722. Sst. 15–16, col. 2.
[128] DAI. 1857. T. VI: 407.
[129] MIYa. Ch. 3: 975–976.

In general, it turns out that part of the Alazeya people died from epidemics, part was killed by the Chukchi, and part of them simply wandered off from the winter quarters, ceasing to pay tribute. It was probably a matter not of some war of the Yukaghir with the Chukchi, but rather about the hunting conflict. The Cossack correctly regarded the situation as unstable, since the Yukaghir, seeing the small number of servicemen at the winter quarters, did not exhibit any reverence for Russian authority, and in addition, there existed a potential threat of attack on the winter quarters by the Chukchi.

From 1670 to 1680, in connection with the impoverishment of hunting furs, particularly sables, on the Kolyma the main directions of movement of the hunting people shifted to the south, and navigation in the north of Eastern Siberia was in decline, as was the fair in Nizhnekolymsk. The garrison has become even fewer in number. In 1676, the Kolyma steward had only 20 servicemen with 25 amanaty in the three winter quarters, and even the Yakutsk governor himself believed it necessary to have 60 Cossacks there.[130] The primary responsibility of the latter now was not bringing the Chukchi into submission, but rather only the collecting and convoying of tribute and, if possible, protecting the tributary Yukaghir.

On 25 May 1676 the Kolyma steward, a boyar's son, Petr Yaryzhkin informed the Yakusk governor A. A. Barneshlev: "In the present 184th year the foreigners were not able to hunt for tribute . . . and the former tributary Yukaghir from to the Lower tribute winter quarters fought with the other Kamennye Yukaghir. And the many Chyukhchi people arrived on the Kovyma River and robbed the food and storehouse of the Yukaghir, and no one to defend them."[131] It turns out that a conflict occurred between the Nizhnekolymsk Yukaghir, who formerly paid tribute to the Russians, and the "Kamennye" (Mountain) Yukaghir, probably the Shoromboi. The Chukchi arrived at the Kolyma, stole the supplies of the Yukaghir acquired in summer, killed the best people and the ordinary people, took captive the women and children, and drove off the reindeer. P. A. Yaryzhkin was not able to protect the tribute payers

---

[130] Miller 1756: 51; Belov 1956: 181; Erunov 1993: 112; Polevoi 2000: 134–135; about the number of servicemen in 1676: DAI. 1857. T. VI: 407.
[131] AII. F. 160. No. 665. Sst. 36; Vdovin 1944: 258; 1965: 72–73; Zuev 2002: 218.

from robbery because of the insignificant number of servicemen available to him at the winter quarters.

In his report to the Yakutsk governor F. I. Bibikov, the Kolyma steward desyatnik Semen Sorokoumov described the situation in a similar way (May 1679): "And in the Lower tribute winter quarters up until now, the servicemen live in seclusion from the non-tributary people the Chyukhchi and from the tributary Yukaghir for lack of people, and they can't catch fish for food or go for firewood. And there are only 10 men servicemen at the Lower tribute Kovyma winter quarters, while previously in this winter quarters lived 30 or more, and at the Upper Kovyma tribute winter quarters there are 8 servicemen, and at the Middle 3 men."[132]

As we see, in fact ten servicemen of Nizhnekolymsk were under seige, since it was dangerous to go to get provisions (to catch fish) and heating material (firewood). This danger came both from traditional enemies—the Chukchi—and from local Yukaghir who turned out to be untrustworthy.

The tributary Yukaghir submitted a petition to the new Kolyma steward, now Pyatidesyatnik Mikhail Kolesov, describing the situation at Nizhnekolymsk in 1680, the content of which the steward thus related to the Yakutsk governor:

> And the unpeaceful Chyukchi people migrated to the Lower tribute winter quarters and live one *dnishche* from the winter quarters, and watch the Russian peoples and tributary people, and if they catch someone, those people are aflicted with all kinds of torments, and they end by receiving a shameful death; and there are no defenders, only eight servicepeople in the Lower tribute winter quarters, and those are old and decrepit; and in previous years twenty-seven servicemen lived in these winter quarters; and the tribute winter quarters have rotted and collapsed, but no one there to replace them with new ones, the people are few, and there is no forest nearby: the forest that was there got cut down.[133]

In the very same complaint of the Yukaghir in the name of the Tsar it was said: ". . . in the past . . . the 187th and in the present 188th years the

[132] KPMGYa: 241; cf.: Vdovin 1944: 254; 1965: 104; Zuev 2002: 220; Nefёdkin 2003: 24.

[133] DAI. 1862. T. VIII: 9; cf.: Gurvich 1966: 49; Zuev 2002: 219 (with an incorrect date of the event of 1678).

Chukhchi pushed us, your servants, from the lower side, and do not allow us to procure fish and reindeer," and along the Anyui, Omolon, and Oloi rivers the Khodyntsy and Chuvan do not permit hunting sable.[134]

In general, in 1679–1680 the Chukchi did not permit the tributary Yukaghir to catch fish or to hunt wild reindeer. Probably in the winter of 1680 the Chukchi approached Nizhnekolymsk to 7 to 15 km—the distance of one migration (based on Yakutsk measure—*dnishche*)—and there set up a camp.[135] Since they considered the Russians and tributary Yukaghir their enemies, those caught were subjected to torture as revenge for their killed and then finished off. The steward was not even able to defend his people and himself, since there was a total of only eight veteran servicemen in the winter quarters. In addition, the structures of the winter quarters themselves were delapidated, and wood for repair was not near—that available had been cut down.

According to the report of the Kolyma steward Pyatidesyatnik Ivan Dorofeev to the Yaktusk governor general M. O. Krovkov of 20 June 1686, when on 26 October 1684 I. Dorofeev arrived on the Kolyma with nine Cossacks; in the three winter quarters were 12 servicemen, of whom six were at Nizhnekolymsk led by desyatnik Ivan Golygin:

> . . . and being in those Kovyma winter quarters I was greatly oppressed by the non-tributary and tributary people from the Chyukchi because of the small number of servicemen; and in this same 193rd year, after the official sending of tribute, the non-tributary foreign enemy of the Chyukhoch clan, gathering 100 or more men, arrived at the Nizhnii Kovymsk tribute winter quarters and at the winter quarters killed four Cossacks. In this same 193rd year the same Chyukhchi foreigners killed 18 men of the tributary old-paying Kovyma Yukaghir on the Kovyma River, and on a sable hunt the non-tributary foreigners of the Chyuvan clan

---

[134] Vdovin 1944: 257; 1965: 73, 104; Zuev 2002: 220.

[135] As A. I. Argentov (1860: 24; 1889: 84) noted, the Chukchi caravan in a day traveled 7 verst [~7 km], but if necessary it was possible to travel even 40 km. In another place, A. I. Argentov (1879: 439) noted that a *proběrd*—the distance traveled by dogsled between rests—amounted to 10 verst on the Kolyma. Among the Indigirka people in the 19th century the distance of a "dnishche" also amounted to 10 verst (AAN. F. 250. Op. 1. No. 112a. L. 192).

killed "six men of the same Kovyma Yukaghir"; "and after their murder the old-paying Yukaghir bowed to the great sovereigns of the Kovyma Lower winter quarters, that servicemen be sent on a campaign to those Chyukchi and Koryak foreigners; because of the small number of people" it was impossible "to send to those foreigners at this time"; ". . . and to go on a campaign to the non-tributary Koryak people and to the Chyukchi and to the Khodyntsy to defend the Kovyma tributary foreigners was not [possible] and because of the small number of servicemen in the three Kovyma winter quarters continue to expect every evil and murder from tributary and non-tributary foreigners." According to an earlier report of I. Dorofeev to the governor of 27 November 1685, after the murder of the Cossacks and Yukaghir an expedition was outfitted against the Chukchi. However, "they were not able to find those Chyukhochi. From their old dwellings in the tundra they, the Chyukhochi, went far to the Kamen' side beyond the Kamen'. And those lower tributary Yukaghir did not take the Cossacks and hunting people and Yukaghir of the Middle winter quarters to find those Chyukhochi far beyond the Kamen', and they themselves did not go, and returned back."[136]

In general, the events of 1685 can be imagined in this way. Since the Kolyma steward had an insignificant number of servicemen he pointed out to the governor the constant threat both from the non-tributary Chukchi and from the unreliable tributary Yukaghir. In the first quoted fragment, the Chukchi are called tributary and not paying tribute, which can be understood both literally (part of the Chukchi began to pay tribute), and simply as, in the general sense, a description of the threat from the local residents to all three Kolyma winter quarters—the Chukchi hardly threatened the Middle and Upper winter quarters. However, the Yukaghir in the document are called "*staroplatezhnye*" [old-payers], as if contrasting them with new payers of tribute, that is, the Chukchi.

If we judge by the number of Chukchi who gathered a substantial group that numbered more than a hundred people, the events most probably occurred in the winter period of 1684/85, when it was possible to move quickly on sleds. The Chukchi carried out a raid on

---

[136] Report of 1686: AAN. F. 21. Op. 4. No. 32. L. 130–131; report of 1685: AII. F. 160. No. 891. Sst. 40; cf.: Vdovin 1944: 254; 1965: 104, 105; Zuev 2002: 223; Nefëdkin 2003: 24.

Nizhnekolymsk. Appearing, evidently unexpectedly, they killed four servicemen, which was substantial for the small-numbered garrison. This group probably attacked the Kolyma Yukaghir and killed 18 of their people. The document does not report whether the Yukaghir were hunting or fishing. Perhaps the Chukchi simply destroyed the camp. The Chuvan killed six people in the forest zone during the hunt for sable for paying tribute, apparently in winter. After these occurrences the Yukaghir, who paid tribute in Nizhnekolymsk, asked in a petition that a campaign be conducted against the Chukchi and Koryak. Since the petitioners were the Yukaghir from the northern part of the Kolyma, it can be supposed that the Reindeer Koryak reached their hunting grounds and/or camps. The Khodyntsy attacked the Omoki from the southeast.

Evidently I. Dorofeev was not immediately able to organize a campaign. He probably did this in the fall of 1685, after which he wrote a report to the governor. Most of the servicemen remained in Nizhnekolymsk as the most dangerous place. Possibly a certain number of Cossacks arrived with the new steward, and recruitment could be held among the Russian promyshlenniki and trading people and it was possible to mobilize the Yukaghir, who were prepared to defend their dwellings and hunting grounds. The detachment moved into the Great Tundra, to places of former camps of the Chukchi, but the latter had migrated east. Having crossed the Kolyma (perhaps on the ice), they in winter went to the mountains, probably beyond Pyrkanai or even to the Rauchuanskii Range. The Yukaghir guides refused to lead the servicemen to the east. In this the campaign ended.

Governor M. O. Krovkov reported the situation in the district to the Tsar, recounting the report of the Kolyma steward Pyatidesyatnik Ivan Tobol'skii of 26 November 1686 in this way: ". . . and the non-tributary Chyukhoch foreigners near the Kovyma lower winter quarters are many, and every year kill servicemen at the fishing grounds." In July 1688 I. Tobol'sk'ii himself, reporting on the collection of tribute on the Kolyma, wrote: ". . . in past years and at present, in the 196th year, many tributary foreigners died and were killed by the unpeaceful non-tributary Chyukoch foreigners, and some have not been to the tributary winter quarters with tribute and have gone to distant unknown places . . . and the non-tributary

Chyukchi, having arrived at the fishing grounds, kill the servicemen."[137]

In the report of another Kolyma steward Tit Bogomolov, son of a boyar, of 9 June 1687 the situation at the Nizhnekolymsk is described in more detail (again in a narrative of the governor):

> . . . at present in the 195th year the crowd of non-tributary Chyukhchi foreigners arrived at the Kovyma lower tribute winter quarters many at a time, and in their villainous fancy wanted to take the tribute winter quarters and to kill all the servicemen, and to release the amanaty, and to rob your great sovereign's treasury; and he, Tit, because of the small number of servicemen, did not go on a campaign after them; and at the winter quarters the Chyukhchi foreigners killed two of the servicemen, and the tributary foreigners are greatly oppressed and murdered by the non-tributary villainous Chyukhchi foreigners every year, and no one can defend the tributary foreigners from the non-tributary foreigners, because the servicemen are much too few.[138]

I. Tobol'skii focused the attention of the governor on the annual attacks by the Chukchi on the servicemen at fishing in summer—the conflict over the fishing grounds, and the hostile relations of the Chukchi and Yukaghir are also mentioned. The report of T. Bogomolov describes events of the winter of 1686/87. Repeating information about the local tributary Yukaghir annually suffering attack by the Chukchi, the steward informs that the Chukchi tried to seize Nizhnekolymsk, repeatedly approaching the winter quarters in substantial number. Based on information of the steward, the Chukchi wanted to seize the winter quarters, to kill the Cossacks, to release the Yukaghir amanaty, and to take away as booty the tribute and other things. The steward did not report why these intentions were not realized. Perhaps these were only threats. However, two servicemen, who for some reason found themselves outside the winter quarters, were killed. The Cossacks were supposed to set off on a retaliatory campaign, but the latter did not happen since the number of servicemen did not permit doing this.

In 1689 the Cossack Mikhail Zinov'ev, sent with a detachment to the three Kolyma winter quarters to collect tribute, explained to the

---

[137] DAI. 1867. T. X: 357; MIYa. Ch. 3: 977–978.

[138] DAI. 1867. T. X: 357; cf.: Gurvich 1966: 49 (with the unreliable date of 1684); Zuev 2002: 224.

Yakutsk governor the shortfall in the current year by the fact that "many foreigners . . . died, and others were killed by non-tributary Chyukochi and Koryak."[139] In general, the report of M. Zinov'ev presents the same situation: part of the tributary Yukaghir died from an epidemic, and part were killed by the Chukchi on the lower reaches of the Kolyma and by the Koryak on its upper reaches.

The last campaign against the Kolyma Chukchi took place in 1708. Desyatnik Ivan Eniseiskii was sent "to the Kovyma by the steward for tribute collection and for a campaign to the unpeaceful Chyukhoch. And, being on the Kovym, collected tribute" from those newly paying tribute—three sable furs and five furs of red fox. And later in the edict of Peter I of 9 January 1711 about the awarding of a noble rank for the desyatnik for service is recounted: "And he went on a campaign to the Chyukochi with the servicemen, and on appeal those Chyukochi did not bring tribute payment to our great sovereign; and he, Ivan, with servicemen twelve yurts of those Chyukochi destroyed in military battle, in order that tributary people there were not offended and robbed by them."[140]

Steward I. Eniseiskii was sent to the Kolyma with a traditional task: to collect tribute and to bring into citizenship the remaining non-tributary people in the Chukchi region. The campaign took place after the collection of tribute, probably in winter, at the end of the year. The reason for the expedition were robberies and murders of the tributary Yukaghir, which probably occurred on hunts. One large or several small camps were discovered during the campaign—a total of 12 yarangas. Through translators they proposed that the Chukchi give amanaty and tribute, which, as usual, was followed by a negative answer. The servicemen then attacked and destroyed the yarangas and took captives. The details incidentally are not known to us. I. S. Gurvich supposes that the campaign was directed against the Nose (that is, eastern) Chukchi. However, to think such, based only on the fact that the Kolyma Chukchi had already disappeared, does not pay, though the campaign actually seems to have

---

[139] MIYa. Ch. 3: 994; cf.: the Kolyma Yukaghir—people, "for the most part killed by the Chukchi (Chulkov 1785. Kn. 1: 486).

[140] AAN. F. 21. Op. 4. No. 34. L. 118–118 ob.: cf.: PSI. Kn. 1: 435–436; Zuev 2002: 241; Nefëdkin 2003:24.

been to the east, beyond the Kolyma. Indeed, in a letter of 30 October 1740 from Nizhnekolymsk, Captain D. Ya. Laptev points out to Count N. F. Golovin that there is a Chukchi woman with his detachment, who 30 years ago was taken captive by the Russians, when she lived on the two large rivers between Kolyma and Chukotski Nos (Cape Dezhneva).[141] Precisely which rivers these are is unclear. If we proceed from the sizes of rivers, then these should be the Pegtymel' or Ekiatap (Ekvyvatap) and the Amguema. In such case it can be supposed that by 1708 the western Chukchi were weakened, and the threat to the Russians on the Kolyma was now presented by the eastern Chukchi. In such case, this campaign to a distance of 500–800 km was directed at the nomadic Chukchi who lived on the rivers of northern Chukotka.

In January 1710 "the villainous Chukchi approached the Nizhnekovymsk winter quarters deceptively; and one person, a Chyukochi man by the name Nitkal', was captured from these Chyukochi."[142] The Chukchi probably wanted to take advantage of the weakness of the Kolyma garrison—part of which was supposed to have gone to Anadyrsk—in order to open hostile activities against the residents of Nizhnekolymsk. The details of the attack and the essence of the trick itself are unknown. The Chukchi possibly wanted to present themselves as a peaceful caravan or even were such, but were taken by the Russians for the enemy. The Russian expeditionary detachment had still probably not left and the servicemen could organize resistance. The Chukchi probably hoping their deception was not revealed, retreated on sleds, not taking up battle, though one man nevertheless was taken captive. This, in essence, is the last winter attack of the Chukchi on the Kolyma people known to us in the documentary (not the folklore) sources that could be connected with the western Chukchi.[143]

Later, in instructions about this, in August 1714, to the Cossack Grigorii Kuzakov, a member of the "Great Kamchatka Squad" who was sent with servicemen to find the islands in the Arctic Ocean, it is noted: ".

[141] Gurvich 1966: 49; Pokorvskii 1941: 328; Okhotina-Lind and Mëller 2015: 776.
[142] PSI. Kn. 2: 527–528; cf.: AAN. F. 21. Op. 5. No. 20. L. 62; Okhotina-Lind and Mëller 2013: 95 (Miller, G. F. *Izvestie o morskom khode iz ust'ya Leny . . .* [News of a Maritime Voyage from the Mouth of the Lena . . .]).
[143] Zuev 2002: 64, 248; Nefëdkin 2003: 24.

. . opposite the mouths of the Kovyma and Khroma and Indigirka rivers, where free foreigners live, and do not pay any tribute to the treasury of the great sovereign."[144] Consequently, at this time local residents who did not pay tribute lived at the mouths of the Khroma, Indigirka, and Kolyma rivers. These evidently were the Kolyma Chukchi, who at this time had already moved to the west, to the Khroma River, in the territory of the Yukaghir who paid tribute to the tsar from the middle of the 17th century.[145]

Thus, after settlement in the Great Tundra, the Chukchi were able to establish relations with the Alazeya Yukaghir, who probably not did not exceed them in number, and therefore were entirely able to impose their terms of coexistence onto the latter. If the Russian administration of Nizhnekolymsk did not try to place the Chukchi under control in the 1640s, having conquered the Yukaghir, from the end of the 1640s the Yakutsk governors required the Nizhnekolymsk stewards to impose tribute on the Chukchi and to take amanaty from them, and if they refused, to take tribute and hostages by force—thus, the usual model of interrelations of the Siberian administration and the local residents was carried out.[146] In 1649–1662 the servicemen conducted several campaigns directed at conquering the Chukchi. However, in distinction from other Siberian peoples, the Chukchi refused to pay tribute, and they simply abandoned amanaty that had been taken, considering them dead. Because of this, a conflict situation permanently existed. The last fueled by Chukchi blood revenge, which was directed not to the direct perpetrators, but against the whole Russian population. In the 1670s–1680s the few servicemen of Nizhnekolymsk turned to passive defense. There are no data about the 1690s, probably because no active actions were taken. Only in 1708 was the Kolyma steward able to conduct a campaign against the Chukchi. And in 1710 the Chukchi in the last raid known to us, as earlier, approached Nizhnekolymsk with hostile intentions. After disappearance of the Kolyma-Alazeya group of Chukchi the constant danger of winter attack also disappeared.

---

[144] PSI. Kn. 2: 41; cf.: 111; Fel' 1960: 77; Zuev 2002: 64.
[145] Dolgikh 1960: 379–392.
[146] Slëzkin 2008: 31.

4. Migration of the Chukchi from the Kolyma Area to the West

Not so long ago researchers asked the question: where did the Kolyma-Alazeya Chukchi disappear to in the first quarter of the 18th century? They explained this disappearance either by the fact that this part of the Chukchi ethnos perished in an epidemic, or, more probably, by the fact that the Chukchi resettled in the east, beyond the Kolyma, sometimes explaining that this happened after a military clash.[147] Both assumptions appear quite plausible, considering a historical parallel: In 1884, a third of the Kolyma Chukchi died from smallpox, and of the remaining, about half migrated to the east, beyond the Kolyma.[148]

Only after the publication of new sources did it become clear that in 1716–1721 the Kolyma-Alazeya Chukchi carried out a genuine "Ice March" to the west, reaching the Yenisei, that is, after approximately five years (and the migrations were carried out only in winter) they made the distance, amounting to more than 2,900 km in a direct line! The primary source of this march are the notes, in essence, the diary of Academician G. F. Müller, in which he entered in the German language various information during his Siberian expedition in 1733–1743, particularly during a stay in Yakutsk from the summer of 1736 to the summer of 1737.[149]

G. F. Müller learned from the boyar's son Prokopii Efimov, formerly in 1734–1735 a Kolyma and Alazeya steward, that "the Chukchi previously lived in the tundra between the Kolyma and the Alazeya and interfered with the Russians on the Kolyma, especially often at the Lower winter quarters. However, they frequently suffered attacks from the Anadyr', or Trans-Kamennye, Chukchi. In 1716, they were brought to such a state by the latter that they were forced to escape along the sea coast

---

[147] Epidemic: Gurvich 1966: 49 (they died from an epidemic of smallpox at the beginning of the 1690s or left); resettlement to the east: Wrangell 1948: 136; A. 1878: 262, 282; Jochelson 1928: 45; Martynov 1957: 108; Vdovin 1965: 105; 1987: 121–122; Kolesov 1991: 42, 79; Erunov 1993: 112 (after a military clash); Mikhailova 1996: 189; Vakhtin, Golovko, and Shvaittser 2004: 45 (after a clash with the Russians).

[148] Bogoras 1899b: 16; 1921: 51, 87; Among the Chukchi. 1899. No. 38: 3; Gurvich 1985: 58; cf.: Argentov 1887a: 3; 1889. No. 5: 77; Popov 1893: 1; Alekseev 1996: 16; Vakhtin, Golovko, and Shvaittser 2004: 48.

[149] Elert 1991: 27–30; 1999: 41, 86–90; see also: Nefёdkin 2013: 80–84.

in order to find themselves a new residence. They crossed the mouth of the Lena River, though in the region of the Anabara River they were all wiped out by the local savage Tungus or Samoyed of the Turukhanskii District."[150] P. Efimov apparently knew vaguely the end of the Chukchi march.

The information of P. Efimov is supplemented by a report obtained by G. F. Müller in Yakutsk in 1737 from *Sotnik* [commander of a hundred.—*Trans.*] Yakov Mokhnachevskii and serviceman Prokhor Anpekov:

> Not far to the left, or to the west of the mouth of the Kolyma, into the sea falls the Chukoch'ya River, called such from the Chukchi. These Chukchi, for distinction from the Chukchi who live on the other side of the mountain called Kamennye,[151] are called Tundra Dwellers. And those and the others were not tributary, but for a long time fought both among themselves and with the Russians. However, in 1724, when Fedot Amosov wanted to go on a campaign against them, they got frightened, and simultaneously with this, they suffered an attack by the Kamennye Chukchi, so that they were saved by fleeing. The Kamennye Chukchi pursued them to the mouth of the Indigirka. From there they fled along this side of the Lena to Olenĕk, where the local Tungus annihilated them, and thus, this clan was eradicated.[152]

This report refers to preparation of F. Amosov's expedition to the Medvezh'i Islands in November 1724, about which the Kolyma Chukchi learned. They also suffered an attack from their eastern fellow tribesmen, who pursued them to the Indigirka. The Kolyma Chukchi continued the campaign to the west: having passed the Lena, they reached the Olenĕk, where the local Tungus killed them. This campaign could possibly be the second wave of flight of the Chukchi from the Alazeya tundra, which can be compared with the secondary attack of the Chukchi on the Russians in the Taimyr in a Dolgan legend.[153] On the other hand, the geography of the

---

[150] RGADA. F. 199. No. 507. Ch. 2. L. 205 ob. (translation from the German by A. C. Elert).
[151] Kamennye Chukchi are Chukchi who live on the rocky shore of the Kolyma River (ABL 2004, 194).—*Trans.*
[152] RGADA. F. 199. No. 507. Ch. 2. L. 69 ob.–71 (translation from the German by A. C. Elert).
[153] Popov 1934a: 120–121.

Chukchi campaign being described corresponds with the preceding description, and it is impossible to exclude the fact that they are the same campaign—simply the time of the event was shifted through chronological error.

During the stay in the Mangazeya District in 1739 G. F. Müller collected additional information on the fate of the Chukchi in the west:

> In 1719, up to 40 Chukchi families arrived at the Khatanga, who had previously fled from the Kolyma River, and began hostile activities against the local Yakut and Russians. However, the latter soon united and killed the foreign guests, so that only seven families of them remained, who in 1720 arrived at the Pyasina. There, however, they were subjected to such a fierce attack by the Avgi Samoyed that only three families could escape to the Yenisei River, where they met their end in the following year of 1721. The men were killed and the women and children, who were successfully caught alive, were taken captive, and sold into slavery in Mangazeya and Yeniseisk. However, captured alive was little, since the Chukchi, when they saw that they had not hope, usually stabbed their women and children to death beforehand so that they would not fall into the hands of the Russians. And most of the captive slaves subsequently hanged themselves.[154]

In general, in 1719 up to 40 Chukchi families went to the Khatanga River in the eastern Taimyr, where the Russians and Yakut crushed them. Only seven families were saved from destruction and continued their way to the west, reaching the Pyasina River in the western Taimyr in the following year. There they were attacked by the "Samoyed-Avgi." The remaining three families continued to move west and in 1721 reached the Yenisei, where the men were killed, and the remaining women and children were sold into slavery in Mangazeya and Yeniseisk, a former large slave market, from where they exported the slaves to Russia, paying customs duty.[155]

Some details supplementing G. F. Müller 's reports we find in the legends of the Nganasan. According to these legends, the Chukchi lived along the Pyasina River; having spotted the Nganasan, they committed

[154] RGADA. F. 199. No. 507. Ch. 2. L. 310 (translation from the German by A. C. Elert).

[155] L. 1901. No. 48: 3.

suicide, and the Nganasan, having taken their reindeer, became wealthy; other Chukchi managed to escape along the sea.[156] The Chukchi probably traveled farther to the west, to the Yenisei.

Overall, the circumstances of the "Ice March" of 1716–1721 can be reconstructed in the following way. The conflict causing the Chukchi to resettle from the Chaun to the Great Tundra of the Alazeya at the beginning of the 17th century was not exhausted, fueled by continuing blood revenge,[157] and the military actions between the eastern "Kamennye" and the western Kolyma-Alazeya Chukchi periodically entered an active combat phase. Since the campaign of the steward I. Eniseiskii in 1708 was directed, judging by everything, against the eastern Chukchi, it is clear that by that time they (and not the Alazeya Chukchi) represented the chief danger for Nizhnekolymsk. The fact that the last time, in January 1710, the Chukchi approached the Nizhnekolymsk stockade with hostile intent, just says that after that time the Kolyma Chukchi could no longer carry out raids onto Russian territory because of active combat activities with their "Kamennye" fellow tribesmen.

The attacking side was the Nose Chukchi,[158] whom their allies—the Anadyr' Chukchi—supported. The allies could press the enemy simply due to their numerical superiority. In 1716, the conflict reached its apogee. The tundra Chukchi decided to migrate farther from their enemies, and since the way to the south was closed to reindeer herders by the line of forests, migration could only be carried out to the west. Also, probably contributing to the decision to resettle was the fact that the Chukchi did not consider the Great Tundra their ancestral territory, still remembering of their arrival there. The nomads moved along the coast of the East Siberian Sea, which can be explained by the presence of reindeer herds and probably an inadequate number of baidary. The eastern Chukchi pursued the enemy to the Indigirka. Such a distant pursuit of the enemy was uncharacteristic for the Chukchi. It can be supposed that they simply occupied abandoned territories or the mutual resentment caused extreme

[156] Dolgikh 1976: 159–160; Labanauskas 2000: 96.

[157] Even in 1842, when several Chukchi arrived on Kolyuchin Island speaking with a strange accent they were considered descendants of the Erruva, who once fled beyond the sea and were killed (Argentov 1857: 103; 1861: 612; 1861a: 21.

[158] The Nos [Nose] Chukchi are Chukchi from Chukotskii Nos or Cape Dezhnev.—*Trans*.

stress, which induced them to such long pursuit of a hated enemy with the purpose of inflicting maximal damage on it.

During their movement to the west the Alazeya Chukchi continued hostile activities against the Russians, with whom peace had not been concluded. Also, relations did not proceed peacefully with the local population since, on the one hand, the local population was allied to the Russians (the Yakut, for example), and on the other, the intention of the Chukchi to stay in foreign territory clearly did not meet the understanding of the local population. Hence arose constant confrontation. Therefore, it would be more advantageous to move just along the sea in baidary, and not on land, where enemies waylaid the Chukchi, but this was impossible because of the reindeer herds. Passing through the thinly settled territory of the Yukaghir, probably without too much loss, the Chukchi passed the Lena and went to the region between the Olenëk and the Anabar, where the local Tungus routed them. These could have been both the Vanyadyry, who later became part of the Vadeev Nganasan and the Yakut who were settled there, and the Olenëk Tungus-Sinigir.[159] In a confrontation with the Chukchi the mastery of the Tungus as archers was probably effective since they could simply shoot the Chukchi from a distance and not engage in a hand-to-hand melee.

Figure 18. A Dolgan settlement of the 19th century. (Reproduced after Middendorff [1878. Pt. 2. Sec. 6: 693]).

[159] See: Dolgikh 1960: 152, 447–449; 1976: 10–11. It should also be kept in mind that at the beginning of the 18th century Tungus from southern regions also arrived in the north of the Turukhanskii District (Gurvich 1966: 85–86).

The remaining 40 families (about 200 people) arrived in 1719 at the Khatanga, where they clashed with the local Russians and Yakut who appeared in the Mangazeya District in 1705, having fled from the heavy tribute in their homeland.[160] These Yakut could possibly have been both real Yakut of the late Nizhnezatundrennyi [literally: Lower trans-tundra.—*Trans*.] clan who lived on the Khatanga, and the ancestors of the Dolgan. According to Dolgan legend, the Chukchi were defeated at the Khatanga by the Dolgan and Russian Zatundrennye peasants [an ethnographic group of Russian old-timers], who used guns for defense of their huts, which they shot through loopholes. The victors took the reindeer of the defeated and captured the women. The remaining Chukchi began to experience hunger.[161] According to Yakut legend, the Chukchi had a clash with the Yakut at the Khatanga "and just as many of them died from arrows as from hunger."[162] From here arose the Nganasan legend that the last Chukchi remaining alive (in one variant of the story—two) reached for bait in the "mouth" of a trap for Arctic foxes, and was pinned down by a deadfall and, not having the strength to get out, simply froze.[163] Probably, in fact, having lost the reindeer, the Chukchi began to experience difficulty finding food during the winter period.

Having suffered defeat, the remaining seven families (about 35 people) arrived in 1720 to the Pyasina. In the vicinity of this river, near the shore, was, according to Taimyr legends, where the Chukchi resided.[164] Here, on hills with stonework, were burials, among which stood sleds and lay yaranga poles, which corresponds completely to the Chukchi burial custom.[165] The ancestors of the modern Nganasan, the Tavgi, discovered the Chukchi in the tundra. The legend localizes Chukotskaya sopka [Chukchi Hill], in the center of the southern part of the Taimyr, as the place of the battle.

The demise of one family is described in an Enets legend, according to which five Tavgi, unexpectedly attacking in the night, shot the husband with a bow and killed his wife and five children. According

---

[160] Dolgikh 1960: 121, 180; Elert 1999: 85.
[161] Popov 1934a: 121; 1937: 8; cf.: Dolgikh 1961: 216–218.
[162] Tret'yakov 1869: 406.
[163] Dolgikh 1938: 22; 1976: 155; Popov 1934: 121; 1948: 11.
[164] Popov 1934a: 120–121; Dolgikh 1938: 22; 1976: 159–160.
[165] Bogoraz-Tan 1939: 1987–188.

to another legend, the Chukchi committed suicide, seeing the approach of the Khatanga Nganasan. Specifically, the Taimyr legends attribute the annihilation of the Chukchi to the ancestors of the Ninonde (Porbiny) clan, which belongs to the western Avam Nganasan. On the Taimyr, the Chukchi still possessed reindeer: the local legends mention that one family had 20 or 30 head, which then passed to the victors. A certain number of Chukchi women were probably taken captive by the Tavgi since the Avam Enets believe that among the ancestors of the Nganasan Chunanchera (Falachera, Kursiminy) clan were sniffing women, that is, Chukchi women.[166]

Three families (15 people) of Chukchi managed to load into baidary (or even into one baidar) and sail to the west. In 1721 they reached the Yenisei, where the men were killed and the women and children taken captive. G. Müller does not indicate who killed the three families of Chukchi on the Yenisei—the local Samoyed (Tidirisy—ancestors of the Nganasan, or Khantai Samoyed [Somatu]—ancestors of the tundra Enets) or the Russians. Since later the prisoners ended up in the cities of Mangazeya and Yeniseisk, the rout was most probably carried out by the Russians. In captivity, a large part of the captives committed suicide. Such was the end of the great "Ice March" of the Kolyma-Alazeya Chukchi, as it can be presented based on the notes of G. F. Müller with some clarifications and additions from the folklore of the peoples of the Taimyr.

5. Relations of the Chukchi with the Shelagi

5.1. Appearance of the Shelagi in Russian Documents

No other people in Chukotkan history have raised so many questions as the Shelagi. Their appearance, ethnic association, and reason for disappearance are not precisely known. After providing the name to a cape in the 18th century, they disappeared without a trace from the arena of history. The data preserved about the Shelagi are fragmentary and do not create a complete picture. This ethnonym is mentioned for the first time in known historical sources in the story of events in the summer-fall of 1709, when the "Shelagi-Chukchi" killed seven Cossacks led by D. Busurmanov, who, on a ship, had slipped by the mouth of the Kolyma,

[166] Dolgikh 1961: 216–218; 1976: 160; Simchenko 1976: 45, 57, 62–63.

where they had been going, and found themselves at the Bol'shaya (Large) River. The Chukchi Nitkal', captured in January 1710 at Nizhnekolymsk, told the Russians about this event.[167] We note that the Cossacks, in their testimonies about this event, reproduce the speech of Nitkal', who probably being a western Kolyma Chukchi, emphasized that the mariners were killed precisely by the Shelagi, that is, in the arrangement of his story for the Russians, the Chukchi who lived to the east, in this case—on the lower reaches of the Rauchua (Large Bol'shaya Baranikha) River.

In Lower Kolyma folklore from the beginning of the 1880s a legend has been preserved about an attack of the Shelagi on the Nizhnekolymsk stockade: "Here, in the tundra, was a people called Shelagi. Now they are dead—their bones still lie there in *saiby* [open-air burials.—*Trans*.], along the edge of the forest. They lived in the tundra, like the Chukchi now (there were no Chukchi then in the local tundra). These Shalagi once attacked the city there, burned all the houses, and killed all the people. Only those who were at seasonal work places [were fishing.—A.N.] remained alive."[168] Thus, the legend connects open-air burials, which are believed to be Omoki (Yukaghir), with the Shelagi, who lived in the tundra. They, in particular, once carried out an attack on Nizhnekolymsk, which could be an echo of some attack of the Chukchi on the stockade in the second half of the 17th–beginning of the 18th century (1656, 1662, 1679–1680, 1685, 1687, 1710). Since the Kolyma people themselves viewed the Shelagi as Chukchi,[169] it cannot be excluded that it is simple contamination of events with a Chukchi attack. However, in the brief notes of the servicepeople about the above-cited sieges and attacks on the stockade, there is no information about setting homes on fire—this is not a Chukchi battle tactic.[170]

5.2. The Expedition of Fedot Amosov (1724)

Regarding the history of Shelagi-Russians relations, it is known that in 1723 the Lower Kolyma promyshlennik Ivan Vilegin sailed by sea to the "Shelagi man" Kopai and took tribute from "several Shelatski

---

[167] AAN. F. 21. Op. 5. No. 20. L. 61 ob., 62; PSI. Kn. 2: 528; Belov 1956: 245; Vdovin 1972: 102; Okhotina-Lind and Møller 2013: 95.

[168] Olyksandrovich 1884: 295.

[169] Smirnov 1991: 401.

[170] Nefëdkin 2003: 149.

clans."[171] Therefore, the Shelagi lived near the coast in summer, and there were several clans of them. In August 1724, the boyar's son Fedot Amosov, wishing to sail to an island in the Arctic Ocean, also sailed to the chief Kopai. The crew of F. Amosov, which consisted of 60 Russians and Yukaghir, sailed in five *shitiki* [boats sewn together with withes or thongs.—*Trans.*]. In a petition, Amosov thus told the Yakutsk governor about his expedition: in 1723, he was instructed to go with service people to search "in a new land, which is opposite the Yana and Indigirka mouth" and extends still farther. In 1724, he was not able to cross the sea to this land because of the ice, but he heard from I. Vilegin that he had been told by "the Reindeer Shalag Kopai, that that land is large . . . it goes by the mouth of the Kovyma and to their Shalatskii land," from where it is also possible to get across to this land. Therefore, F. Amosov sailed from Nizhnekolymsk to Kopai,

> . . . and went near the land with much difficulty, on August 7th reached the tributary Shalagi, Kopai and relatives, from whom the aforementioned Vilegin took tribute of His Imperial Majesty in the past 723rd year. And when I arrived at their dwellings and upon invitation by this Vilegin and the interpreter Ivan Chernyi of those Shalagi, at his arrival he, Kopai and his relatives, on August 15th for this 724th year, paid the tribute of His Imperial Majesty in walrus tooth for each of fifteen people to the service people without any delay or excuses; and on the 14th day his relatives paid for six people in tooth each, in sum: twenty-one teeth. And at their payment Kopai and relatives were asked: Whether in their land were sables and foxes, and other fur things, but they themselves told nothing of this, except walrus bone [tusks], and of that they had a little, and even then it was imported . . . .

The Shelagi as gifts received iron blocks, axes, and coral beads.

> And he, Kopai, according to his foreign belief and by *sherti* [oath] exchanged with the sun, to be from that day persistently under His Imperial Majesty's hand and to pay tribute from his relatives for all the years without scolding, and promised to bring his other non-tributary relatives with tribute. And he,

[171] AAN. F. 21. Op. 4. No. 33. L. 130 ob.; Miller 1996: 33–36; Okhotina-Lind and Møller 2013: 84; Bogoraz-Tan 1934: 7; Efimov 1948: 227–228; Sarychev 1952: 81; Dolgikh 1960: 435; Vdovin 1965: 18.

> Kopai, gave his native son as an amanat without any delay and controversy and he left another of his relatives with his son.
>
> He did it and had a crafty intent and in order for his evil intention not to be revealed, he, Kopai, not appreciating noble words and gifts, and affection, and leaving his son and relative, on August 17 betrayed, and those who willingly wanted to go to the new land one serviceman and seven Cossack children, and one Yukaghir person who went with him by their will, Kopai killed; on August 16 their friends had wanted to give them clothes for winter: they had a great need, so that in winter without clothing—in the service of His Imperial Majesty not to freeze to death. And having killed these people, he, Kopai, leaving the yurts, fled with relatives. The body of the newly baptized interpreter, who was called Ivan Chernyi by the Shalagi and was at the tribute payment, did not appear with those slain bodies: it is not known whether he was taken away or killed with him, Kopai, betrayed together. And the relative of Kopai, who was left with the mentioned amanat, not knowing the evil and cunning murderous intent and anger, was sent to him, to Kopai, on August 21st, in order with them, the Russian people who went from us to him, soon went; but he, sent, at the departure, did not return to us. And the brother of the killed Yukaghir, having learned of the slaughter of his brother, stabbed suddenly not telling anyone this mentioned amanat, son of Kopai.
>
> And on August 23rd at the mentioned Shalagitskie dwellings, servicemen and Russian people of any rank and tributary foreigners of His Imperial Majesty submitted a petition in which was written: 'For this traitor Kopai to go on foot is incomprehensible, because they are nomadic reindeer people. And for us to winter over in their dwellings is impossible: there are no standing forests, and driftwood is very sparse; a winter quarters and stockade cannot be made from this and there is nothing to heat them. And for such obstacle to wintering over and to get them, the traitors, and to visit the new land opposite their Shalatskie dwellings is impossible.'

On 25 August F. Amosov returned and sailed to the Nizhnekolymsk winter quarters only on 9 September as a consequence of strong winds.[172]

In general, according to the petition, the following picture of the expedition is obtained. In the campaign were sent 60 Cossacks, Cossack

[172] AAN. F. 21. Op. 4. No. 33. L. 130–131 ob.; cf.: Okhotina-Lind and Møller 2013: 84–85.

children, and Yukaghir, and the traveler I. Vilegin and Yukaghir I. Chernyi appeared as interpreters. The goal of the expedition was not in conquering the Shelagi, but rather in reaching the northern land, which F. Amosov was not able to reach in summer of 1724 from Nizhnekolymsk by sea because of the ice. On the advice of I. Vilegin they went to the camp of Kopai, in order to sail from it to the sought-after land. On A. F. Shestakov's map this Great Land is shown, which was "discovered in 1723 by a Shelagskii prince," that is, it concerns a northern land that Kopai told I. Vilegin about only in 1723. Since it is shown on the map that in this land are "many residents, a multitude of sable, foxes, beavers, martens, and forest," it is clear that it is a matter of the western part of North America, the extremity of which was placed on maps of the 18th century to the north of northeastern Siberia.[173] Having given tribute in 1723, Kopai was already formally considered, from the Russian point of view, a subject, which he confirmed by an oath of allegiance.

Upon the arrival of the ships, I. Vilegin and I. Chernyi were sent to the camp of Kopai to invite him to pay tribute in the current 1724. A week after the arrival of the Russians, six Shelagi from the camp paid tribute at the standard amount: one walrus tusk per person. On the following day, 15 August, Kopai himself also arrived with his relatives and paid 15 tusks. Kopai with the primary part of the men probably came to the camp from summer pasture. The tributaries were 21 men, that is, the total in Kopai's camp was about a hundred people. Having received as a gift valuable iron items, Kopai gave an oath of allegiance and provided as hostages his son (probably a boy) with one accompanying relative.

After two days, 17 August, Kopai "betrayed." The son left as a hostage did not stop Kopai's activities. The reason for the "betrayal" in the petition was not indicated—the servicemen probably did not know it either. Kopai, apparently for some reason unknown to the Russians, did not want to lead the Russians to the northern land they were searching for. Nine members of the expedition were killed—eight Russians and one Yukaghir, and the interpreter I. Chernyi went missing. These were people who had no winter clothing with them. On 16 August, they went to Kopai to obtain these clothes in his camp for the intended continuation of the

---

[173] Zubov and Badigin 1953: 11; cf.: Breitfuss 1939: Pl. between 90–91; Efimov 1964: No. 68.

expedition to the north. Kopai fled from the camp, even abandoning his "yurts"—evidently, summer yarangas. This flight could have coincided with the migration to winter pasture in the interior regions. On 21 August, F. Amosov, trying to establish a dialog with Kopai, sent to him the Shelag left with Kopai's son, but he did not come back. At this same time, a Yukaghir from the expedition, learning that his brother had been killed, based on the law of blood revenge, unexpectedly stabbed Kopai's son, who had been left as an amanat. On 23 August members of the campaign, both Russian and Yukaghir, submitted a petition to F. Amosov, in which they pointed out the impossibility of continuing the expedition. It was noted in the document that pursuit of the Reindeer Shelagi by servicemen on foot was useless, which probably attests to the appearance on 17 August (the 28th in New Style) of snow cover, on which the Shelagi left on reindeer sleds. Therefore, F. Amosov did not follow the Shelagi, which he should have done in this situation, but rather returned back by sea. This petition also served him as a basis to return, not waiting for the onset of winter, though not having completed the tasks of the campaign—reaching the northern land. In general, the campaign was a geographical expedition, during which the troop acted toward the local residents in the usual way: collected tribute and brought them into citizenship.

A copy of this petition, preserved in the "Müller Portfolios," is practically the only known authentic document that tells in sufficient detail about the relationships with the Shelagi. The document permits making some observations about the Shelagi from the words of an eyewitness who stayed among them for more than two weeks, from the 7th to 25th of August. First, the residents of the camp are referred to in the document precisely as Shelagi, and not Chukchi. Second, the Shelagi appear as reindeer people, nomads, and not settled. We note that also on the ethnographic map of this time (1729–1730) also written is: "the Shelagi wander." Consequently, the assumption of W. G. Bogoras, that the Shelagi were settled coastal residents, is incorrect.[174] Third, the Shelagi were not occupied, like the Chukchi in the time of M. V. Stadukhin, with walrus hunting, going out in summer to the sea shore, but were getting ivory by means of trade or gifts. Fourth, some customs of the Shelagi are similar to

---

[174] Map: Efimov 1964: No. 65; opinion: Bogoras 1921: 58.

those of the Chukchi: the oath by the sun and neglect of their amanaty. Finally, fifth, it pays to note that as interpreters in the expedition, as usual, were Yukaghir, who incidentally were able to speak both Russian and Chukchi, as well as their own language.

According to G. F. Müller 's notebook, in 1724 the servicemen led by F. Amosov went on an expedition against the Chukchi, who, as the academician noted, were "frightened." Regarding F. Amosov's expeditions in 1724, it is known that in July he was not able to sail by sea to the northern island because of the ice, in August the conflict with Kopai did not develop into military actions, and in November of this year F. Amosov and 16 companions reached the Medvezh'i Islands on dog sleds, where he discovered only old abandoned pit houses.[175] Probably Kopai's flight, who fled because of his fear of the Russians, should be understood as information about "intimidation" of the Chukchi.

Based on information obtained by the French cartographer J. N. Delisle in St. Petersburg, Kopai was taken captive on the island. The main source of the information for J. N. Delisle was one of the maps of the Cossack leader A. F. Shestakov. Though Shestakov himself was illiterate, the map was clearly composed based on his words (1725–1726). On this map, by an island northeast of the mouth of the Kolyma, is written: "Island, occupied by Shelagi, subjects of a prince named Kopai." Kopai's dwelling, based on the estimates of F. Amosov who told G. F. Müller about this in Yakutsk, was on the coast no more than 200 *verst* from the Kolyma mouth and "is there on a small island near the seashore." A member of the expedition, Cossack Yakov Oshlapov, called this island Kagin, and between the island and the mainland was shallow water. A distance of 200 *verst* (213 km)—overlain on a modern map—gives us the region of the Rauchua River, in the mouth of which in fact there are small islands, where Kopai's people could have summered over. This island at the mouth of the Rauchua is probably shown on the French map, but in magnified form. G. F. Müller states directly that he has no information about the capture of Kopai on the island.[176] The cartographer received the information itself in

[175] Miller 1996: 33–36; Elert 1991: 29; Zuev 2002: 271–272; Okhotina-Lind and Møller 2013: 83–86; petitions of F. Amosov: AAN. F. 21. Op. 4. No. 33. L. 130–132.
[176] Miller 1996: 35–36; Vize 1933: 10; Berg 1946: 97–98; Okhotina-Lind and Møller 2013: 85. Map of A. F. Shestakov, see: Breitfuss 1939: Pl. between pp. 90–91

St. Petersburg, where he was located in 1725, from Shestakov, who could hear something of the events in Siberia. If this information about the capture had a historical base, this could have been done only by the crew of F. Amosov, who, however, sailed away, and did not hurry off in pursuit of Kopai to the island. There is no information about another Russian expedition to the Shelagi.

5.3. Information on the Shelagi in the 18th to the Early 19th Century

Information on the Shelagi can be found in cartographic materials of the first half of the 18th century. On a map of Kamchatka, composed in 1699–1715, we read: "The *nos* [cape] of the unpassable Shalatskii, and on it Chukchi with both a special language, and do not [pay] tribute"; a lengthier description is on the Yakutsk map of 1710–1711: "[On] Shalatskii Nos live people of the Shalatskii clan, and their language is different and especially the warriors, and they fight with the bow, and to suppress them is not possible, because any taken from battle as prisoner, they sacrifice themselves, slaughter with knives." On the map of the Yakutsk nobleman Ivan L'vov there is also written: "Live the Shalagi, a different clan from the Chukoch," and on a map of eastern Siberia (ca. 1726) is indicated: "On Shalatskii Nos it is known, live the unpeaceful Chuktskie of the Shalatskii clan, do not pay tribute."[177]

Judging by the legend on the first map, it is a matter of the territory of Chukotka proper, the population of which had its own language and was not tributary. The second map, composed in Yakutia, is more detailed: on it the author distinguished the non-tributary Chukchi, who lived north of the Anadyr', from the Shelagi, who were situated east of the Kolyma. The map also relates some ethnographic details: about the Shelagi's own language, about their militancy, about archery battle as the primary kind of warfare, and about the suicide of prisoners. These characteristics, like those among the Chukchi, can attest to the similarity of the two ethnic groups or about the simple transfer of qualities of the better-known Chukchi to the somewhat lesser known Shelagi. I. L'vov indicated on his

---

(Russian); Golder 1914: 111 (French); Zubov and Badigin 1953: 11. Testimony of Ya. Oshlapov (1739): Okhotina-Lind and Møller 2015: 370.

[177] Maps, see: Efimov 1964: 34: No 49; 38: No. 54; 39: No. 55; 41: No. 60; cf.; Miller 1996: 38.

map only that the Shelagi were a separate people. The last map, composed on the basis of information from Shestakov, constitutes the place of residence of the people and calls the Shelagi a Chukchi clan. As we see, on maps of the first half of the 18th century, the information about the ethnic association of the Shelagi is contradictory; it often corresponds with a description of the Chukchi, and it is possible to speak of similarity of these two ethnic groups and of some genetic connection, though on the Yakutsk maps the Shelagi appear as a people separate from the Chukchi with their own language.

In 1743, the translator Jakob Lindenau (1710–1795) advanced such hypothesis about the origin of the Koryak: ". . . based on my assumption, the Koryak were formed from a mixture of the Chukchi, Shelagi, Monoki, and Kamchadal as a result of internecine wars, in which the weakest had to retreat before the strongest and went to some other place for their safety. If one believes the reports, even part of the Shelagi not so long ago could be found on the Yenisei."[178] In this case, J. Lindenau advances his supposition regarding the ethnogenesis of the Koryak from a mixing of different peoples of the northeast (Chukchi, Shelagi, Omoki, and Itel'men) and notes that with wars, those who lost were forced to migrate to new places of residence, as did part of the Shelagi, who not so long ago could even be found on the Yenisei. Based on Alexey A. Burykin's asumption, the report refers to the Evenk-Shilyagiry, who were sometimes called Shilyagi. In fact, the Shilyagiry-Shilyagi (Kureiskie Tungus) were well known to the Russians, but they lived on the Nizhnyaya (Lower) Tunguska and the Lena. J. Lindenau, who traveled throughout Siberia, knew rather well the Tungus-speaking peoples, on whom, incidentally, he wrote two works and could hardly have taken the Chukotka Shelagi for the Tungus Shilyagi.[179]

In a description of the Anadyr' River J. Lindenau noted that ". . . the Foot Chukchi live by catching polar bears near Shalaginskii Nose [Cape Shelagskii], which is called this way because people named Shalaginskie lived near that place, and from this Shalaginskii Nose around

[178] Lindenau 1983: 105.

[179] Burykin 2000: 48; 2013: 303; about the Shilagir, see: Dolgikh 1960: 167–172; works on the Lamut by Ya. Lindenau, see: Lindenau 1983: 53–101.

by sea to the Nerpisha River have places of residence."[180] Consequently, as J. Lindenau supposed, "the Foot Chukchi" live along the coast of the Chukchi Peninsula from Cape Shelagskii to Kanchalan. And formerly (before the 1740s) on Cape Shelagskii lived the Shelagi, who gave the cape its name. In 1742 J. Lindenau marked on his map by an oblique red line the boundary of settlement of the Chukchi, beginning from the coast between the mouth of the Anadyr' and Kresta (Cross) Bay and ending at "Shalaginskii Nos" inclusive; the Shelagi themselves are not noted on the map.[181]

In 1777 Captain T. I. Shmalev, in a letter to G. F. Müller, briefly commented on the Shelagi, that "this people became extinct or moved from the Russian settlement on the Chaun River," that is, the captain, who was serving then in Gizhiga, knew essentially nothing about the Shelagi. In July of 1787 Captain G. A. Sarychev (1763–1831), telling about his excavations of pit houses on the northern part of Baranov Kamen' [Baranov Rock] east of the Kolyma mouth, noted: "Regarding the residents of these places, who without doubt must be Chukchi, the Kolyma Cossacks told that they were called Shelagi. Because of the settlement of Russians in their neighborhood, they moved to the east and established their residence near the cape that stands out farther than others to the north, which from that time became called Shelagskii."[182]

As we see, G. A. Sarychev compared the Chukchi with the Shelagi. He obtained this information from Kolyma people, who informed him that the Shelagi, because of the Russian threat, resettled from Baranov Kamen' in the northeast to Cape Shelagskii. Folkloric memory of events was probably still fresh then. This migration, based on this description, can be dated to the middle–second half of the 17th century, when from the

---

180 Lindenau 1983: 162.

181 Efimov 1964: 83: No. 122; Lindenau 1983: 2, 15, 162; cf.: Vdovin 1943: 53; Berg 1946: 54, note 6.

182 Quotation from T. I. Shmalev: RGADA. F. 199. No. 528. Ch. 2. D. 9. L. 35; quotation: Sarychev 1952: 83–84. During the course of excavations on Cape Baranov by the Kolyma Expedition in 1946, led by A. P. Okladnikov, pit houses were discovered that belonged to the Neo-Eskimo culture, dated by dendrochronology to 1690 ±165 years ago (Okladnikov 1947: 181–182; Okladnikov and Beregovaya 1971: 16–18, 22–26, 138–141). However, it must not be forgotten that local residents used the ancient abandoned dwellings as storehouses (Nordenskjöld 1879:222).

end of the 1640s the campaign of the Cossacks began for the purpose of obtaining tribute from the local Chukchi. The historicity of these events is attested by the flight in 1685 of the Alazeya Chukchi to the eastern bank of the Kolyma and farther due to the approach of the Nizhnekolymsk servicemen and the possible campaign of the Cossacks in 1708. In this case, the Shelagi could have become an ally of the Alazeya Chukchi in the second half of the 17th century and, under threat of ruin of their camps by the Russians, were forced to go to the more peaceful distant region of Cape Shelagskii. However, probably not all Shelagi migrated (or, at least they later returned) since in the first quarter of the 18th century the Reindeer Shelag Kopai will reside to the east of the Kolyma.

Based on Kolyma legends that were still preserved in the mid-19th century, the missionary A. I. Argentov compiled this story about the first acquaintance of the Russians with the Chukchi:

> . . . we have been acquainted with the Chukchi for more than 200 years. Russians saw them for the first time at the mouth of the Kolyma River. A wandering group of our hunters (volunteers), sent for a survey of the country and peoples, here met people accustomed to attack. But audacity and courage, bone-tipped arrows and guns were not equal. After the first attempt to engage in flight seized by secret terror, the savages, scaring each other with stories of hairy-faced fire-breathers, fled deep into their country. Saving themselves, the Chukchi with their herds distanced themselves in Chavan country and left the Omoki in peace, hitherto oppressed by them.[183]

Viewing the historical core of the story, it can be supposed that after the arrival of the Russians in the Kolyma region and several clashes with the servicemen, that part of the Reindeer Chukchi migrated from the Alazeya tundra to the Chaun region. This in fact could, as A. I. Argentov suggests, even have occurred soon after the beginning of Russian-Chukchi confrontations in the region in the mid-17th century. In any case, this legend shows us the connection of the Kolyma-Alazeya and Chaun regions, signifying preservation among the Kolyma people of an oral tradition similar to that told to G. A. Sarychev. However, A. I. Argentov changed the ethnonym of the Shelagi to the better understood ethnonym Chukchi.

---

[183] Argentov 1886: 7 (about the sources), 11 (quotation); cf.: Wrangell 1948: 180; Dioneo 1895: 145; Jochenson 1895: 161.

In June 1810, the traveler M. M. Gedenshtrom [M. M. Hedenström] (1781–1845), wishing to receive information from the Chukchi about northern islands, sailed to the upper reaches of the Bol'shoi (Large) Anyui, where at the camp of the "Yukaghir Chuvan prince" Kobelev he met an elderly Chukchi woman translator, widow of the Anadyr' Cossack Krasnoyarov. She told the investigator the following: "The Shelagi (a tribe of the Chukchi), who lived in the Chaun and on Shalatskii Nose, being in a long-term feud with the Chukchi, were forced to leave their homes. Some part of them, going through the Kolyma, Indigirka, and Yana, were killed by the Lena Tungus. A large part went to America. The Americans of Bering Strait, trading with the Chukchi, notified them of the escape and wealth of the Shelagi consisting of reindeer. The Chukchi, themselves looking for prey, going along the shore of America, found traces of them at the Arctic Ocean, but were not able to find them."[184]

This concise story by the Chukchi woman was clearly built on legends (Chukchi, Nizhnekolymsk Russians, and perhaps Yukaghir) known to her. And an explanation of the ethnic attribution of the Shelagi could be assigned to M. M. Hedenström himself, who inserted explanations in the text for the readers. The enmity between the Chukchi and Shelagi in the story is long-term and not limited to the one last war. The Shelagi suffered defeat and had to leave their territory. They went in two groups: one to the west, the other to the east, which can be explained by simple geographic division of those departing: the western group moved toward the sunset, and the eastern—toward the sunrise. The way to the west is depicted quite accurately and corresponds with the route taken earlier by the Alazeya Chukchi. Arriving at the Lena (even to the Yenisei—according to J. Lindenau), the Shelagi were killed in a struggle with the more numerous Tungus. A larger group went to the east, to America. During trade the Eskimos of Alaska told the Chukchi about the route of flight of the Shelagi. Quickly assembling, the Chukchi went after the fugitives, but they did not find the Shelagi. This pursuit probably occurred in the same winter as the flight: otherwise traces of the Shelagi

---

[184] Hedenström 1822. Ch. 19. P. 149–151

would have disappeared. But at the onset of spring the pursuers had to return back.

Some 18th century sources in fact assert that there was reindeer herding in North America that can be juxtaposed with the Reindeer Shelagi. In 1711, the Nose Chukchi man Makachkin, who repeatedly visited Alaska, told the serviceman P. I. Popov that the American Eskimos ". . . keep large herds of reindeer." In 1730 two Eskimos from Ratmanova Island, telling serviceman A. Mel'nikov about the Great Land (Alaska), mentioned that there "the reindeer foreigners and foot foreigners are rather numerous." It turns out that the Eskimos of the Diomede Islands, being themselves in Alaska, spoke of the existence of some reindeer people in Alaska. According to I. Skurikhin's chaotic description of the return of the boat *St. Gavriil*, the ship dropped anchor at one of the Diomede Islands (Ratmanova Island?) in 1732. M. S. Gvozdev approached the shore in a sloop with servicemen, where he obtained 20 reindeer and took from three local dwellings reindeer furs and walrus ivory.[185] It can be supposed that the reindeer, which pastured on the shore by the village, were domestic, though this is not specifically stated in the source. Since there is no mention of wild reindeer in the description of the fauna of Alaska, it can be assumed that wild [*dikie*] turned into domestic [*domashnie*] in the translation. On the other hand, it is impossible to exclude that these mentions of reindeer people relate to the fugitive Shelagi, who later disappeared.

In 1765 N. I. Daurkin also reported that, according to the Chukchi, there were herds of domestic reindeer among northern peoples. Even Academician P. Pallas remarked, concerning this testimony, about reindeer in Alaska: "Up to now travelers have not yet noted this about any North American people."[186] This comment of the academician remains relevant even at the present time—domestic reindeer in Alaska were not recorded until the end of the 19th century, when the animals were brought there from Chukotka. The information of N. I. Daurkin was evidently

---

185 Testimony of 1711: PSI. Kn. 1: 458; KPTs: 157; 1730: Sgibnev 1868: 138–139; Efimov 1950: 154; Okhotina-Lind and Møller 2001: 180; cf.: KPTs: 158; testimony of I. Skurikhin: Polonskii 1850: 402; absence of deer herding in Alaska: Bogoras 1927: 44; Vdovin 1964: 1.

186 Testimony of N. Daurkin: Pallas 1781: 248, Note "d".

taken from oral tradition of the Chukchi about the flight of the Khrakhai reindeer herders. Indeed, on his 1765 map the land north of Chukotka, believed to be an extension of America, is designated: "Land of Tikegen . . . reindeer people live in this land, and the Chukchi call them Khrakhai, their language is like the Chukchi, and they have sufficient red copper: copper spears, knives, and pots." In the oral report of N. I. Daurkin to Lieutenant Ya. M. Peresypkin, in explanation of the map, similar information was given: "There are still in the northern land reindeer people, whom they, the Chukchi, call Khrakhai and these people have copper spears and knives, and their copper is red; they speak Chukchi." And in relation to the Great Land (Alaska) proper, N. I. Daurkin mentions only reindeer hunting by the local residents, but not the presence there of reindeer herding. Tikegen itself is viewed by researchers as Kitigen, that is, the Alaskan village of Kygmin; M. Krauss proposes that this is Hotham Inlet in Kotzebue Sound, and that the name Tikegen is derived from the Eskimo name of Point Hope—Tikiġaq. Agreeing with the information of N. I. Daurkin about Alaskans is a story by a captive Alaskan Eskimo woman Itteni (1763) about the absence of sled reindeer among her fellow tribesmen and the testimony of the Koryak woman Yutta (1772) about the fact that Americans practice only hunting drives of wild reindeer [caribou].[187] Describing the voyage to the shores of western Alaska in 1732, M. S. Gvozdev mentions wild reindeer only on the peninsula. Later, the Chukchi Pangao told G. Maydell that among the Kergauly (Alaskan Eskimos) the reindeer "were very few and then only those that they traded from the Chukchi."[188] All this in general is key to explaining this phenomenon—the existence of reindeer herding in Alaska. Knowing the legend of the Khrekhai, the Chukchi simply retold it to the servicemen.

In 1821, the British captain J. D. Cochrane (1793–1825), who conversed with the Chukchi through the translator A. I. Kobelev at the

[187] Krauss 2005: 173; cf.: Breitfuss 1939: 96; Berg 1946: 109; Medushevskaya 1954: 120; Alekseev 1961: 34; Efimov 1964: 87, no. 114; report of N. I. Daurkin: Gel'mersen [Helmersen] 1876: 462; information of Yutta: RGADA. F. 199. No. 528. Ch. 2. D. 3. L. 43. A. A. Burykin (2006: 197–206; 2013: 445) suggests that the name Kitigen (Chukchi *k'ytryn*—"spit, cape") refers to Cape Blossom on Wrangel Island.

[188] The absence of traditional reindeer herding in Alaska: Bogoras 1921: 29; 1927: 44; testimony of Itteni: KPTs: 187; evidence of M. S. Gvozdev: Efimov 1948: 241; information of Pangao: Maidel' 1894: 215; Burykin 2006: 203.

Anyiui fair, noted: "They have no knowledge nor tradition of any nation called the Shelages, but they recognize the word "kopai," as applicable to the name of a person in their language."[189] The Scot evidently tried to clarify whether the Chukchi knew of people by the Russian name "Shellagi." However, the last ethnonym is not Chukchi, therefore the Chukchi did not understand what people were talking about. Indeed, at the same time, F. P. Wrangell directly mentions the Shelagi "who lived on the coastal tundra that lies east of the Kolyma. They were pushed out of there by the Chukchi. Cape Shelagskii received its name from the Shelagi. The Chukchi call this people Chauadzhan, or Chavacha, and together with that, the river and bay Chaun or Chavan." In March 1823, the expedition met the sixty-year-old Chukchi elder Kamakai at Cape Shelagskii. "The old man presented himself as a descendant of the ancient Shelagi, or as the Chukchi usually call them, the Chavany, who many years ago moved west along the seacoast and never returned. A river and a bay received the name Chavan, or Chaun, from the name of this people."[190] F. P. Wrangell dealt with the local ethnonyms and found from the Chukchi the name of the Shelagi. Based on folkloric data, he reconstructed events of the past by this method: the Chuvan lived in the tundra east of the Kolyma; some time ago they were pushed out of there by the Chukchi and went west along the coast never to return. This information generally corresponds with the evidence of the Chukchi woman, the widow of Krasnoyarov.

A member of F. P. Wrangell's expedition, Doctor A.-E. Kiber, received this description of the Shelagi:

> Here I can't help but mention the Shelagi. This people, who completely disappeared from Northern Asia, once inhabited the vicinity of Cape Shelagskii and Chaun Bay. The Chukchi, who still remember them, tell that they spoke a foreign language, that the clothing and way of life was similar to that of the Lamut, Tungus, and Yakughir who live in the tundras, and finally, that these Shelagi went to the east from Chaun Bay; after that nothing was known about them. The Chukchi call the Shelagi Chavany and Chaun Bay Chavan; by contrast we call

---

[189] Cochrane 1824: 202.

[190] Wrangell 1948: 310, 295; cf.: Argentov 1861a: 18; Maydell 1894: 273; Berg 1946: 55; Burykin 2006: 104. In February 1792, the Chaun toion [chief] Khamakhai signed a petition for renewal of the Anadyr' fort (KPTs: 190)—this is probably the chief that was with F. P. Wrangell.

> Shelagskii—Ekriem—words that lead to the assumption that Chaun Bay has some relation to dwellings of the Shelagi. Evidence of the Shelagi is also preserved in the vicinity of the Kolyma, albeit obscure and confused. They tell that they were in uninterrupted wars with the Tungus and Yukaghir and that they were finally completely destroyed by them. The Shelagi perhaps paid such tribute to the Yukaghir. . . ."[191]

The doctor cites two traditions that existed in the region in his time: the Chukchi and the Kolyma Russian. It turns out that in the beginning of the 1820s the western Chukchi themselves still remembered the Shelagi. The Shelagi (Chavany) were viewed by the Chukchi as a people with a different language, who lived in the tundra around Chaun Bay, and who were evidently occupied, like the Tungus and Yukaghir, with small-herd reindeer herding, from which emerged their comparison with the Tungus and Yukaghir. For some reason the Shelagi went to the east, and not to the west, after which the Chukchi heard nothing more about them. The author cites the adopted Chukchi name of Cape Shelagskii, *Erri* (Chukchi *Ytrin*),[192] which he learned from the local residents. According to the legends of the Kolyma people, the Shelagi carried out constant warfare with the Tungus and Yukaghir and were simply annihilated by the latter, which generally corroborates the information about the annihilation of the Shelagi in the west by the Tungus or the latter were simply mentioned anachronistically. On the whole, the Shelagi disappeared, having migrated or being destroyed.

Later, in the 1840s–1850s, the Chukchi missionary A. I. Argentov specially gathered information about the Shelagi from various local peoples:

> The legend preserved among them [the Chukchi] that they arrived on the Chaun from the east and forced from there the original inhabitants, who, not being a tribe of the Chukchi, spoke a special language and were known among the Chukchi by the word "Chavacha." The Chavachi, who lived on the Chaun and on Cape Shelagskii, having been conquered by the Chukchi, went to the west. The Chaun in Chukchi is called Chava, and Cape Shelagskii—Erren. The Chavachi, according to the interpretation of some Omoki Yukaghir, were one tribe

[191] Kiber 1824: 93–94.
[192] Leont'ev and Novikov 1989: 418.

> with the Omoki, and the Shalagskie Omoki were allegedly called Shalagi. The Chukchi recognize the relationship of the Chavachi with the present-day Omoki Yukaghir. The Nizhnekolymsk Russian old-timers recognize in their stories the Shalagi as Chukchi, adding that in the old days they sailed on the Kolyma in baidary and killed the original inhabitants; the Omoki old-timers credit this circumstance to the Russians. The Alazeya Tungus in testimonies about the old days recognize the Shalagi as a Russian people, adding that they were not baptized, perhaps were cultists. The Chukchi say that only the Chavacha people lived on the Chaun and on Cape Shalagskii.

In another place, A. I. Argentov writes:

> Regarding the tribe of the Shelagi, the Russian inhabitants of the Nizhnekolymsk Region . . . recognize this people as Chukchi. Based on the search of Baron Wrangel, these peoples were the same tribe as the Tungus; the Alazeya Tungus in their legends about the old days call the Shelagi baptized peoples of the Russian clan, the Chukchi consider them a tribe of the Omoki and call them by the word Chavach or Chavachi. At this time I readily believe the Chukchi, whose opinion about the brotherhood of the Shelagi with the Omoki is supported by the story of Anyui Yukaghir, that the Shelagi and Omoki gathered together at folk festivals, occupied themselves with games, wrestling, running, mock battle—fencing with spears. . . . The Shelagi, based on stories of the Chukchi, lived in the vicinity of Chaun Bay and on Cape Shelagskii. On Aĕk Island, 50 verst west of the cape, there are seven mounds, known there by the name Kostyanye yurty [Bone Yurts]; they consist of reindeer antlers and were built, they think, by the Shelagi. . . . The Shelagi quarreled with the Chukchi, were defeated, and went north beyond the sea. So the Chukchi told me all this during my stay among them on Cape Shelagskii. To my question, in what direction the Shelagi went, the storyteller, by the name Rina Tevlyaorgi, son of Kymakai, descendant of one of the Shelagi captured by the Chukchi, waved his hand to the northwest.[193]

As we see, A. I. Argentov gave different versions of the stories about the Shelagi that he collected from four local peoples. The Russians of the lower Kolyma definitely considered the Shelagi as Chukchi, who

[193] First quotation: Argentov 1857: 89–90; second quotation: Argentov 1861: 601–602; 1861a: 18–19; cf.: Argentov 1861: 601; 1879: 440, note 1; 1887a: 2; 1889: 71–72.

carried out sea raids on the Yukaghir; in particular, in the legends a raid by the Chukchi in baidary on the Omolon River is mentioned.[194] The missionary himself did not succeed in establishing the sources of legends of the Tungus from the Alazeya about the Shelagi as of some Russian old believers. The Tungus, who appeared in this region after the western Chukchi and Shelagi had already disappeared, brought some of their ideas about the past, juxtaposing the Shelagi with some Russian old believers. The point of view of the Chukchi, who occupied the territory of the Shelagi, seemed to A. I. Argentov as the most authoritative. According to the latter, the Shelagi were Omoki, that is, Yukaghir. Therefore, they had festivals in common with the Omoki, which the Anyui Yukaghir told about, and even the Omoki themselves were called Shelagi, which probably corresponds to the Yukaghir name Chuvantsy [Chuvan people]—the Sholilau, provided by V. I. Iokhel'son [W. I. Jochelson]. The Shelagi lived on Aion Island, where shrine piles of reindeer antlers were preserved (which among the Chukchi themselves symbolized the herds of the ancestors). The shrines are connected with the Shelagi, which suggests that they were reindeer herders. He cites in passing the history of the disappearance of the Shelagi, without details. In July 1848, he talked about this on the sea coast west of Chaun Bay with the son of the elder Kamakai, whom F. P. Wrangell met in 1823:[195] the Shelagi fought with the Chukchi, were conquered, and were forced to leave to the northwest beyond the sea. The last assertion could have emerged as a result of certain contamination with stories on the flight of the Khrekhai. Based on evidence obtained from other Chukchi, the Shelagi went to the west, as told in other stories.

At the boundary of the 1860s and 1870s, the head of the Chukotka Expedition, G. Maydell, tried to find out some information about the Shelagi from the local Chukchi, but the latter said that "their elders, of course, knew something, but that generation had died, and they themselves know entirely nothing." In another place, he noted: "In spite of my careful questioning, I was not able to learn anything about the Shelagi from the

---

[194] Bogoras 1902: 162–164; Shelagi = Chukchi: Smirnov 1991: 401; Burykin 2013: 453.

[195] Argentov 1860a: 185–187. About the archaeological sites, including the piles of antlers, on Aion Island: Dikov 1977: 206–210; deer herds of the ancestors: Vdovin 1977: 156.

Chukchi who live on Cape Erri. They said that this word was Russian, and therefore its meaning is not known to them; in their words, the cape has always been called Erri and on it, as on the island, have always lived the settled Chukchi and Chukchi reindeer herders. . .. On the Anadyr' as well they knew nothing about the Shelagi."[196] Consequently, like J. Cochrane, G. Maydell did not succeed in finding out anything about this people—this is clearly connected with the complexity of ethnic attribution in the translation, because indeed the baron did not know the Chukchi language.

That legends about the Shelagi were preserved among the Chukchi is attested by the more careful investigations conducted by W. G. Bogoras. On 5 March 1896, he recorded in his notebook: "Legends of the Chaacheny are so serious that, for example, individual families preserved legends about the origin of the Chaacheny, who lived at Chaan before their resettlement in the tundra. For example, Aichinvat says that, based on his father's line, he came from a Chaachen girl. When the Koryak defeated the Chaacheny at Chaan, one of the girls was taken captive and from her came his ancestors—he said it was great-grandfather, but of course this is incorrect. But they still tease him about being a descendant of a slave. . .. Such is the origin of Araro. Along the seashore as far as Chaun there are also Chaacheny *tyimai*, though more rarely, . . . in the tundra."[197]

In general, it turned out that W. G. Bogoras's primary informant, the Reindeer Chukchi Aivanvat (Aichinvat), descended in the fifth generation from the Chaacheny. The researcher himself believes that the genealogy of Aivanvat must be more ancient. However, it seems the Chaucha was not mistaken. If we accept one generation as 25–30 years, then the defeat of the Chaun people can be dated to 1750–1770, which corresponds to the campaign to the Koryak in the Chaun region at the end of the 1760s. This genealogy can completely confirm the reality of the events. And if these Chaun people were not Chukchi, who resettled from the west, then the Chaacheny can be the remains of the ancient population of the region. We note that, in W. G. Bogoras's opinion, the Chaacheny resettled from the Chaun into the tundra, probably to the western Alazeya. The piles of antlers (*tyimai*), made, it was believed, by the Chaacheny,

---

[196] The first quotation: Maidel' 1871: 69; cf.: Burykin 2000: 47; the remaining quotations: Maidel' 1894: 274.
[197] AAN. F. 250. Op. 1. No. 112a. L. 116 ob.

were on the coast of the ocean west of Chaun, which completely corresponds to the territory where the Shelagi lived.

If we turn our attention to modern Chukchi folk tradition, then we see that in it the Koryak in fact are active in the region of the north coast of Chukotka. Folk etymology translates the name of the cape and later the town of Pevek (Peek) as "pakhuchii" ["odorous"], since there were decomposed corpses of Koryak there, from which came a stench. Though linguists suggest that the toponym emerges from the name of neighboring Mount Peekuņei—"Swollen Mountain." However, even if the folk etymology is secondary, it is important to note that the folklore connects the locality on the northeast coast of Chaun Bay with the Koryak. The Chaun legend specifically localizes the place of the death of the hero Yurg'eyu from the Koryak on Vuskenmeem Mountain at Billings, that is, east of Chaun. East of Chaun were the three mountains of Ryrkalyaut, where the Chukchi shaman Tinyl defended himself from the Tannit, that is, Koryak. A slain Koryak hero was buried on Mount Rym`einech'yn (near Komsomol'skii), named in his honor.[198] G. Maydell suggested that still at the beginning of the 19th century the Koryak lived along the southern slopes of the mountains stretching from the upper reaches of the Anyui and Anadyr' to Vostochnyi [East] Cape (that is, on the Anadyr' plateau and Chukotka upland). It can also be mentioned that in 1850 there were 242 Chuvan people and Khodyn people in the Nizhnekolymsk parish of A. I. Argentov, as well as 35 Koryak, who did not have permanent dwellings; which also speaks of the presence in the region of Koryak, who arrived from the south.[199]

The above-cited passage about the Chaacheny in W. G. Bogoras's notebook is not unique. In another place, he noted: "Chaachen is the name of that Chukchi people who lived in the western tundra in ancient times. The Chukchi find in the tundra many sepulchral remains that belong to this people; the difference is that in these remains there are also shed reindeer

---

[198] Information on the traditional translation of the word "Pevek: I. G. Gyrgol'naut and V. N. Nuvano, Anadyr', October 2013; Mount Rym`einech'yn: Tyneru 2009: 51; translation of the name of the mountain: Leont'ev 1969: 107; Menovshchikov 1972: 141–142; Leont'ev and Novikova 1989: 302, 312; Yurg'eyu: Tyneru 2008: 9; about Vuskenmeem Mountain: Leont'ev and Novikova 1989: 109; Tinyl: Leont'ev and Novikova 1989:330.

[199] Maidel' 1925: 23; Argentov 1857: 89; 1857b: 19.

antlers, which present-day Chukchi do not put into tyimai. This people left because of conflicts with the Russians. They went beyond the sea, first to islands, then it is not known where; they went on reindeer. . . ."[200]

In this case, it turned out that the Chaacheny is the Chukchi people who lived in the Great Tundra west of the Kolyma. However, whether W. G. Bogoras in this case transmits his thoughts or a local tradition is unclear. The first suggestion looks most probable. The Chaachen were reindeer herders and placed antlers in piles not only of slaughtered reindeer, like modern researched Chukchi, but also shed ones. The reason for the disappearance of the people was war with the Russians, as a result of which they left on reindeer, first to islands, and then farther. If we superimpose this information on that known to us in other sources, we see that in the Great Tundra in fact lived reindeer Chukchi, who fought with the servicemen, and then, in the first quarter of the 18th century, they left. The fact that they left to the (Medvezh'i) islands, could be the episode of the flight of the Alazeya Chukchi.

Another legend was told to W. G. Bogoras by the Chuvan chief Vaevna at the Anyui fair on 30 March 1897: ". . . the Chaacheny—precisely Chuvan people, and not Chukchi—and the *val'kar* [dwelling] on the islands—a structure of the settled maritime Chaacheny, just as now there are settled people among the Chuvan. These Chaacheny went beyond the sea, fleeing from the Chukoch during a conflict."[201] Thus, the Chuvan people believed the Chaacheny were their relatives, who built the semi-subterranean *val'kary* on the Medvezh'i Islands. The Chaacheny were forced to flee from the Chukchi beyond the sea, which can correspond to the migration of the Omoki or even the Shelagi.

In a later unpublished article, "The Limits of Settlement of the Chukchi Tribe in Antiquity and Modern Times," W. G. Bogoras directly separates the Chaacheny from the Shelagi who lived west of them, enumerating the peoples of the region: "The Chaacheny, who lived on Chaun Bay, and some other special Chukchi, who lived to the west of the Kolyma River and soon after the arrival of the Russians disappeared without a trace, the Shelagi, who lived along the seacoast between the

---

[200] AAN. F. 250. Op. 1. No. 112a. L. 53; cf.: Bogoraz-Tan 1934: 6–7.
[201] AAN. F. 250. Op. 1. No. 112a. L. 344 ob.; cf.: Bogoras 1921: 56.

Kolyma and this same Chaun Bay."[202] This enumeration is obviously fruit of scholarly work; it was built on the addition to the oral information about the Chaacheny of written evidence about the Shelagi. W. G. Bogoras now speaks of three peoples: the Chaun-Chaacheny, the western Chukchi who vanished without a trace, and the Shelagi.

In another place in this article W. G. Bogoras provides in more detail folkloric data about the Chaacheny:

> The Chukchi legend, however, does not recognize this tribe as Chukchi. It calls them Chaachen (Čáačên). The name is derived from the name of the Čáan River (Chaun on the maps) and means "chaunskie zhiteli" [Chaun residents] are 'chauntsy' [Chaun people]. The Chaacheny lived on the coastal tundra along both banks of the Kolyma and farther east to the Chaun itself. The Chukchi, of course, encountered them on the Chaun. From this comes their Chukchi name. Chukchi legend recognizes them as a related people, but still different from the Chukchi, who had their own language. Incidentally, the Russian form of this Chukchi name, *chauntsy* [Chaun people] or *chavantsy* [Chavan people], almost matches the name *chuvantsy* [Chuvan people].

Further, W. G. Bogoras, on the basis that in Chukchi the Shelagi are called Chavan, concludes: "Incidentally, all the information on the Shelagi, that has come to us, define them as an exclusively coastal tribe," and the Chaacheny, in Chukchi legend, were reindeer people. "If, in fact, we accept the identity of the Shelagi with the Chaacheny and the western 'Chyukhchi,' then it should perhaps be accepted that the Chyukhchi-Chaacheny were a reindeer division of the tribe, and the Shelagi a coastal division. ... The name Shelagi is of unknown meaning and even the sounds of it are foreign to the Chukchi-Koryak language, as well as the allowable combination of consonants and vowels in it." Based on Chukchi legends, the Chaacheny and Shelagi "emigrated beyond the sea."[203]

W. G. Bogoras gives his scholarly thoughts, based on information about the Chaacheny parsed folklorically. The place of habitation of the Chaacheny the author now determines as from the Alazeya tundra to the Chaun (where their name was derived from), that is, including the

[202] Bogoras 1921: 31; cf.: Bogoraz-Tan 1934: 6–7.

[203] The first quotation: Bogoras 1921: 54; cf.: Bogoraz-Tan 1934: 7; the remaining quotations: Bogoras 1921: 57–58, cf. 59; Anuchin 1918: 41.

territories of habitation of the western Chukchi and Shelagi. With reference to the Chukchi, the scholar suggests that the Chaacheny had their language (which the Chukchi told A. I. Argentov about), but at the same time they were related to the Chukchi people, which can be understood not literally, but as a description of peaceful relations between ethnic groups. The opposition in claims, on the one hand, that the Chaacheny were reindeer herders, as follows from descriptions in Chukchi folklore, and that, on the other hand, the Shelagi lived on the coast, W. G. Bogoras suggests resolving by the division of the people into settled and nomadic. Such identity is also supported by the legend about resettlement beyond the sea of both the Shelagi and the Chaacheny.

As we see, the ethnic association of the Shelagi remains unclear. In the 18th century the Shelagi were believed to be Chukchi. Academician G. F. Müller, on the basis of inquiries in Yakutsk of the mariner F. Amosov (in 1736–1737), who himself visited the Shelagi Prince Kopai, directly names the Shelagi "a clan of the Chukchi," evidently from the words of the traveler. In 1739, another member of F. Amosov's expedition, Cossack Yakov Oshlapov, told of a voyage to the Chukchi, not mentioning the name of the Shelagi. This established view was included in the *Geographic Lexicon* of F. A. Polunin (which, incidentally, G. F. Müller himself edited), and also in the work of M. D. Chulkov about commerce.[204] However, it should be noted that the Russians identified as "Chukchi" not only proper representatives of this ethnic group, but also Eskimos, which can be explained by the bilingualism of the latter and their interaction with the servicemen in the Chukchi language. There is no data about the presence of such bilingualism among the Shelagi. At the beginning of the 19th century, M. Hedenström continued to believe the Shelagi were a Chukchi tribe. However, later, in the middle of this century, the Shelagi began to be viewed as the Yukaghir or as a people related to them (A. I. Argentov). From this, possibly, emerged the legend of the flight of the Shelagi beyond the sea—in this way the Omoki vanished, who departed beyond the sea, probably escaping an epidemic of smallpox in the

[204] Miller 1996: 34; Polunin 1773: 448, 453; Chulkov 1785. Pt. III. Bk. 1: 486; Bk. 2: 390, note 2 (G. F. Miller's edition of the *Lexicon*: Kosven 1961: 192); "Something about Siberia. . .," 1806. No. 8: 94; Shchekatov and Maksimovich 1809: 171, 225; information of Ya. Oshlapov: Okhotina-Lind 2015: 370.

1690s.[205] Juxtaposition with the Yukaghir on the basis of linguistic data at the end of the 19th century was supported by W. I. Jochelson, pointing to the fact that the Kolyma Yukaghir called the Chuvan people Sholilau, from which he derives the Russian ethnonym *chuvantsy*. However, as I. S. Vdovin noted, the Yukaghir themselves, fishermen and forest hunters, did not live on the seacoast.[206] A. A. Burykin, developing the assumption of the scholar of Chukchi studies V. V. Leont'ev, proves that the naming of the Chuvan people comes from the ethnonym *chaal'yt*—the name of the Chukchi residents of the Chaun, whereas the name of the Shelagi was formed from the ethnonym *chul'yt*—the name of the Chukchi residents of the Chulek River, a left tributary of the Malyi [Little] Chaun River; and the very topography of the Chaun, with one exception, according to the author, is also Chukchi.[207] This idea is not new: as early as the end of the 18th century C. Merck indicated that some of their fellow tribesmen called the Chaun Chukchi *Tschawa-ramkin*, that is, "those who live on the named river."[208] Another supposition emerged on the basis of excavations and the finding of permanent dwellings on Cape Baranov—the place of habitation of the Shelagi—the latter began to be juxtaposed with the Eskimos.[209] There, in fact, were found remains dated to Old Bering Sea, Birnirk, and Punuk times. These cultures are connected with the Eskimos, who occupied the coast in an earlier time. The Shelagi themselves were not a settled people who lived in pit houses, but rather were nomadic reindeer herders. In general, the question of the ethnic association of the Shelagi remains open. A substantial part of modern researchers continues to

---

[205] Opinion: Hedenström 1822: Pt. 18: 150; about the flight of the Yukaghir beyond the sea: Tugolukov 1986: 94–100; date of the departure of the Omoki: Hedenström 1830: 97; Argentov 1861a: 17–18; Gurvich 1975: 15; Tugolukov 1986: 96; reduction of the Yukaghir population at the end of the 17th century by 44%: Gurvich 1985: 16; about the demise of the Omoki from smallpox in 1694: Slëzkin 2008: 42.

[206] Jochelson 1897: 161, note 1; Anuchin 1918: 41–42; Ogorodnikov 1922: 258, note 1; Antropova 1957: 112; Dolgikh 1960: 433–434; Tugolukov1986: 98; Gurvich 1992: 76; Burykin 2000: 47–50; 2006: 103; about the Yukaghir: Vdovin 1972: 111.

[207] Burykin 1998: 87, note 37; 2000: 47; 2006: 97–105, 185; 2013: 299–305; cf.: Vdovin 1972: 102; Leont'ev and Nikolaeva 1989: 418

[208] Merck 1978: 150.

[209] Juxtaposition with the Eskimos: Beregovaya 1954: 310–312; Ŗaushenbakh 1969: 48; Okladnikov and Beregovaya 1971: 4, 142 (cf.: Okladnikov 1947: 179–182); Belov 1973: 82; Krupnik 1989: 187; Kir'yak 2012: 124; contra: Dolgikh 1960: 434; cf.: Kondratov 1983: 160; Mikhailova 1996: 178 (ascertaining of ethnic attributes).

consider the Shelagi as Chuvan-Yukaghir people. However, somewhat more probable is the assumption of those researchers who believe the Shelagi were Chukchi, which corresponds with the opinion that existed in the 18th century. Indeed, at that time the Russians had already been in the region for more than half a century and could have sorted out the ethnolinguistic situation.

5.4. War between the Chukchi and the Shelagi

Thus, as we see, the folkloric data collected by M. Hedenström, F. P. Wrangell, A. I. Argentov, and W. G. Bogoras explain the disappearance of the Shelagi as a result of a conflict with the Chukchi. The conflict of the Chukchi and the Shelagi was long, as the Chukchi woman translator told M. Hedenström about this. At least, A. I. Argentov, citing one of the variants of the story about the flight of the Krekhai, names the latter a captured Shelagi raised by the Chukchi,[210] that is, taken captive in infancy. And this is an event of the first third of the 17th century.

The details of the conflict are unknown to us. However, it is possible to assign to this war several folklore subjects, in particular, the legend of the hero Vel'vynelevyt and the local war between the reindeer herders at Ryrkaipii, where the mountain of reindeer antlers made by the characters was preserved, as well as the subject of the "Yanranai story" about the war between the western and eastern Chukchi and about the adopted child of the latter; the action of these stories occurs in the region of Mount Yanranai in Chaun Bay. The complicated relations in the region are revealed in the story of trade of the northern settled and nomadic residents with the rich western reindeer herders that was based on certain established rules; this annual fair could escalate at any moment into fighting, which was what happened, and during the flight of one family from enemies a child was lost, who was raised by western residents. In other stories the name of the character Talë—an adopted Tangitan foster child—is also mentioned. In 1927, the Settled Chukchi man Tynaargen (probably from Ryrkaipii) told about the origin of the custom of smearing the face with the blood of a reindeer brought to sacrifice. This custom, in the words of the storyteller, emerged with the reconciliation after the battle of the settled residents and the newly arrived reindeer herders, who spoke

---

[210] Argentov 1861: 612; 1861a: 19–20.

a different language.[211] These legends were memories about war in the northern part of Chukotka and could entirely be an epically reinterpreted episodes of the conflict of the eastern Chukchi with the reindeer herding Shelagi.

If we accept the suggestion of W. G. Bogoras about the connection of the Alazeya and Chaun Chukchi, then the latter could have supported the first in the war with the Nose Chukchi in the second half of the 1710s. As in the case with the Alazeya Chukchi, the Shelagi lost the war to the eastern Chukchi and were forced to migrate. The recorded legends give several directions for resettlement of the Shelagi: to the west, to the northern islands, and to America. All these directions of migration the ethnographer V. A. Tugolukov uncritically accepted as real, in addition juxtaposing the Omoki, Shelagi, and Krekai's people.[212] In 1810, a Chukchi woman told with details about the flight of the Shelagi with reindeer to America. The details of this narrative about the unsuccessful pursuit of the Shelagi by the Chukchi suggests the idea of some historical base for the narrative. Though the flight itself over the ice with reindeer herds could be simple contamination with the more ancient flight of Krekhai to America, history could also have been repeated. On the other hand, the fact of the flight to the west over the Kolyma, Indigirka, and Yana to the Yenisei is supported by information obtained by J. Lindenau and M. Hedenström. Since the Shelagi fled with reindeer, it is clear that they were saved from complete defeat and captivity. The Chukchi, wishing to obtain booty, pursued them (to the shores of Alaska). The flight, analogous to the migration of the Alazeya Chukchi, was carried out along the coast to the west, where the Shelagi perished in a battle with the Lena Tungus and Yukaghir (M. Hedenström, A.-E. Kiber). The flight to the northern islands (Medvezh'i) could be both a peripheral episode of the flight of the Shelagi, and also a simple contamination with the flight of the Yukaghir, Alazeya Chukchi, or even Krekhai. Possibly the primary directions of the migration of the Shelagi were two: the western part of

---

[211] About Vel'vynelevyt: Menovshchikov 1974: 307–310; the "Yanranai story": Baboshina 1958: 213–214; the legend about trade: Voskoboinikov and Menovshchikov 1959: 542–546; Menovshchikov 1958: 159–163; 1964: 122–126; 1969: 218–222; 1988: 303–308; Vakhtin 2000: 225–240; about Talë: Bogoras 1900: 359–365; Tynaargen: Kavelin 1931: 98.
[212] Tugolukov 1986: 98–100.

them fled along the coast to the west, and the eastern part to the east toward Alaska. However, part of the Shelagi remained in place, which is attested by the lineage of elder Kamakai—maybe it was just the captives (usually—women and children). Indeed, the Reindeer Chukchi man Aivanvat considered himself a fifth-generation descendant of a Chaun woman who had been taken captive by the Koryak. In such lineages were simply meant descendants of Shelagi women and children captured by the Chukchi.

The time of disappearance of the Shelagi, as already indicated, is not precisely known. In legends on Russian maps of the first third of the 18th century, representing northeastern Siberia, the Shelagi are constantly listed. For example, if the master map of the Russian empire of 1734 shows "narod shelatov (Schelati populus)," then the map of 1745 now shows only Cape Shelagskii.[213] In the beginning of the 1740s J. Lindenau mentions the Shelagi in the past tense; over several years the campaign of D. I. Pavlutskii on the Chaun (1746) was directed against the Chukchi, and not against the Shelagi, who are not mentioned in the description of the expedition. In 1763, the merchant N. Shalaurov had already placed the Chaun Chukchi in a special cultural-economic group of reindeer herders who lived permanently on the coast. In 1775 T. I. Shmalev explicitly stated that only Chukchi lived on the Chaun.[214] Thus, it turns out that the terminus ante quem for the disappearance of the Shelagi is the beginning of the 1740s, whereas the terminus post quem is absolutely clear—1724, the expedition of F. Amosov to the Shelagi of Kopai. If J. N. Delisle's information is reliable, that Kopai later fled to some island, then this could be Aion,[215] where piles of reindeer antlers were located. This flight could have occurred when the strait was frozen, that is, at least in the winter of 1724/25. However, Kopai's flight to the island was probably the flight of only one group of Shelagi, and not from their region but within it. We note that between 1733 and 1738 some calm was observed in the Anadyr'

[213] Shelagi on maps, see: Efimov 1964: No. 49 (1699–1715), 54 (1710–1711), 55 (ca. 1710), 60 (ca. 1726), 65–66 (1729–1730), 71 (1733), 79 (1736–1743); maps of the Russian empire 1734 and 1745: Burykin 2006: 102.

[214] Campaign of 1746: KPTs: 168–169; RGADA. F. 199. No. 528. Ch. 2. D. 3. L. 16 ob.; N. Shalaurov: Belov 1954: 181; T. I. Shmalev: RGADA. F. 199. No. 539. Ch. 2. D. 6. L. 63.

[215] Vize 1933: 10.

theater of warfare. Could it possibly be connected with the revival of warfare of the Chukchi in the western direction, in particular against the Shelagi?

According to the investigations of Dr. F. M. Augustynowicz (1809–1891), who worked in the region in 1874–1876, the Nose Reindeer Chukchi arrived on the Chaun only at the beginning of the 19th century in search of pastures for the reindeer, but this could have been a matter of a mass resettlement of the Chukchi, which appeared on the Chaun even earlier, in the middle of the 18th century, which is clearly attested, for example, by the maps of J. Lindenau and F. Plenisner, where Cape Shelagskii and Chaun (in the second case) are shown as Chukchi territory.[216]

In general, as we see, the sources of information are completely inadequate for building any coherent history of the Shelagi and their wars with the Russians and Chukchi. Probably with the consolidation of the Russians at the mouth of the Kolyma, the Shelagi who lived to the east, by the seacoast, in the second half of the 17th century moved because of conflicts with servicemen farther to the east, to Chaun Bay. This distancing can explain why the Shelagi appear in Russian documents only in 1710. On the Chaun, among the Shelagi, hunting conflicts began to appear with the Chukchi penetrating there from the east, which J. Lindenau mentions, telling of Chukchi there hunting for polar bears. Blood revenge led to maintenance and prolongation of conflict, which ended in a series of military actions and flight of the Shelagi in the second quarter of the 18th century. They fled both in a western direction and to the east toward Alaska. The further fate of the fugitives is unknown; this permits supposing that they did not long survive their flight.

6. Warfare in the Second Quarter of the 18th to the Early 19th Century

Warfare on the Kolyma in the 18th century is a subject that has not been studied extensively. In order to demonstrate this, it is sufficient to examine the relevant parts of presently published works. For example, in the book about the Kolyma region by the regional historian M. I. Kolesov there is generally no information about this. In the most complete

---

[216] Avgustinovich 1878: 50; maps: Efimov 1964: No. 126; Gol'denberg 1984: 86.

monograph of A. S. Zuev, about the relationships of the Russians and natives in the region, only events up to 1726 are analyzed. And in another monograph, dedicated to the joining of Chukotka to Russia, the author examines primarily events in the Anadyr' theater of warfare.[217]

In November 1720 two Kolyma promyshlenniki, Ivan Vilegin and Grigorii Sannikov, went on sleds from the mouth of the Chukoch'ya River to one of the Medvezh'i Islands (probably to Krestovskii). In November 1724, 18 men did this on sleds, led by the Yakutsk boyar's son Fedot Amosov, who set off on their way from the tundra west of the Chukoch'ya River.[218] No Chukchi or obstacles for this expedition were reported, which is evidently not by chance—there were now none in the area.

The situation in the region at the boundary of the 1730s and 1740s is rather clearly outlined by statements of recruits from Anadyrsk who arrived in Selenginsk; the Irkutsk provincial chancellery sent a copy of these statements to the senate in February 1741: "And they [the Chukchi] arrive in the fall from Shalatskii nos [Cape Shelagskii] at the mouth of the Kovyma River to the Nizhnekolymsk stockade, and to other rivers in a multitude and in the best places hunt every kind of animal, and they cannot be prohibited because of the small number of Russians and there is always fear of them, and where they converge with the Russians and the tributaries, then they kill them and take them captive, and also arrive at the Anadyrsk stockade along the Anadyr' River and do the same."[219]

It is worth noting that in this "story" only the Chukchi are spoken of; there is nothing about the Shelagi, though the Kolyma legends attributed similar maritime raids on the Omoki from Cape Shelagskii precisely to the Shelagi.[220] The Chukchi arrive from Shelagskii Nos, in substantial number for the sake of their security, to the mouth of the Kolyma and the surrounding rivers (evidently including the Chukoch'ya) for hunting, where they are competitors of the Russians and tributary Yukaghir, whom on occasion they kill or take captive. It says nothing about attacking settlements or the stockade.

---

[217] Kolesov 1991; Zuev 2002; 2009.

[218] AAN. F. 21. Op. 4. No. 33. L. 130 (I. Vilegin); 132 (F. Amosov); Miller 1996: 34–36; Zubov and Badigin 1953: 82; Belov 1956: 247; Lebedev 1957: 202.

[219] KPTs: 162.

[220] Argentov 1861a: 16.

The lower Kolyma legend connects a certain stage of confrontation of Russians and Chukchi with the name Iogann, a servant of some anonymous gentleman baron. The regional historian from Irkutsk, I. S. Sel'skii (1808–1861), retells the legend in the following way, adding in the name of the baron: ". . . at this time the Chukchi made frequent raids on these winter quarters [Nizhnekolymsk] and local living places, but by his orders the baron weaned the savage neighbors from attacks forever. As soon as Mengden learned about a mob of Chukchi, he sent with the people as leader his man Iogan, who was tall and extremely strong. Finding himself always in front with the banner in his hands, he mercilessly killed Chukchi, and after many fights, which ended with large losses for the Chukchi, they finally did not dare to attack the residents of Nizhnekolymsk." In a variant of this legend, transmitted by F. G. Safronov, Iogann headed the militia of the Kolyma Yukaghir, whom the Chukchi attacked.[221]

In order to understand the period concerned, we turn to the documents. In January 1742 the chamberlain, president of the Commerce College, Baron Karl Lyudvig Mengden (1706–1761), based on accusation of removal of Elizaveta Petrovna from power and raising to the throne Anna Ioannovna, was sent to Siberia, in this case to Nizhnekolymsk, together with other nobles (S. Kh. Minikh, A. I. Osterman, and others). The baron arrived at the stockade in September 1742. Judging from the documents, his spouse, son, daughter, sister-in-law, two servants (Rein Kurmen and Egon Pudal), and two female servants accompanied him. K. Mengden was under guard and did not have the right to leave the stockade, which in the 1750s was commanded by Lieutenant Aleksei Parshukov. The baron died on 23 February 1761 and was buried there. His daughter and wife had died earlier, and those left alive departed Nizhnekolymsk for the European part of Russia.[222]

In general, the picture drawn by the legend can be presented in the following way. The Chukchi conducted offensive warfare, attacking the population of the lower Kolyma, which consisted largely of Russified Yukaghir. How the raids were carried out—overland in winter or by water in the summer—is not specified. Since the baron himself was not able to

[221] Sel'skii 1861: 18; Safronov 1975: 35; cf.: Maksimov 1900: 384.
[222] Sel'skii 1861: 18–19; Zuev and Minenko 1992: 160; Puzyrev 1994: 157–160.

leave the stockade, being under guard, he sent in the campaign his servant Iogann—probably the Baltic German Egon Pudal, whom they called Iogann, having changed it from the rarely encountered German name Egon. He was tall and possessed great strength, which could not help but command the respect of the Chukchi, who assessed men for their physical qualities. Evidently Iogann also possessed some military experience, since the baron placed him at the head of the local militia.

The period of warfare, in which the servant of K. Mengden participated, is dated to 1743–1760 (probably earlier than 1752), that is, to the period of active battle actions against the Chukchi on the Anadyr' and on Chukotskii Nos in general. It is known that in the summer of 1759 A. Parshukov feared attacks from the coast, since only one sergeant and six veterans remained in the stockade—the other residents of Nizhnekolymsk had gone to catch fish.[223] Consequently, the lieutenant had a basis for an fearing attack—such had happened previously. Since an attack was expected from the coast in summer, it was believed that the Chukchi would sail up in baidary. It was not fortuitous that in the lower Kolyma legends the Chukchi conducted raids by water. The legend describes Iogann as a man that stopped the raids of the Chukchi. But this information, if it is not just simple epic exaggeration, can attest to the fact that raids were stopped for some time or that their intensity dropped during negotiations with the Russians on the Anadyr'. At least, in 1763 Colonel F. Plenisner, mentioning an attack of the Chukchi on the Russians in 1752, believed that the Chukchi would no longer attack Nizhnekolymsk.[224] And so it happened that between 1752 and 1763 there was no attack. At least in the summer of 1760, on the Anadyr' the Nose Chukchi told the Russians that the Chaun people were going to go to Anadyrsk and kill the Russians and Koryak there, especially Ivaka Lekhtelev, who had persuaded his fellow tribesmen to kill the Chukchi of Tatymkin and Arelpuga.[225]

In this period of 1740–1742, Major D. I. Pavlutskii was governor in Yakutsk. In July 1742 the major received instructions to transfer command of Yakutsk and go to Anadyrsk to confront the Chukchi. However, based on a Siberian story delivered by the Yakutsk merchant I.

---

[223] Puzyrev 1994: 158.
[224] AAN. F. 3. Op. 10. No. 137. L. 65 ob.–66, 173–174.
[225] RGADA. F. 199. No. 539. Ch. 2. D. 6. L. 39 ob.–40.

S. Moskvin, from Nizhnekolymsk arrived a "report with an exaggeration of the Chukchi threat. He [Pavlutskii] asked to transfer command and go to the Chukchi." Pavlutskii "prepared to set off to Kolymsk," but the Chukchi, "after looting of the Nizhnekolymsk stockade, went to Gizhiga, and therefore Pavlutskii was also directed there."[226] Of course, the legend interprets the events in its way, replacing the major's assignment from the then-non-existent Anadyrsk to Gizhiga—a village that later became the center of the region, but the information itself about the threat to Nizhnekolymsk at the beginning of the 1740s and even about the looting of the stockade is notable. It can definitely be based on a real political situation on the Kolyma, which in this case the legend connected with the different campaign of D. I. Pavlutskii in 1744 (see inset).

The Kolyma theater of military action is connected not with this, but with the following (1746) campaign to the region of Chaun Bay. The primary source for the campaign of 1746 is the report of D. I. Pavlutskii himself to the Irkutsk provincial office, which then forwarded to the senate the document with an appendix showing a map of the campaign made by Timofei Perevalov. We will let the major speak for himself, he who created the report, in his dry impartial military-bureaucratic language:

> In this year of 1746, March 18th, from the Anadyrsk stockade, taking a detachment of my regular soldiers and servicemen, I marched to the Kolyma Sea in a campaign against the unpeaceful Chyukoch and went with a convoy in all possible haste.
>
> On April 15th, having taken from the command 141 soldiers and servicemen and Yukaghir and Koryak foreigners, I went lightly ahead of the train to the sea to search for the enemy Chukoch, and Corporal Fedor Taksheev of the Yakutsk regiment remained with the command. He was ordered to follow behind the wagon train carefully and with caution and all the rest.
>
> On the same April 17th, light foreign Yukaghir marched along both sides, in front of us for scouting out good signs of the enemy.

[226] Moskvin 1864: 186.

And on the right side of our way the Yukaghir searched for the enemy Chyukoch trail, along which I hastily followed.

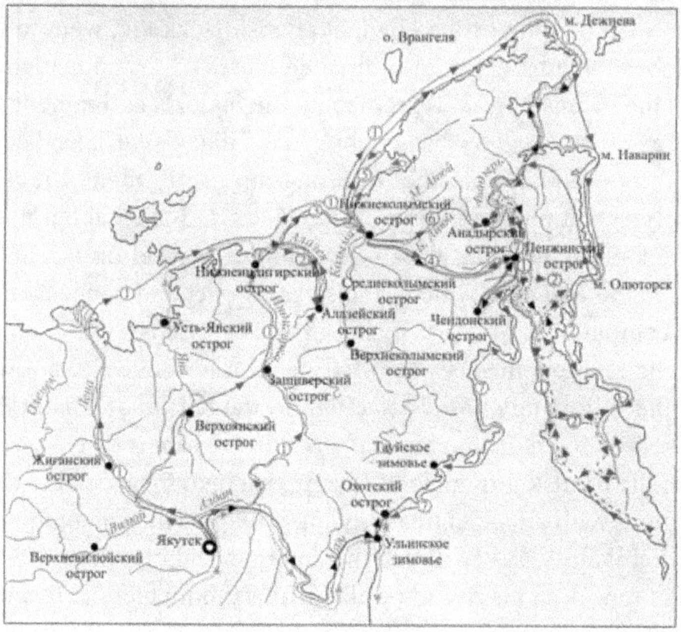

Рис.1. Походы русских землепроходцев XVII в.

Иван Ребров (1636–1641 гг.)
Постник Иванов (1638–1639 гг.)
Иван Москвитин (1639–1640 гг.)
① Михаил Стадухин (1641–1642 гг.)
Дмитрий Зырян (Ярило) и Иван Ерастов (1642 г.)
② Михаил Стадухин 1642 г.
③ Михаил Стадухин Дмитрий Зырян (1643–1644 гг.)
Исай Игнатьев (1646 г.)
Семен Шелковников (1647 г.)
④ Михаил Стадухин (1648 г.)
Семен Дежнев и Федот Алексеев-Попов (1648–1649 гг.)
⑤ Михаил Стадухин (1649 г.)
⑥ Михаил Стадухин (1650 г.)
⑦ Михаил Стадухин (1651–1657 гг.)
Федор Чюкичев (1657–1658 гг.)
Иван Камчатый (1658–1660 гг.)
Курбат Иванов (1660 г.)
① Иван Бакшеев (Рубец) (1661–1662 гг.)
② Иван Бакшеев (Рубец) (1662–1663 гг.)
Иван Голыгин (1683 г.)
① Лука Старицын (Морозко) (1693–1694 гг.)
② Лука Старицын (Морозко) (1696 г.)
Владимир Атласов (1697–1699 гг.)

Figure 19. Campaigns of Russian explorers in the 17th century. (Reproduced based on Zuev 2009: Fig. 1).

On the 18th day at sunset we ran into five yurts of the Chyukochkie that stood along the edge of the river, named Chaun, having run and attacked, out of which came sixteen men of the enemy Chyukoch to battle, and with the help of almighty God we killed them all (on our side in this battle one foreigner

> was wounded, the best Koryak), and the other Chyukchi enemies, themselves in the yurts, not wanting to go into captivity, cut themselves down, also their women and children stabbed and hanged, though two male teenagers were taken captive, two wives and a Koryak man and wife formerly captured by the Chyukchi, who were given to their relatives the Koryak. With this, five herds of sled and non-sled reindeer were taken, for example, six hundred, which reindeer were given to the servicemen, the sled reindeer to ride and the *karginy* [reindeer for slaughter] for food.
>
> On the 19th I sent the Sotnik Semen Krivogornitsyn with a detachment of sixty men and 17 Yukaghir and Koryak foreigners beside the sea to search for the Chyukoch enemy.
>
> On the 27th the aforesaid Sotnik Krivogornitsyn arrived at camp safely with the detachment and he, the sotnik announced: The enemy Chyukochi, seeing us, went escaping from one yurt, the people few in number on their light standing reindeer, after which we pursued three days, not able to catch them, besides, after a long distance the reindeer got tired, having taken only fifty of the non-rideable reindeer, and they, the Chyukchi, abandoned the yurt and property and they, the enemy, stabbed their children. Then the small number of people in the detachment pursuing the enemy, the indicated Chyukchi, and also the absence of food supplies for the soldiers and servicemen, as well as because of the reindeer being tired, were able to go no farther.
>
> Major Pavlutskii
>
> June 20, 1746.[227]

As we see, the source says nothing about the purpose of the campaign. The senate document notes only: "for the search for enemy Chukchi." The campaign in this region was evidently dictated, on the one hand, by the fact that the Chaun Chukchi appeared as allies of the Nose and together conducted warfare against the Koryak and Russians (it was not by chance that captive Koryak turned up among them), and on the other—the threat of the Chukchi to the population of the lower Kolyma. The forces involved in the campaign were significant—250 soldiers and Cossacks and 150 Yukaghir and Koryak. They moved on sleds harnessed

[227] KPTs: 168–169; cf.: RGADA. F. 199. No. 528. Ch. 2. D. 3. L. 16 ob.; Senate archive. Vol. VII. 1895: 133. Briefly about the campaign, see: Zuev 2009: 125; maps of the campaign route: Efimov 1964: Nos. 124, 127.

to reindeer collected from the Koryak and Yukaghir. Leaving Anadyrsk on 18 March, the detachment passed up along the Anadyr' and Malyi Anyui and went to Chaun Bay. On 9 April, in the region of the middle course of the Chaun, they discovered a camp of Chukchi who were fishing, whom they killed. Then the weather deteriorated, and because of a blizzard they had to stop for five days. On 15 April, the major himself took 141 fighters, leaving the remaining forces in a convoy with Corporal F. Taksheev, and with an accelerated march moved ahead along the left bank of the Chaun to the mouth, sending forward a reconnaissance of quick Yukaghir trackers, probably not burdened by heavy defensive weapons. From the text, it is clear that they were scouts and not an avant guarde that had to engage in battle. At sunset on April 18 (that is, between 9:43 pm and 12:15 am, based on data of 28 April 2014, in the Pevek region) the detachment discovered five yarangas of Reindeer Chukchi on the river. The major, without any negotiations, attacked the opponent. The Chukchi, who apparently did not expect the attack, took up battle, presenting about three warriors from each yaranga. In an unequal, obviously fleeting, battle 16 warriors against 141 of the major's fighters, all perished before successfully wounding one Koryak. Those in the yarangas, seeing the defeat of their warriors, killed their families and committed suicide. As booty, the major took four Chukchi captives, freed two captive Koryak, and seized five herds of reindeer—600 head (120 reindeer per yaranga). Memory of this battle is possibly preserved in the explanation of the hydronym of the Rytkucha River, according to which a battle was there between the Chukchi and some riders, who went then to Markovo (that is, to Anadyrsk).[228] The horsemen in this case could only have been Russians, who generally used horses, but in the region these animals were used only by merchants for transporting wares and not for military purposes.

On 19 April, having set up camp, the major sent the Cossack Sotnik S. Krivogornitsyn with 77 soldiers along the coast (and, judging by the maps representing the route of the campaign, farther east toward Kolyuchinskaya Bay) to search out the enemy. The sotnik succeeded in finding only one yaranga of Reindeer Chukchi, who, seeing the Russians, managed to get on their sleds and go, abandoning property (yaranga and

[228] Leont'ev and Novikova 1989: 200–201, 330; Nutekeu 2005: 93.

garno—reindeer hides) and killing their children. In the words of the sotnik, he pursued the enemy three days. Several more days passed in returning from the chase, and after eight days S. Krivogornitsyn rejoined the major, having seized a herd of 50 reindeer.

Not finding the Chukchi, the major decided to return. In the report, he explained his decision to return as the lack of people, supplies, and reindeer. If it is possible to agree with the second and third reasons, a detachment of 400 fighters could hardly be considered insignificant for conducting warfare in Chukotka. The expedition returned along the same route to Anadyrsk on 21 May, having spent on the campaign two months, of which they battled the Chukchi two weeks. That a total of only three camps of Reindeer Chukchi were discovered during the campaign can be explained by the fact that the population of the region was extremely sparse, or by the fact that the Chukchi, having learned of the Russian campaign, migrated to secluded places, or even by the fact that the nomads had not yet gone to the seacoast on their summer migration. At least, in 1771, moving forward along the shore of the sea from Baranov Kamen' to Shelagskaya Bay (= Chaun) from the 18th to 26th of March and returning to Nizhnekolymsk from the 28th of March to the 6th of April, the small expedition of Ensigns Ivan Leont'ev, Ivan Lysov, and Aleksei Pushkarev did not encounter anyone on their way.[229] If, in 1763, the merchant N. Shalaurov believed that the Chaun Reindeer Chukchi lived settled on the coast, occupying themselves with hunting walruses and whales, then later A. I. Argentov noted that to the west, on the coast in the region of the Rauchua River, the Chukchi lived only during the summer, whereas in winter, from September to April, they spent in the mountains.[230] In this case, it turns out that D. I. Pavlutskii poorly knew the cycle of the migration of the Chukchi.

On 23 November 1752, the head of the guard of Baron K. Mengden, Ensign Aleksei Parshukov, reported to the senate the situation at Nizhnekolymsk:

> 1752 October 25, at the request of one prisoner, to the Chukotskaya River were sent his, P[arshuko]v's, troops, an officer's servant, a worker, and 4 servicemen to bring fish oil

[229] Zubov and Badigin 1953: 61, 113–115.
[230] Belov 1954: 181; Argentov 1887: 56.

> and fish as food for themselves, and these sent to the river were attacked by unpeaceful Chukchi and everyone killed who went to the river, which is, for example, over 100 *verst* from the stockade; and at this same stockade the servicemen do not have food for themselves, besides food of fish, and every year this is satisfied from the Chukotskaya River. And since near this river the unpeaceful Chukchi presently set up a camp and took all the food and, what is more, that no harm would be perpetrated on them, if they attack this stockade, and in this stockade are 15 servicemen and there is no one else besides them around the stronghold of the stockade, and they have fear from the above-described enemy.

Then, the Yakutsk office ordered that five *pudy* [1 *pud* = 36.11 lbs] of lead and 66 fusils for the Cossacks be delivered to Nizhnekolymsk, and a decree followed about repairing the fortifications of the stockade, which was done by I. S. Shmalev in 1754. The son of the latter, captain T. I. Shmalev, thus briefly described the occurrence in 1775: "In the 751st and 752nd years, more than 100 Chukoch arrived in winter on reindeer from the Chaun River to the vicinity of the Nizhnekolymsk stockade and at the river of the name Malaya Chukoch'ya, where the Nizhnekolymsk inhabitants live in summer and store fish, and go in winter to bring it and to hunt different animals, and killed several Cossacks who were there at the time."[231]

It is obvious that before 1752 the Chukchi did not appear on the Chukoch'ya River for a rather substantial time (the Anadyr' theater of warfare distracted them), and the appearance in the region of the enemy was a surprise for the Nizhnekolymsk people. More than a hundred Chaun Reindeer Chukchi crossed the frozen Kolyma (according to T. I. Shmalev) to the Malaya Chukoch'ya River, where they killed two servants of the ensign and four Cossacks and took the food supplies. Colonel F. Plenisner learned that the events on the Chukoch'ya River developed somewhat differently than described by A. Parshukov: ". . . and in reality his, Ensign Parshukov's, report is not true (as Ensign Shmalev said when I was in Nizhnekolymsk), that he, Parshukov, sent declared people from himself

---

[231] Senate archive. Vol. IX. 1901: 204 (quote); 291–292 (decree from Yakutsk); RGADA. F. 199. No. 528. Ch. 1. D. 17. L. 15, note; No. 539. Ch. 2. D. 6. L. 77, note (quotation of T. I. Shmalev); AAN. F. 3. Op. 10. No. 137. L. 66, 173–174; Vdovin 1944: 254; Gurvich 1966: 49–50; Nefëdkin 2003: 27; Zuev 2009: 144, 159.

for Arctic fox hunting, in which case these people left from this Nizhnekolymsk to a far distance and empty places without any danger; why then do the dogs that they had, having felt the reindeer smell among the Chyukoch brought them to the Chyukchi, where because of a small number of them were killed."[232] However, the details of the case described by F. Kh. Plenisner do not change the essence of the matter. The Chukchi, probably knowing the weakness of the Russian garrison (15 Cossacks and 5 soldiers of the baron's guard), not especially fearing an attack by the Russians and thus, establishing winter quarters on the Chukoch'ya River, threatened in this way a possible attack on the stockade, which was not fortified. Probably the Chukchi viewed their hostile actions against the Russians as revenge for D. I. Pavlutskii's campaign.

In a report of 20 September 1753 to the senate, A. Parshukov noted that he, "with the troop, is in great fear of the not peaceful Chyukchi, because these arrived this September 25 for surveillance at the stockade, at a distance of 10 verst, and with this, the stockade has no fortification, and the servicemen are few in number." Consequently, on 25 September 1753 the Chukchi approached to a distance of 10 km to reconnoiter Nizhnekolymsk. Again in 1756 several Nizhnekolymsk fishermen were killed on the Chukoch'ya River, after which Ensign Zykov prohibited going "to those places for fishing." In a report to the senate the merchant N. P. Shalaurov noted that, on the Chukotskaya *viska*, "a few years this these Chaun Chyukhchi arrived in this place and killed several people." Though it cannot be excluded that for N. P. Shalaurov it was the matter of the same attack on the fishermen, but the Chukotskaya *viska*—a river flowing from a lake into the ocean—was most probably the Chukotskaya channel, located 19 *verst* from the Pokhodskaya River. In 1763, he reported to the senate: on the "Kolyma River there were no clear attacks or insults to the Russian villages on the Kolyma River or the loyal tributaries living nearby," except on the hunting grounds. In general, it turns out that the situation was again similar to that at the boundary of the 1730s and 1740s. On the eve of the expedition of 1771 Ensigns I. Leont'ev and I Lysov received a warning from the Nizhnekolymsk residents that the

[232] AAN. F. 3. Op. 10. No. 137. L. 66, 174.

Chukchi "are sometimes at the mouth of the Kolyma River near the sea."[233]

The most important observation of N. P. Shalaurov was that the Chukchi from the Chaun carried out the attack, and not the locals. This explains exactly why the promyshlenniki fished so fearlessly in the places of former Chukchi occupation. The events of 1752 attest, on the one hand, to the fact that Ensign A. Parshukov thought that fishing on the Bol'shaya Chukoch'ya River was rather safe for a small group of people—indeed, there had been no attacks there by the Chukchi for a long time—and on the other, that in the old place of settlement were Chukchi who had evidently sailed there in summer or moved over the ice of the rivers in fall, the Chaun Chukchi and not the remnants of the local Kolyma Alazeya. In fact, somewhat later, in 1769, the storehouses and huts for fishermen were located along the water arteries west of the Kolyma, beginning from the Krestovaya River 19 verst from the old Nizhnekolymsk stockade and ending at the mouth of the Krestovka River, also including both the Malaya and Bol'shaya Chukoch'ya rivers.[234] All this attests to the relative safety of fishing in former Chukchi possessions. At least, in October 1763 F. Plenisner proposed withdrawing the primary forces from Nizhnekolymsk, believing that the 30 Cossacks there led by a boyar's son would be quite sufficient, because the Chukchi would not dare attack the stockade since they lived far away and did not have firearms.[235]

In winter, passage in the region was rather safe. The Cossack F. S. Tatarinov and baptized Omok Yukaghir E. F. Konovalov, having arrived from Nizhnekolymsk on the Krestovaya River in May 1755, lived there until March 1756, and then, after the campaign to the Medvezh'i Islands—before the winter season of that same year. At the end of November 1762 N. P. Shalaurov with three companions passed to the Chukoch'ya River and farther to the Krestovaya. From 4 March to 2 May 1763 and from 16 March to 8 May 1764 the small expedition of Sergeant S. Andreev went

---

[233] Events of 1753: Senate Archive: Vol. IX. 1901: 291; Zuev 2009: 144; 1756: Zubov and Badigin 1953: 26, 84, 116 (quotation of the order); report of N. P. Shalaurov to the Senate of 1763: Belov 1954: 181; location of the Chukotskaya channel according to the expedition log of 1769: Helmersen 1876: 463; Zubov and Badigin 1953: 102; news of the ensigns based on their report of 1771: Zubov and Badigin 1953: 113.

[234] Helmersen 1876: 463–466; Zubov and Badigin 1953: 100–104.

[235] AAN. F. 3. Op. 10. No. 137. L. 65 ob.–66, 173–175.

by dogsled from Nizhnekolymsk to the Krestovaya River to investigate the Medvezh'i Islands and returned without incident. From 1 March to 7 April 1769 Ensigns I. Leont'ev, I. Lysov, and A. Pushkarev, with a small detachment, went safely from Nizhnekolymsk to the Bol'shaya Chukoch'ya River and farther to the Krestovka River, from which they set off to the Medvezh'i Islands, and then returned by the same route; and 27 February to 9 March 1771 they proceeded safely from the mouth of the Chukotskaya channel and Laptev lighthouse to the fifth Medvezh'i island.[236]

In June 1787 Captain G. A. Sarychev saw on the Omolon *letov'e* [summer pasture], at the confluence of the Omolon and Kolyma, a cross set-up in 1718, on which could be seen traces left, according to local peasants, by Chukchi arrows during an attack on Russian settlements.[237] Since the village itself was 20 verst above on the Omolon, and fishing occurred only in summer at the letov'e, it can be supposed that the Chukchi entered the empty letov'e in winter. However, based on a Yukaghir legend told to W. G. Bogoras in 1895, a hundred years ago the Chukchi arrived in baidary to the settlement on the Omolon River, from where the Yukaghir had time to evacuate, leaving the *karbasy* [sail-rowing boats] with punctured bottoms. The Chukchi, who became stranded in these boats, drowned, and those who sailed in their baidary could not cope with the current of the river. As a result, the whole flotilla sank.[238] Of course, the events described by G. A. Sarychev and W. G. Bogoras could belong to different raids of the Chukchi on the Omolon, but the parallelism of them is nevertheless striking. At the time of G. A. Sarychev's voyage the attack on the letov'e was already distant history, but it happened after 1718. The dating of the raid by an informant of W. G. Bogoras as the 1790s is, of course, rather tentative—it is entirely probable that the event could have happened even earlier.

Legends of the Kolyma people recorded at the end of the 19th century preserved some information on Chukchi raids on the residents of

---

236 About F. S. Tatarinov, see: Zubov and Badigin 1953: 82–84; about N. P. Shalaurov, see: Belov 1954: 171; documents of the expeditions of S. Andreev in 1763 and 1764: Sokolov 1852a: 104–118; Zubov and Badigin 1953: 86–98; expedition log of 1769 and 1771: Helmersen 1876: 463–471; Zubov and Badigin 1953: 99–115.
237 Sarychev 1952: 82, cf.: 74.
238 Bogoras 1902: 164.

the lower Kolyma. These legends are connected to geographic toponyms. According to folklore, some Russian settlements on the Kolyma were simply destroyed by the Chukchi. Pogromnoe—a *zaimka* [settlement] of fishermen three verst below Nizhnekolymsk—was destroyed by the Chukchi in retaliation for the destruction of their camp. The attackers arrived in baidary, penetrated the cross with spears, killed the men, led the young women away captive, and burned the houses (about ten). In the Duvannoe zaimka, where there were many houses, the Chukchi attacked in the fall. The residents, noting the gathering of the Chukchi, were able to sail away in the fog, leaving three karbasy punctured and plastered with clay. The Chukchi launched the karbasy in pursuit, but drowned. In revenge, those remaining on shore burned the houses. In another version of the legend, it says that the Chukchi attacked Duvannoe at night, annihilated the residents, while the rest drowned trying to flee in karbasy punctured by the Chukchi. The place received its name due to the fact that the Chukchi divided the booty there. On a fall night, the Chukchi attacked the village of Chukoch'ya—a Russian settlement on the western mouth of the Kolyma 53 km from the East Siberian Sea, which did not exist at the end of the 19th century. In the settlement, there was a high observation tower facing the tundra, from which the enemy was seen, but no significance was attached to the report. The Chukchi attacked the residents and killed all those running out from the dwellings. News of the massacre was brought to Pokhodsk by a fellow who had pretended to be dead. The Chukchi, having crossed the Kolyma on the ice, wintered west of the river; the retaliating campaign of Nizhnekolymsk Cossacks against the Chukchi ended in nothing: they allegedly did not find the enemy. Pokhodsk, itself on the western mouth of the Kolyma, was also destroyed and burned by local western Chukchi.[239] Attesting to clashes with the Chukchi, in the opinion of local residents and, following them, researchers, are the names of the localities and rivers near Nizhnekolymsk: the rivers Ubiennaya [Slain], Krovavaya [Bloody], Razboinich'ya [Predatory], and the

[239] Pogromnoe: Olyksandrovich 1884: 295; Dioneo 1895: 116; Bogoras 1902: 164; 1902a: 71; Duvannoe: Olyksandrovich 1884: 296; Shklovskii 1892: 96; Bogoras 1902: 162–163 (second variant of the legend); 1902a: 71; village of Chukoch'ya: Bogoras 1902: 161–162; Bogoraz-Tan 1934: 48; Pokhodsk: Bogoras 1921: 53; Bogoraz-Tan 1934: 6.

settlements of Pogromnoe [Massacre], Duvannoe, and Tomilino, where, according to tradition, a girl, wounded by an arrow, languished [*tomilas'*].[240]

Of the numerous clashes between the Russians and Chukchi during fishing and on hunting trips, a rather detailed description of an attack by the Chukchi on the Russians during a hunt for molting geese in mid-July was preserved in the legends. During this hunt, the old man Portnyagin discovered the Chukchi, who shot at the hunters from hiding places. Portnyagin, evidently a historic person with whom the name *toni*—a certain expanse of the river for the fishing artel'—was connected.[241]

The precise dating of these attacks of the Chukchi is not clear. The folklore clearly tells of diachronic raids on different settlements. At the time of F. P. Wrangell's expedition (beginning of the 1820s) there were "legends . . . about frequent devastating raids of the Chukchi on the primitive Russian settlements along the left bank of the river. Some say it was then when Pogromnoe and Ubiennoe localities got their names." Based on the information of Olyksandrovich, the raids were before the campaigns of D. I. Pavlutskii, whereas W. G. Bogoras, by contrast, suggests that the routs occurred after these campaigns; he dates the attack on Chukoch'ya village to the first decades of the 18th century. Based on the information of I. V. Shklovskii, the Chukchi destroyed Pogromnoe a hundred years ago, that is, in the 1790s.[242] Of course, all the attacks belong chronologically to the second half of the 17th to the 18th century. Raids on land can be assigned primarily to legends, telling about the actions of the Kolyma-Alazeya Chukchi, whose military actions were directed at annihilation of Russian settlements on the Kolyma (Duvannoe, Pokhodsk, and the village of Chukoch'ya); from this the burning of Russian settlements also emerges. All these events can be dated between the 1650s

---

240 Shklovskii 1892: 96; Bogoraz-Tan 1934: 48–49; cf.: Wrangell 1948: 310 (the localities of Pogromnoe and Ubiennoe); Bogoraz-Tan 1934a: 6–7 (Tomilino, Pogromnoe). Incidentally, according to A. I. Argentov (1861a: 16), Pogromnoe and Duvannoe received their names due to the rout of the Omoki Yukaghir by the Chukchi.

241 Bogoras 1902: 162–163; Bogoraz-Tan 1934: 48.

242 Wrangell 1948: 310; Olyksandrovich 1884: 295; Bogoraz-Tan 1934a: 6 (dating of the attacks after D. I. Pavlutskii's campaigns), 6; 1934: 47 (Chukoch'ya); cf.: Bogoras 1921: 40 (attacks on Nizhnekolymsk—end of the 17th–beginning of the 18th century); information of I. V. Shklovskii: Dioneo 1895: 116.

and 1710. If the Chukchi, according to the legends, arrived in baidary, then these raids were conducted by the eastern Chaun Chukchi for the purpose of enrichment (Pogromnoe, the Omolon River). These raids can be dated to the time after the departure of the western Chukchi and disappearance of the Shelagi—1730s and later, up to the end of the 18th century.

According to memories of an eyewitness to the events, a 92-year-old woman, in the Korytovo zaimka near Nizhnekolymsk in 1891, the Chukchi attacked the Russians for the last time 85 years ago, that is, in 1806. Here is how this raid is described, in her words, by the exiled *narodovolets* [Russian revolutionary] I. V. Shklovskii: "When all the men were out fishing the Chukchi sailed up in boats to Nizhne-Kolymsk, killed the old women, and took the young away captive."[243] This raid was carried out in baidary in July-August, when the Kolyma was free of ice, and most of the residents of the village, as usual, were dispersed to the *zaimki* [open lands] for fishing. In Nizhnekolymsk during this period, an insignificant part of the population remained, chiefly the incapacitated or those intended to be there by the nature of their service. What is more, private homes (in 1787—33 homes) stood outside the wooden stockade.[244] The Chukchi, who were probably conducting the usual raid for booty, took advantage of this. Since the old people were of no interest to them, they were killed, and the young, evidently girls, were taken captive. This raid was probably little different from earlier ones, which were conduced not for the purpose of revenge on the Kolyma people for any insults (though this also cannot be excluded), but rather simply to take booty and captives.

[243] Shklovskii 1892: 99; Dioneo 1895: 26; cf.: D'yachkov 1992: 196.
[244] Sarychev 1952: 74.

CHAPTER 3

RUSSIAN ATTEMPTS TO CONQUER THE CHUKCHI IN THE MIDDLE OF 17TH TO THE FIRST THIRD OF THE 18TH CENTURY

From the middle of the 17th century Russian explorers tried to develop the natural resources of the Chukchi Peninsula and, performing state service, sought to impose taxes on its residents. Initially the Russians were interested in the acquisition of walrus tusks and, after the founding of Anadyrsk in 1649, both fishing and hunting wild reindeer in Anadyr'. In the second half of the 17th century the primary vector of the foreign policy of Russia in the region was directed toward conquest of the Yukaghir and Koryak; the Chukchi lived far away in this period and encountered the Russians only in summer during hunting and fishing in Anadyr'. In general, the Russians who arrived on ships had contact mainly with the settled residents on the coast of Chukotka (that is, primarily the Eskimos); they did not try to penetrate into the peninsula.

Russian mariners sailed east of the Kolyma for the first time in the summer of 1646. The crew of one koch with nine men led by promyshlenniki Isai Ignat'ev Mezenets and Semen Alekseev Pustozerets, having left the Kolyma, reached after two days some bay where the Chukchi lived. In the absence of an interpreter, mute trade was conducted with them; through this the promyshlenniki received valuable walrus ivory. It is believed that the mariners reached Chaun Bay.[245] B. O. Dolgikh suggests that the mariners found Chuvan people, since at this time the Chukchi had not yet moved so far to the west, whereas I. S. Gurvich

[245] Belov 1952:110; 1956: 159; cf.: Miller 1996:21; Krasovskii 1894: 156; Zuev 2002: 195. They reached Chaun: Samoilov 1945: 66; Belov 1952: 111, note 2; Dolgikh 1960: 550; Gurvich 1966: 53; Vdovin 1972: 102 (Chukchi); Zuev 2002: 195; 2009: 33.

completely accepts that the Russians discovered precisely a camp of Chukchi, who had migrated into Chuvan territory.[246]

We will turn attention to the fact that the mariners passed along open water between the shore and the massed ice ("po zal'du" in the document). Accepting as the average speed of the Siberian koch in open water as 5.0–6.5 knots,[247] it is possible to calculate that in two days the mariners could have gone on the order of 430–560 km along the shore. If we accept that the mariners reached southern Chaun Bay along the shore, then, judging by the program "Yandekskarty" [Yandex.Maps], their coasting voyage took only about 350 km. If it is assumed that the promyshlenniki sailed past Chaun Bay from the north, they could have gone much farther to the east, for example to Nol'da Bay, making an approximately 460 km voyage from the mouth of the Kolyma, where there could quite possibly been "sitting Chukchi," who engaged in the sale of walrus tusks.

This expedition fueled interest in the business of walrus tusks in the east. On 20 June 1648, the well-known expedition of S. I. Dezhnev set off from the Kolyma to the Anadyr'. On 20 September during a landing a clash occurred with coastal residents who wounded the trader Fedot Popov.[248] A dispute probably occurred with the Eskimos during a landing on Chukotskii Nos [Cape Dezhnev] to replenish supplies and for trade, which developed into an armed confrontation. M. I. Belov suggests that in this battle a girl was taken captive, who then became a concubine of the promyshlennik Foma Semenov Permyak.[249] This was the first military conflict of the Russians with residents of the Chukchi Peninsula known to us.

In July 1649 two kochi led by M. V. Stadukhin set off from the mouth of the Kolyma to the east in search of the Pogycha River. They passed seven days under sail without rest, but did not reach the river. After this, M. V. Stadukhin sent a group led by the promyshlennik Ivan Fedorov Kazanets and the Cossack Ivan Baranov to capture people. They went two

---

[246] Dolgikh 1960: 550; Gurvich 1966: 53.
[247] Vladimirov and Tsiporukha 1992: 202–203.
[248] Belov 1952: 130; Miller 1996: 22.
[249] Belov 1952: 269, 272–273, notes 2, 13; 1956: 164–166; 1973: 121; cf.: Samoilov 1945: 69.

days along the shore of the sea and, as I. F. Kazanets reports in his petition, having found "two yurts of the Koryatskie people," "with them [the Koryak] fought a deadly battle and in this fight" he was wounded "with a mortal wound," but they nevertheless took people. The latter, under torture, told that in 1648 their relatives killed people from two shipwrecked kochi, which were part of S. I. Dezhnev's expedition.[250] Researchers, on the basis of their understanding of the route followed by M. V. Stadukhin's expedition, suggest that these captives were either Chukchi, thought to be Koryak, or actual Koryak who migrated to the Arctic Ocean from the south.[251] And S. Dezhnev and N. Semenov, challenging the priority of the discovery of the *korga* [shoal] with walruses not far from the mouth of the Anadyr' (modern Russkaya Koshka) by Yurii Seliverstov, a traveling companion of M. V. Stadukhin, reported to the Yakutsk governor in 1655 that Yu. Seliverstov's information about the discovery of the korga with walruses was false, since he did not even reach Bol'shoi Kamennyi Nos (believed to be modern Cape Dezhnev).[252]

In order to understand how far it was possible to sail the kochi of M. V. Stadukhin, it is simply necessary to turn to the average speed of this type of vessel, which amounted to an average of 5.0–6.5 knots under favorable conditions.[253] Thus, for seven days of uninterrupted travel the

---

[250] Belov 1952: 116, 127, 246; Miller 1996: 25; Krasovskii 1894: 156; Okhotina-Lind and Møller 2013: 88. The petition of I. F. Kazanets: Belov 1952: 245–247 (with lacunae); AAN. F. 21. Op. 4. No. 31. L. 121–123 (complete).

[251] Chukchi: Bogoras 1921: 55; Dolgikh 1960: 550 (Chukchi between capes Billings and Ryrkapiya); Koryak: Krasovskii 1894: 56; Gurvich 1966: 52; Polevoi 2000: 71; Zuev 2002: 199; 2004: 60. A. A. Burykin (2000: 54) suggests that at this time coastal residents, independent of national association, were simply called Koryak (from the Chukchi *ak'orakyl'yt*—"people without reindeer") (cf.: Burykin 2004: 39; 2006: 192, 203; 2013: 432–434). Though, of course, it must be noted that experienced Cossacks understood rather clearly the ethnic situation of the region (Gurvich 1966: 52); the name "Koryak" itself is not a self-name, but is believed to have originated from the word *korak* (*qorak*)—in the southeastern dialect of the Koryak language: "being with reindeer" (Stebnitskii 1934: 49; Leont'ev 1983: 7).

[252] Belov 1952: 127, 131–132; 1973: 115; Miller 1996: 25.

[253] S. I. Dezhnev himself believed that from the cape, subsequently named after him, to the mouth of the Anadyr' was no more than a three-day journey (Belov 1952: 132), that is, on a coastwise voyage with a distance between points of approximately 820 km, it was necessary to sail 273 km in a day; correspondingly, the speed of the ship would be 11.4 km/hr, or 6.3 knots. According to the estimates of M. I. Belov (1956:

kochi could sail 1,500–1,950 km, to which must be added about 50 km spent for two days of the party on foot. If this distance is placed in a line along the coast of the Chukchi Peninsula, the distance of 1,930 km will lead to the southern shore of Kresta Bay. M. V. Stadukhin could have met Koryak who had migrated there. Moreover, in 1654 S. I. Dezhnev, in the campaign from Anadyr' along the seacoast, recaptured the Yakut wife of Fedot Popov (Alekseev) from the Koryak. Fedot, like I. Ankudinov who also sailed with him, survived a shipwreck, and at a later date they both died from scurvy.[254] On the Remizov-Atlasov map of Kamchatka (ca. 1700), "Chukotskie Koryaki" is indicated on the coast north of the Anadyr'. And based on later information of the Anadyr' resident Sotnik I. Kobelev (1779), from Serdtse-Kamen' Mountain "begin the Settled Koryak, that is, those not having reindeer," that is, Koryak dwellings started from Priskal'naya Mountain on the north shore of Anadyr' Bay.[255]

In 1649 S. I. Dezhnev established the Anadyr' winter quarters, which for more than a century became the easternmost settlement of the Russians. The stockade, constructed there 10 years later, served not only for control over the tributary Yukaghir of the Anadyr' valley, but was also a base for further campaigns for the purpose of subjugation and collection of tribute from the Koryak and Chukchi. S. I. Dezhnev himself did not go on campaigns against the Chukchi: the first decade of existence of the winter quarters passed in clashes with the Yukaghir and Koryak, besides the force at the winter quarters was inadequate—S. I. Dezhnev's successor, the boyar's son Kurbat Ivanov, who was assigned as steward (commandant) of Anadyrsk, reported to the Yakutsk governor in spring of 1659 that there were only 5 servicemen and 32 promyshlenniki in Anadyrsk.[256]

---

212), during the voyage from the Lena to the Kolyma, the speed of the kochi fluctuated within the range of 150–250 km per day.

[254] Belov 1952: 132; cf.: DAI. 1851. Vol. IV: 12, 22; Efimov 1948: 222; Dolgikh 1960: 550; Gurvich 1966: 52–53.

[255] Efimov 1948: 228 (quotation of I. Kobelev); 1964: No. 48 (map); cf.: Chernenko 1957: 121; localization of the mountain: Zuev 2009: Fig. 2. According to A. A. Burykin (2013: 314), M. V. Stadukhin sailed only to the region of Cape Kamen'-Serdtse.

[256] Belov 1952: 272, note 1.

In 1660 Kurbat Ivanov completed the first maritime campaign to the Chukchi Peninsula. The primary goal of the campaign, as before, consisted of procurement of walrus ivory, and not imposing tribute on the residents. When the Anadyr' was released from the ice, a koch with 22 Cossacks and promyshlenniki led by K. A. Ivanov sailed from Anadyrsk to the mouth of the river, and then to the north along the shore of the sea, where a storm broke up the boat and the mariners were forced to spend 10 days repairing it, and then they pulled the koch with a towrope. During the rebuilding of the ship the Chukchi appeared.

> And we encountered foreigners, Chukhoch'i people, ten boats, and they took the shore from us and began to fight with us. And they battle with stone and bow. And they began with stone to beat and shoot and fought with us from noon until the evening, pierced plank and copper shields. And we went out from them to anchor and, having stopped in the morning, we made strong shields and from anchor went to the shore. And they started to strike with stone and to shoot from bows at the shields, and penetrated the shields and wounded four of our men. And we reached a stream at the shore, started to shoot them with guns. And they ran away to escape and they left their boats behind. And their boats are skin, walrus hide, and will hold twenty or thirty men each.

Next, the mariners reached Bol'shaya Bay (modern Kresta Bay), going with paddles or towing the boat with a rope. On 10 August, they found empty "four yurts of Chukhoch'i men" in the bay, where they discovered a store of 2,000 geese. Then they went under sail "and ran upon many yurts. And many men gathered and put up a battle with us. And we went to the shore and killed men from the yurts and there took a hundred pudy and more of reindeer meat for food." Having entered into negotiations with the "Chukhchi," they learned of a walrus korga that, with the aid of a guide, they reached after 15 August. The haulout at Chukotskii Nos turned out empty. On 25 August K. A. Ivanov decided to return and on 24 September reached Anadyrsk, not bringing any walrus ivory with him.[257]

---

[257] Orlova 1951: 405–408; Belov 1952: 269–270; cf.: Belov 1956: 166–168; 1973: 121; Dolgikh 1960: 550; Vdovin 1965: 108; Gurvich 1966: 51; Zuev 2002: 213; Nefёdkin 2003: 23.

The character of K. A. Ivanov's expedition being predominantly about the hunt, and not war, is attested in certain degree by its small number—22 men—and also that the Anadyrsk steward did not especially have military forces. Since the korga closest to Anadyr' was abandoned, the steward would have to go to a new, more northern korga, about the existence of which he knew only from the testimony of a captive Chukchi woman. The date of the beginning K. A. Ivanov's expedition is not indicated in his first report to the governor, but it is possible to suppose that this was the beginning of June, when Anadyr' normally became ice-free.[258] The detachment's first encounter with the "Chukchi" occurred on the north shore of the Anadyr' Estuary: during the long, ten-day repair of the koch the detachment was noticed, and the length of the stop made it possible for the maritime residents to gather a militia from neighboring villages. As a result, nine large fighting baidary arrived at the place of portaging the koch, each of which, in K. A. Ivanov's own words, held 20 to 30 men, that is, a total of 180 to 270 men, and possibly including some women. It is also possible that this was an encounter by the Russians with a "Chukchi" hunting expedition sailing to the mouth of the Anadyr' to hunt reindeer at the end of July. At least, the Russians did not notice the settlements in the vicinity. There were no attempts at negotiations; the "Chukchi" immediately engaged in battle, trying to drive the Russians into the sea. The skirmish lasted half a day, forcing the Anadyr' people to get on board and go into the sea. On the following day, the Russians made shields and under cover of these began to approach the shore, not opening fire from the muskets. Upon reaching the stream at the shore, the Anadyr' people opened fire, turning the enemy to flight. The "Chukchi," abandoning the baidary, departed overland. K. A. Ivanov examined and described the construction of the abandoned boats. It should be noted that accurate shooting with stones from slings indicates that the enemies of the Russians were coastal residents, primarily Eskimos. This is also attested by the long-range battle, one not evolving into a melee traditional for the Chukchi.

Further, in Kresta Bay the Russians discovered a coastal settlement of four yarangas. The residents quickly hid, having learned of

[258] D'yachkov 1992: 169; Olsuf'ev 1896: 9; Gondatti 1897a: 85.

the approach of the Russians and knowing about their hostile intentions, and since they themselves probably participated in the preceding battle. The supply of goose meat found by the servicemen attests to the time of the hunt for molting geese when the birds were slaughtered in mass, that is, in June-July. Then, it is believed, the detachment reached a large settlement in Provideniya Bay—K. A. Ivanov does not give the number of dwellings, though he spent five days there. Whether this settlement was of Reindeer or Settled Chukchi ultimately remains unclear. Here the Russians found more than 1,640 kg of reindeer meat. This was possibly meat of reindeer killed by the nomads approximately in mid-August. If these were reindeer slaughtered on the summer hunt, during the northward migration of the animals in June-July, it could have been done by the settled residents. The local residents came out in front of the dwellings and engaged in battle. However, the Russians, having gone down to the shore, with the aid of firearms forced the residents to retreat and seized the village, near which they stayed five days, bringing themselves in order. In this, the military part of the expedition was concluded. It did not lead either to any taxation of the residents of Chukotka nor did it acquire any new data about them.

Approximately in 1667–1668 the hunter Taras Stadukhin, brother of Mikhail, commanding kochi with a crew of 90 men, set off from the Kolyma to Chukotskii Nos, but because of difficult weather conditions could not pass the Nos. Turning back, they crossed the Nos on foot "from the other side" and, having gone to the sea, constructed kochi, on which they sailed along the shore to the mouth of the Penzhina River. During battles with "non-tributary foreigners" the expedition lost 81 men, after which the remainder returned to the Kolyma.[259]

Information about this expedition is quite scarce and incomplete—in 1710 the Yakutsk serviceman Nikifor Malgin only told in his words what he heard once from T. V. Stadukhin in Nizhnekolymsk. It is possible to suggest that the overland route of the expedition from Nizhnekolymsk was the standard one, known as early as the 1650s: they went along the Anyui to the upper reaches of the Anadyr', along which they moved to the mouth of the Anadyr'. There they built kochi, on which

---

[259] PSI. Bk. 2: 500–501; Miller 1996: 29; Efimov 1948: 223; Belov 1952: 63, note 2; 94–95; 1956: 169; Polevoi 2000: 83; Zuev 2002: 212–213.

they set off south to Penzhina Bay, rounding Kamchatka. Other routes of the expedition were also possible: southwest from Anadyr', as M. V. Stadukhin had already gone in 1651, to the region of Gizhiga, or up along the Omolon, then to the Penzhina River, and from there to the Gizhiga River, as F. A. Chukichev did in 1657;[260] then T. V. Stadukhin's detachment must have gone into Shelikhova Bay, where kochi were built for this route. In any case, the primary opponent of T. V. Stadukhin's expedition were the Settled Koryak, and not the Chukchi.

For more than a quarter of a century, to 1688–1689, the Anadyr' servicemen and the Chukchi did not conduct active warfare. More important for the Russians was the southern direction, where they planned to carry out the acquisition of tribute from the Koryak, and also the Chukchi still had not reached the middle course of the Anadyr'—the nearest Chukchi lived at Kanchalan. In addition, the force at Anadyrsk was clearly inadequate for active military campaigns: part of the Cossacks guarded the amanaty, and the others were occupied with hunting and fishing, providing food for themselves. Thus, in 1676 there were only 16 servicemen at Anadyrsk, and they were looking after 12 amanaty, whereas, based on the estimate of Yakutsk governor A. A. Barneshlev, there should have been 30 Cossacks there.[261]

Together with that, the policy of the authorities, directed toward acquisition of tribute from the local population, was nevertheless carried out. In 1676 the Cossack desyatnik Ivan Rubets, who in 1663–1666 was steward of Anadyrsk, reported to A. A. Barneshlev: "On the Anadyr' River live non-tributary Chukhchi foreigners close to the sea, and many servicemen before this caught those Chukhoch and took from them children and brothers as amanaty and they, the Chukchi, give up these amanaty and the amanaty don't make them accountable."[262] It is evidently the matter of the Chukchi who lived on the lower reaches of the Anadyr'

---

[260] DAI. 1851. Vol. IV: 120, 147.

[261] DAI. 1857. Vol. VI: 407 (report of governor A. A. Barneshlev, 1676); cf.: AAN. F. 21. Op. 4. No. 35. L. 404; Vdovin 1965: 109.

[262] Vdovin 1965: 110; cf.: Dolgikh 1960: 435–437; Zuev 2002: 65, 67. I. S. Vdovin suggests, on the basis of this report, that the Chukchi had paid tribute from the 1670s, but the text itself calls the Chukchi non-tributary, and also, in the report of A. A. Barneshlev, only the Yukaghir are mentioned in the lists of tributary subjects of Anadyrsk (DAI. 1857. Vol. VI: 407).

and who came in contact with the Russians during hunting for reindeer and marine resources. The Cossacks took children and brothers from the Chukchi as amanaty, for whom, as later on, they would not bring tribute. These acquired Chukchi were not held in the amanaty hut. Since there are no Chukchi in the lists of Anadyrsk *zalozhniki* [hostages] for 1670–1680.[263] For this reason G. F. Müller noted: "Though they took amanaty from the Chukchi so that the rest would be concerned by this and make steps toward paying tribute for these amanaty, few examples of this are known. They believe these amanaty are already lost, and though the Yakutsk Cossacks in Anadyrsk once tried to hang those amanaty for whom tribute was not forthcoming so some Chukchi could see, but this nevertheless did not bring any results."[264] As we see, the academician even provided an explanation why the Chukchi did not pay tribute for the amanaty: they considered them already dead.

Desyatnik Vasilii Tarasov, who served as steward at Anadyrsk in 1685, describes the circumstance at Anadyrsk to the Yakutsk governor in this way:

> And being in the Anadyrsk winter quarters, I, Vaska, lived, with great security. And the tributary foreigners pay the great sovereigns tribute when they want, because seeing the servicemen very few in number while going into the Anadyrsk winter quarters, they kill many servicemen from the winter quarters on the road; the tributary and non-tributary Khodyndy and Chyuvandy foreigners, and the Koryaki, and Chyukhchi of different clans arrive at the winter quarters both for catching fish and for feeding the amanaty, and do not let go out from the winter quarters, and they kill the servicemen and promyshlenniki (trappers) at the fish camps. And those foreigners who pay the great sovereigns tribute, and those among the tributary speak of payment: you ask tribute from us, and you do not defend us from the non-tributary Koryak and Chyukchi; those Koryak and Chyukchi kill many of their relatives on the sable hunt, and take captive wives and children and their relatives, and steal the reindeer, knock their dwellings down, and do not allow sable hunting for the great sovereign's tribute; and there is no one to defend against these non-tributary

263 Vdovin 1965: 110; Zuev 2002: 67.

264 Miller 2009: 156; cf.: Elert 1998: 125; 1999: 95; Zuev 2002: 63; 2007: 156–157.

> Koryak and Chyukchi, the servicemen are few, there are no guns and powder, and lead."[265]

In general, it turns out that because of the inadequacy of servicemen and fighting supplies the steward himself had to travel outside the stockade with great care. Indeed, outside the winter quarters "they beat" the Russians on the roads and at the fish camp, which fed both the Russians themselves and the hostages. The nomads fearlessly approached the winter quarters, not seeing a threat from the servicemen. Both the tributary and non-tributary Khodyn Yukaghir did this, and the Chuvantsy, as well as the Koryak and Chukchi from various places. It should be noted that the last two peoples acted simultaneously against the Yukaghir and Russians, which, however, does not indicate an alliance between them. Even the tributary Yukaghir began to pay tribute randomly, since they were not able to catch sable. Of course, when complaining of obstacles by the non-tributary on the sable hunt, the Yukaghir primarily had in mind the Koryak who lived south of the Anadyr', but the Chukchi are also mentioned, which can attest to their movement in migrations to the southwest. From the Yukaghir, they took women and children captive, drove off the reindeer, and destroyed the dwellings. Consequently, the Chukchi initially took the domestic reindeer as booty from the Yukaghir, who incidentally did not have a significant number of reindeer, and those only sled reindeer.

For several years, the situation on the Anadyr' did not improve for the Russians. The next steward of Anadyrsk, desyatnik Andrei Tsypandin, thus in 1688 explains to the governor the shortfall of tribute: ". . . in past years many tributary foreigners by the will of God died, and others were killed on sable hunts and in their dwellings by unpeaceful non-tributary foreigners of the Chyukhochi and by the Koryak foreigners, and others . . . went to distant unknown places . . . and for the small number of servicemen it is fearful to live at the Anandyrsk [sic] winter quarters because of the non-tributary unpeaceful Koryak and Chyukhochi foreigners, and the Koryaki and Chyukhochi arriving at the winter quarters kill the servicemen at the fish camps."[266]

[265] DAI. 1867. Vol. X: 351; cf.: Zuev 2002: 66.
[266] MIYa. Ch. 3: 978.

As we see, the reason for the reduction of the tribute collection is explained by the fact that part of the Anadyr' Yukaghir died from smallpox, others simply migrated from the winter quarters, saving themselves from the epidemic, and some were killed on hunts and in their camps. The Koryak and Chukchi did this. Near the very winter quarters, they killed servicemen who went out to replenish food supplies by fishing. The Russian presence in the region was not frightening for the non-tributaries—the garrison at Anadyrsk was too small in number.

At the same time, the Anadyrsk administration did not sit idly on their hands. The Cossack G. Fedorov, who served at Anadyrsk from 1682 to 1688, indicated in his petition, that "he went to the unpeaceful non-tributary foreigners, the Koryak and the Chyukhchi, with servicemen, and participated in many battles."[267] The lists of prisoners at Anadyrsk attests to the clashes with the Chukchi: in 1681 there were among the Cossacks and promyshlenniki three Chukchi women (a serf, a housekeeper, and a baptized woman who became a wife), and the boyar's son Rodion Kobelev, who was the Anadyrsk steward, had two boys, one of whom had been baptized. Now in 1685 at Anadyrsk there were seven Chukchi women and one boy—R. Kobelev had left the stockade and gone to Yakutsk.[268]

In the second half of the 1680s some of the coastal Chukchi even consented to pay tribute with walrus tusks. On 27 May 1687, the steward Pyatidesyatnik Vasilii Permyak sent the clerk Ivan Ankidinov "down along the Anadyr' River to the non-tributary foreigners in Chyukhotskaya land to the Chyukhchi." A little later the Cossack Vasilii Borisov was sent in a koch to collect tribute as well. The collectors managed to have the coastal "Chukchi" swear an oath of allegiance and take from the "best of the princes" Kopeichka and from his relatives one tusk each—for a total of ten. Kopeichka and his brother Tasirei stated: "we will pay the great sovereign tribute along with the other many Chukochi'." Though the historian of Siberian studies B. P. Polevoi preferred to believe that the tribute was collected from the Chukchi who lived near the mouth of the Anadyr', they are called "morskie kamennye" [maritime stone] in the

---

[267] Vdovin 1965: 111–112; Zuev 2002: 66.

[268] About prisoners: KPMGYa: 193 (1681); Vdovin 1965: 110; about R. Kobelev: Belov 1952: 301, note 1.

document, that is, those who lived on the Kamen' [Stone], the Chukchi Peninsula.[269]

In the spring of 1688 the Anadyrsk steward Pyatidesyatnik Vasilii Fedorov Kuznetsov sent by water to the Chukchi this same Ivan Ankidinov, the serviceman Sidor Ivanov, and promyshlennik Efim Kirilov to collect tribute—walrus ivory. The tribute was collected from the same Prince Kopenko and nine relatives, who probably lived in one village, as well as from two other Chukchi, Elmo and Tintega. I. Ankidinov did not return on time to Anadyrsk.[270] If I. Ankidinov's group of collectors arrived to Kopenko, then another collector—Cossack Vasilii Borisov, who sailed somewhat later in a koch for the same purpose—could get tribute from the other Chukchi. The tribute was collected from the "Chyukhoch'ya clan of the maritime Stone Chyukhoch," evidently coastal residents of the Chukchi Peninsula. As B. P. Polevoi noted, the name of the Kopeikina korga in Anadyr' Bay suggests the name of the prince, who probably moved there.[271]

In this same year of 1688, after the Anadyr' opened, V. F. Kuznetsov himself set off on an expedition. In an earlier reply to the Yakutsk governor it says that the steward "sailed in the sea to the korga in three kochi, having taken with him 52 Cossacks and hunting people . . . and as he, Vasilii, arrived on the sea at the korga in three kochi, and from the korga went in two kochi along the sea toward the Koryatskaya side, and with himself he, Vasilii, took 12 Cossacks and 18 hunting men, and one koch he left with the remaining people on the korga, and those remaining people arrived from the korga in the sea to the Anadyr' winter quarters in good health, while he, Vasilii, from the sea with comrades at present in the 197th year was still not in Anadyrsk on December 29th."

In 1690 the steward Pyatidesyatnik Andrei Tsypandin told the Yakutsk governor this about the campaign of V. F. Kuznetsov: ". . . in the past 196th year the Cossack Pyatidesyatnik Vasilii Kuznetsov went from the Anandyrsk outpost in kochi to the sea for a visit of Koryatskaya land and to summon them under the royal hands, and to call the Koryak into

---

[269] Polevoi 1976: 220–221; 2000: 135; Zuev 2002: 67, 224.
[270] AAN. F. 21. Op. 4. No. 32. L. 159 ob.; Vdovin 1965: 112; Polevoi 2000: 59, 137–138; Zuev 2002: 67, 224.
[271] Polevoi 1976: 222; 2000: 136.

tribute payment, with his, Vasilii's, kochi with servicemen and hunting people, wrecked in the sea, and he, Vasilii, and Cossacks and hunting people going on sleds in the 197th year on December 6th went to the tributary Chyukhchi and those Chyukhchi committed wickedness, betrayed the great sovereigns, and killed him, Vasilii Kuznetsov, and thirty comrades."

The Yakutsk governor P. P. Zinov'ev, on the basis of A. I. Tsypandin's account, thus reported to the tsar about what happened:

> . . . in the past 196th year, the Cossack Pyatidesyatnik Vaska Kuznetsov, with servicemen and with hunting people, forty men, went from the Anandyrsk winter quarters in kochi to the sea to visit the Koryatskaya land and call those Koryak under your great sovereign royal autocratic high hand, in tribute payment; and in this same 196th year, by the will of God, the kochi of Vaski and comrades wrecked on the sea; and after the wreck, in the past 197th year on December 6th, they, Vas'ka and comrades, went from the sea on sleds to the tributary Chukhchi, and those tributary Chukhchi betrayed you, great sovereign, and killed Vaska and comrades, twenty Cossack people and twenty hunting people."[272]

In general, the campaign of V. F. Kuznetsov can be represented in the following way. The steward set off on an expedition from Anadyrsk with a dual purpose: to obtain walruses on the korga and bring into tribute payment the coastal Koryak. In three kochi were 52 Cossacks and promyshlenniki—a rather large detachment based by local standards. Remaining on the korga to hunt walruses, was one koch with 20 men, probably primarily promyshlenniki, while thirty men in two kochi went to the south, to the Koryak. However, the kochi wrecked in a storm, though all 30 men survived. At this time, fall was coming on and snow fell. The passage of such a long period of time most probably attests to the fact that the kochi were already returning from the expedition to the Koryak. The Koryak probably did not show active resistance, since the same 30 men were in the kochi. Having acquired or made sleds, the Russians reached the coastal tributary Chukchi on the shore of the Gulf of Anadyr' on 6

---

[272] Quotations, see: AAN. F. 21. Op. 4. No. 32. L. 158 ob., 162 ob.; report of P. Zinov'ev: Akty istoricheskie [Historical Acts] 1842: 354; about the campaign, see: Belov 1956: 184; 1957: 99; cf.: Vdovin 1965: 111; Leont'eva 1997: 56; Zuev 2002: 224–225; Nefëdkin 2003: 24.

December 1688. Why a conflict developed with the "Chukchi" is unclear, but it can be supposed that it was because of requisition of draft animals and sleds. All the Russians were killed during a surprise attack. B. P. Polevoi writes that the "Chukchi" attacked the sleeping, whereas M. I. Belov and G. A. Leont'eva believe that the detachment was ambushed.[273] Thirty Russians were killed; the figure of 40 in P. P. Zinov'ev's report is erroneous. This attack led to the activation of warfare by the Russians against the coastal Chukchi.

In the following year of 1689 the Chukchi themselves continued hostile actions. A. I. Tsypandin, probably not yet knowing of the fate of V. F. Kuznetsov, sent out four tribute collectors, including the returning I. Ankidinov. He thus writes to the governor: "And in the past 197th year I, Andryushka, sent along the Anadyr' River to the sea to the Chyukhchi for tribute bone [ivory] collection Ivashka Ankudinov, Sidurka Grigor'ev, Stepan, Ivanov, and Kharka Samoilov, and those villainous traitors the same tributary Chyukhchi killed those tribute collectors."[274]

In this same year of 1689 a detachment consisting of servicemen, hunting people, and trading people was sent from Anadyrsk to search for the expedition of V. F. Kuznetsov. A. I. Tsypandin thus reports to the governor about this campaign in his statement: "And in the past 197th year I, Andryushka, sent the Cossacks Ivashka Kotel'nik and comrades, and with him Ivashka, to sea to find Vasilii Kuznetsov with companions. . .. Those villainous traitors Chyukhchi killed Aleshko Shchukin, and shot the Cossacks and the traders, and the hunting people, with muskets they took from Vasilii Kuznetsov, and they continue to boast that the Anadyrsk outpost and tribute winter quarters will be taken and the Cossacks and every rank of people killed, and speak of their boasting and of their villainy, murder the captive women who were in captivity among those villainous Chyukhchi, and ran up to the Anadyrsk outpost."[275]

Of course, the Chukchi, having destroyed the detachment of V. F. Kuznetsov, did not count on leniency from the opponent and engaged in

[273] Polevoi 2000: 164; Belov 1957: 99; Leont'eva 1997: 56.
[274] AAN. F. 21. Op. 4. No. 32. L. 162 ob.; cf.: Vdovin 1965: 112; Polevoi 2000: 138; Zuev 2002: 225.
[275] AAN. F. 21. Op. 4. No. 32. L. 162 ob.–163; cf.: Vdovin 1965: 111; Polevoi 1972: 152, 155; 2000: 138; Zuev 2002: 225.

battle. This was the first time the Chukchi used firearms, which became widespread among them much later, from the middle of the 19th century. In this battle one of the servicemen was killed—Aleksei Shchukin. Since there is no information about the outcome of the battle, it hardly brought victory to the Russians. Indeed, after this the Chukchi boasted that they would take Anadyrsk and kill the Russians. These reports, as well as the news of V. F. Kuznetsov's death, brought to the outpost captive women purchased from the Chukchi: Grun'ka (Agrafena), wife of Kh. Samoilov and daughter of Savin; Odotitsa (Avdot'ya), the baptized captive of V. F. Kuznetsov; and Oksinitsa (Aksin'ya), wife of Cossack V. Borisov, who accompanied the steward on the campaign, and daughter of Afanas'ev with her son. The Khodyn Motora had bought them all from the Chukchi and brought them to Anadyrsk.[276] The threat of attack by the Chukchi was perceived as entirely real considering that in 1690 there were 13 Cossacks in the stockade, who guarded 27 amanaty, thus the strength of these was clearly inadequate for rendering opposition to a large raid. In the stockade in 1688 only the tribute hut and storehouses were surrounded by a palisade.[277]

In the summer of 1690, modern historian B. P. Polevoi suggests, a new maritime campaign to the Chukchi took place, during which the Chukchi Tygagin was taken captive. He then called from a Russian koch to his relatives to pay tribute for him, but the Chukchi refused.[278]

In the following year, the Anadyrsk steward, the boyar's son Semen Chernyshevskii, went to the mouth of the Anadyr' in search of the Chukchi who had killed V. F. Kuznetsov's people, but the Chukchi were not found.[279] This was probably the summer river campaign to the mouth of the river, where it was planned to find the Chukchi out hunting, but the latter were able to elude the Russians or did not appear, fearing the enemy's actions.

In the summer of 1692, servicemen led by S. Chernyshevskii again went "to the unpeaceful Chyukoch at the sea, on a fighting

[276] Polevoi 2000: 139, 241, notes 50–51.
[277] The number of the garrison at Anadyrsk: AAN. F. 21. Op. 4. No. 32. L. 163; fortification: AAN. F. 21. Op. 4. No. 33. L. 246; No. 35. L. 412 (No. 70); Vdovin 1960: 36; Zuev 2003: 138.
[278] Polevoi 2000: 139; Zuev 2002: 226.
[279] Leont'eva 1997: 59–60; Zuev 2002: 226.

campaign, and destroyed 16 yurts."[280] The campaign was a fight and did not have as its goal the collection of tribute from the Chukchi. The Russians sailed to the mouth of the Anadyr' against the coastal group of the Chukchi who had taken active part in the battle actions in the preceding years. Since there is no information indicated in the report about a battle or about losses or about captives, then the residents of the village (or even two or three camps) probably managed to flee, and the Cossacks simply destroyed the abandoned dwellings.

In subsequent years, the Anadyrsk servicemen were engaged in a struggle with the Koryak and Itel'men in Kamchatka. In 1701, the Khodyn Yukaghir Nekras with relatives filed a petition to steward S. Chernyshevskii with a complaint about the Chukchi, who prevented hunting wild reindeer on the lower reaches of the Anadyr', with request to send servicemen with them to Anadyrskii Nos. In 1702 Aleksei Chudinov was sent on a campaign with a detachment of 24 Anadyrsk servicemen and townsmen and 110 tributaries, mainly Yukaghir, as well as Koryak.

Three Cossack members of the campaign—Timofei Daurtsov, Fedor Portnoi, and Petr Mungal—later, in 1710, described the course of the expedition to the Yakutsk provincial office in the following way:

> And in this campaign at the Anandyrskii Nos, found near the Anandyrskii Sea, were Foot Chyukhchi with thirteen yurts and they were brought under the royal high hand of the great ruler to tribute payment and tribute requested; and those Chukchi did not give them tribute but hid in the yurts. And the steward Aleksei Chudinov ordered them to attack the yurts, and upon the attack of the male sex in those yurts 10 men were killed and their wives and children taken captive and many of them among the captured strangled themselves and stabbed each other to death. And other Chukchi, seeing them, the servicepeople, ran from the yurts and went to the Nos to gather in one place. And they, the servicepeople, going from the above-described slaughter, found them, the Chukhchi, three hundred and started to fight with them and killed two hundred and more of their people, and others fled. And on the next day they encountered them, the servicepeople, and they, the Chukchi, in a multitude of Reindeer and Settled, and there were three thousand and more of them at this time in the gathering, and they fought with

[280] AAN. F. 21. Op. 4. No. 34. L. 116; cf.: PSI. Bk. 1: 433; Belov 1957: 96; Polevoi 2000: 139, 164; Zuev 2002: 227–228; 2009a: 84; Nefëdkin 2003: 24.

> them from morning to evening, and many people were killed, and of them, the servicepeople and tributary Yukaghir, they, the Chukchi, wounded 70 people, and these Chukchi moved away from them but stayed near them. And they, the servicepeople, sat 5 days under siege by these Chukhoch, then went from them and escaped to Anandyrsk.

Academician G. F. Müller possessed some other document (or a copy) of these testimonies of the Cossacks in 1710, it differs from the above-cited in some details.[281]

G. F. Müller writes that the campaign lasted eight weeks, from April to June, whereas the "story" itself points out that the expedition lasted 28 weeks, beginning in April, that is, to October. The expedition by local standards was rather large: 134 men, almost 20% of whom were Russians, the remaining being tributary Anadyr' Yukaghir and Koryak. Judging by the time of the beginning, the campaign was overland, and not by sea. They moved on sleds and/or skis. The detachment reached the coast of the Anadyr' Sea (that is, the Gulf of Anadyr'), where it came to a settlement of 13 "yurts" of coastal residents, who had probably already moved into the summer yarangas. Through an interpreter the residents were ordered to pay tribute and to provide amanaty, to which, as usual, a positive response did not follow. Then the commander of the expedition, A. Chudinov, gave the order to storm the dwellings, in which their inhabitants sat. The Chukchi probably saw that their forces were not equal—the enemy detachment being more numerous—and did not engage in open battle. The Cossacks, bursting into the dwellings, killed a dozen men and took the women and children captive. Many of the captives taken ended their lives by suicide (in G. F. Müller 's version, evidently erroneous—only the men). Some men of the village, seeing the approach of the Russians, fled north to the Nos and let the enemy's arrival be known. Forces began to gather on the Chukchi Peninsula, the Chukchi evidently arriving rather quickly on sleds, using the last snow cover. Pressing on, the Russians encountered a group of Chukchi superior in number, with which they gave battle. The clash was probably primarily shooting—otherwise the number would have crushed the Russians. In the course of battle the

---

[281] PSI. Bk. 2: 524–525; KPTs: 155–156; Miller 1996: 38; Bogoraz-Tan 1934: 44–45; also see: Bush 1872: 238; Gurvich 1966: 112; 1985: 54; Zuev 2002: 235–236; Nefёdkin 2003: 24.

Chukchi suffered large losses: based on the report of the Cossacks—2/3 of the total number. The Chukchi remaining alive fled to other forces. Already on the next day A. Chudinov's detachment encountered this overall militia of the Chukchi—both the nomadic and settled. This fact most probably indicates that on the preceding day the Russians had battled a detachment that was on its way to join the militia. The Cossacks' estimate of the total number of the Chukchi force at more than 3,000 warriors is clearly exaggerated (the enemy is always many!). Indeed, in 1711 the serviceman Petr Popov, in his "story," estimated the total number of Nose Chukchi at only 2,000 archers.[282] The Chukchi militia, not due to the circumstances, but intentionally attacked the Russians in the morning. Fighting at a distance continued all day. It probably did not reach hand-to-hand combat, to which attest the losses of the servicemen. The losses of the Chukchi were substantial, but the Cossacks, since they left the field of battle before the enemy, were not able to count the latter, as they had in the preceding fight. A. Chudinov's detachment lost 70 men wounded, while based on G. F. Müller —a total of only 20, Russians and Yukaghir. If we try to align these two numbers, perhaps in the second case it is a matter of seriously wounded. In general, more than half of the combatants of A. Chudinov received wounds during the course of the day-long battle. In the evening the Chukchi moved away and set up camp nearby, threatening the opponent with an attack (from G. F. Müller : "the Chukchi enclosed the Russians and Yukaghir"). It is unclear if this was a real siege of the Russian camp during the course of five days. It is possible that the servicemen and allies were simply not able to leave the camp because of the permanent threat of attack by the nearby enemy. However, this time was clearly well spent on doctoring wounds. The Chukchi also did not undertake any action since they had suffered substantial losses as well—otherwise there would have been information. Since the Chukchi did not keep careful observation of the enemy, the servicemen were able to escape the dangerous situation, probably at night, after which the detachment safely returned to Anadyrsk.

On the whole, the campaign was to no avail: tribute was not obtained, the Chukchi did not accept Russian citizenship, and they were

[282] PSI. Bk. 1: 459; KPTs: 158; Miller 1996: 40.

not defeated, though they suffered substantial losses. However, more reconnaissance information about the Chukchi was obtained. This campaign was the first and only large combat operation of the Anadyrsk garrison against the Chukchi in the first quarter of the 18th century up to the expedition of Shestakov-Pavlutskii. More important for the Russians now was the southern direction, military actions against the Koryak and Kamchadal.

In general, the situation at Anadyrsk at the beginning of the 18th century was described to the governor in Yakutsk in 1710 by servicemen who were at the stockade thus: ". . . and from this Anandyrsk stockade the servicepeople and tributary Yukaghir sail down along the Anadyr' to hunt reindeer; and those servicepeople and tributary Yukaghir are killed by the unpeaceful Chyukchi and the wives and children taken captive. And at present there are Cossack children in captivity among those Chyukchi; in various years those Chyukchi approached the Anandyrsk stockade in large numbers. They wanted to take the Anandyrsk stockade and to kill the servicepeople, and throughout all those years they boast of destroying the stockade and because of those Chukchi the Anandyrsk servicepeople and tributary Yukaghir live in great danger."[283] As we see, the hunting conflict on the Anadyr' between the Russians and tributary Yukaghir, on the one hand, and the Chukchi, on the other, continued. Precisely there, in the reindeer hunts on the river, clashes occurred between the hunters of both sides when they met. The Chukchi groups approached the vicinity of Anadyrsk and threatened to take the stockade. This possibly occurred in baidary in summer, as well as on sleds in winter. In general, the residents of the stockade lived under constant threat of attack.

In 1709, the Anadyrsk steward Pyatidesyatnik Efim Petrov sent the newly baptized Yukaghir Ivan Tereshkin with the Khodyn Pelech to the Chukchi for next calling them to tribute payment. Returning, I. V. Tereshkin reported: ". . . having gathered, they, the Chyukchi, in a multitude, went for battle to the Anandyrsk stockade; and he, Ivan, meeting them on the Anadyr' above the Chukoch'ya River near a Yukaghir reindeer hunting camp and talked to them, and the Chyukchi he turned back; and on faith they, the Chyukchi, sent one man from

[283] PSI. Bk. 2: 524; KPTs: 155; Zuev 2002: 69.

themselves to Anandyrsk, and with this Chyukchi a red fox was taken to Anandyrsk to the treasury of the great sovereign, and with him, Ivan, they sent the captive Cossack son, captured in the early years in the lethal year of the Anandyrsk steward Vasilii Lokosov." In a personal report of 1710 to the Yakutsk governor, E. Petrov laid out some other details of the campaign, as well as the events that followed later: the serviceman Ivan Lokosov was sent to the Chukchi for acquisition of tribute and amanaty; he also brought the Chukchi, who paid tribute, and the captive Cossack son Ivan Ankidinov, who had lived among the Chukchi twelve and a half years. And then the steward sent I. Ankidinov and I. Lokosov again to the Chukchi for the collection of tribute, as well as the Chukchi who arrived at the stockade.[284]

The events probably developed in the following way. In the summer of 1709 E. Petrov sent the detachment of serviceman I. Lokosov with two Yukaghir interpreters for acquisition of tribute from the Chukchi, who were hunting reindeer on the lower reaches of the Anadyr'. Evidently, these were Anadyr' Chukchi. At the end of the 19th century the river Chukchi lived on the left bank of the Anadyr' Estuary; they had insignificant herds of reindeer and were for the most part fishermen and maritime hunters.[285] Possibly, the Chukchi actually wanted to sail up to the stockade, and it was no idle boast. The interpreter I. V. Tereshkin met the Chukchi west of the Chukoch'ya River (Kanchalan) and managed to find arguments to frighten the Chukchi and not allow their campaign on the stockade. As a demonstration of his good will, one Chukchi was sent to Anadyrsk, who had never been there; he delivered tribute. This was possibly a scout or simply a curious person, or both. Also, the Chukchi released the Russian captive I. Ankidinov who had accompanied them on hunts. Ankidinov had fallen captive in 1697, probably when the detachment of the official Afanasii Balushkin, which was delivering tribute from Anadyrsk, was killed on a mountain pass by revolting Yukaghir, [286] from whom I. Ankidinov then fell into the hands of the Chukchi. In general, this was the first case in which the Chukchi clearly

---

284 PSI. Bk. 1: 405–406; Bk. 2: 530; cf.: Andrievich 1889: 122; Bogoraz-Tan 1934: 45; Gurvich 1966: 112; 1985: 54; Zuev 2002: 68, 245.
285 About the Anadyr Chukchi, see: Gondatti 1897b: 170; Bogoraz-Tan 1934: 47.
286 Zuev 2002: 232.

demonstrated to the Russians their inclination toward peace, stopped a military campaign, and even gave up a captive Cossack's son.

In the spring of the following year, 1710, I. V. Tereshkin was sent to collect tribute in the same place. And now five river Chukchi, Nokon and relatives, gave tribute, probably arriving at the Anadyrsk stockade for this.[287]

On 7 October 1710 the Yakutsk governor D. A. Traurnikht, examining the prevailing situation on the Anadyr', issued an edict that instructed the new Anadyrsk steward Pyatidesyatnik Matvei Skrebykin, together with the Yakutsk and Anadyrsk Cossacks and tributary peoples, to go on a campaign to the Anadyrskii (that is, Chukotskii) Nos and urge the Chukchi to pay tribute and to provide amanaty, and with their nonagreement, to bring them to obedience by force, for which a pud of powder and lead was issued.[288]

In 1711, an expedition from Anadyrsk took place with the purpose of obtaining tribute for the current year from the river Chukchi along the Anadyr', from the same Nokon and his relatives. Further, the expedition was supposed to set off to Chukotskii Nos to summon the Chukchi into payment of tribute and—which was no less important—to gather information and create a map. Sent in the campaign was the detachment of serviceman Petr Popov accompanied by two translators: promyshlennik E. V. Toldin and the same Yukaghir I. V. Tereshkin. When P. I. Popov's detachment passed Bol'shaya Bay (Kresta Bay), near Matkol' Mountain (modern Bol'shoi Matachingai), the Chukchi refused him tribute, reminding him that earlier (probably in the 1680s) the Russians had demanded tribute from them, but they had not given it, and now they were doing the same. No forced actions of the servicemen in answer followed. P. I. Popov evidently visited a village of the Settled Chukchi, the head of which was Makachkin, who had participated in many expeditions to Alaska, and informed the servicemen about that land. On the return route, at the end of July, the servicemen received from relatives of Nokon five fox pelts as tribute. This was in spite of the fact that in all the river Chukchi on the Anadyr', according to the estimates of P. I. Popov, there were more than 50 archers. The brother of Nokon, Kapochila, sailed to Anadyrsk

[287] PSI. Bk. 1: 406, 456; KPTs: 156; cf.: Gurvich 1966: 112; Zuev 2002: 68.
[288] PSI. Bk. 2: 526; Miller 1996: 39; Zuev 2002: 69–70.

together with Cossacks for delivery of tribute. In general, P. I. Popov brought little tribute and did not enter into conflicts, but he obtained reconnaissance information on the geography and way of life of the population of Chukotka.[289]

At the end of 1718 five Nose Chukchi led by chief Idoga, according to the summons of the Khodyn Sopina Yukol'nikov, voluntarily arrived at the Anadyrsk stockade for delivery of tribute, in order to become subjects of Russia. They informed the Russians in adequate detail about the geography of Chukotka, the way of life of the people, and about the Eskimos of Bering Strait and Alaska.[290] We note that these Chukchi were from the Chukchi Peninsula, where the Khodyn Yukaghir arrived to call the population into Russian citizenship. This was a kind of excursion for the Chukchi who arrived at the stockade, for which they paid tribute.

At the same time, the situation on the Anadyr', especially in summer during the period of hunting, was not safe. In December 1716, the Yakutsk governor Colonel Ya. A. Elchin was assigned as head of the "Great Kamchatka Expedition"—an expedition whose task included investigation of Kamchatka and adjacent territories (including Chukotka), the discovery of new islands, establishment of trade with the local residents, and bringing the residents into Russian citizenship. However, because of the colonel's arrest in 1718 and the organizational confusion that followed, the tasks of the expedition were in substantial degree unattained. A member of the "Expedition," promyshlennik and mariner Prokopii Nagibin, wanted to go in 1720 on a ship from Anadyrsk to America, but he received no support from the local authorities. In 1725, he outfitted a ship on his own but, it is believed following the work of A. S. Sgibnev, he was killed by the Chukchi on the Anadyr'. However, in D. I. Pavlutskii's order of 9 November 1730 to navigator Ya. Gens and assistant navigator I. Fedorov, regarding the order for the boat *St. Gavriil* to arrive at the mouth of the Anadyr', it is mentioned that it was necessary to take from Kamchatka with them "the sailor, Prokopii Nagibin, sent from us." This is probably the same man who remained alive and was sent from

[289] PSI. Bk. 1: 456–459; KPTs: 156–158; Miller 1996: 39–40; cf.: Bogoraz-Tan 1934: 46–47; Berg 1946: 50–54; Fel' 1960: 75; Polevoi 2000: 58; Zuev 2002: 248.
[290] AAN. F. 21. Op. 4. No. 33. L. 134–135; cf.: Miller 1996: 40–41; Efimov 1948: 224; 1971: 196; Vdovin 1965: 12; Zuev 2002: 268.

Anadyrsk, and not his namesake.[291] On the whole, the expedition of P. Nagibin was the first attempt (even before V. Bering) by Russians to navigate to America.

After the unsuccessful expedition of P. Nagibin, the serviceman Afanasii Mel'nikov arrived in Anadyrsk in October 1725 carrying out the task of the "Expedition." Before him was the mission of investigating Chukotka and Alaska, as well as "searching out and bringing into citizenship and tribute disobedient foreigners." In 1728, a *shitik*[292] commanded by A. Mel'nikov and V. Shipitsyn and with a crew of 30 men left Anadyrsk, but, entering the sea not far from the mouth of the Anadyr' in a storm, sprang a leak. The crew was forced to return to Anadyrsk by land (in February 1729), in passing collecting tribute from ten Chukchi.

On 26 November 1729 A. Mel'nikov set off overland to Chukotka for the collection of tribute from the residents and investigation of the islands in Bering Strait. In December, the expedition arrived to the maritime "Chukchi," among whom it peacefully wintered over, taking tribute from four people. Then, on 15 March 1730, members of the expedition went on dogsleds to Chukotskii Nos. A. Mel'nikov reached Vostochnyi Nos (Cape Dezhnev), where the serviceman took tribute of fox furs from two people. In April two "toothed" Eskimos arrived there from the Islands of Gvozdev [the Diamedes]. They gave the serviceman gifts of 14 valuable white pebbles strung on a thread (possibly pearls), and the coastal residents gave five blue stones. Those who arrived, having told A. Mel'nikov about the islands and the Great Land (Alaska), refused to take the serviceman with them, pleading insufficient dogs and food. A. Mel'nikov returned to Anadyrsk, bringing with him two of the "best men" with families. In particular, the Chukchi Nekann came and left three of his sons as amanaty, but did not pay tribute for them. On 19 November 1730 A. Mel'nikov gave a report about his trip to D. I. Pavlutskii, who thus

291 Polonskii 1850: 393; Sgibnev 1868:131, 138; 1869: 27; 2005: 49; Efimov 1950: 145–146, 153; Belov 1956: 249; Fel' 1960: 79; cf.: Gol'denberg 1984: 80, note 3 (P. Nagibin was not killed by the Chukchi). A. S. Zuev (2009: 71) in one place mentions the delivery of D. I. Pavlutskii's order by P. Nagibin, and in other passages he has the sailor die in 1725 (Zuev 2002: 70, 272; 2009: 45). D. I. Pavlutskii's order to Ya. Gens, see: Pokrovskii 1941: 74–75; Fedorova 1984: 91–92.

292 Shitik—a flat-bottomed river boat 10 to 12 m long with stitched hull that traveled along the Anadyr under leather sails (Merck 1978: 144).

received the necessary information for his campaign the following year, 1731.[293]

Figure 20. Chukchi yaranga (1881). (Reproduced after Gil'der 1886: 83).

The peaceful expeditions of A. Mel'nikov were somewhat successful; tribute was obtained from a few people during them, first on the shore of the Gulf of Anadyr', and later on Cape Dezhnev. In the first case, these were most probably Chukchi, and in the second, Eskimos. However, it is clear that the bulk of the coastal population, as earlier, was not going to pay tribute. For three months, spent among the Settled Chukchi, A. Mel'nikov managed to acquire tribute from only two people. It was possibly they who went with him to Anadyrsk for the purpose of further establishing relations. In April 1730 two residents of Ratmanov [Big Diomede] Island arrived in the coastal village on dogsleds, probably for trade. Perhaps they were arriving for the fair in Uelen or Naukan, which was still remembered at the turn of the 19th to 20th century.[294] We note that life in Chukotka ran its course, and in spite of a clash with Shestakov's detachment in March 1730, no special hostile actions occurred from the Chukchi. This conflict was for them only the isolated episode of one raid

[293] Lindenau 1983: 163 (about the maritime part of the expedition); Sgibnev 1868: 138–139; Efimov 1948: 212–213; 1971: 196; Gol'denberg 1984: 82; Fedorova 1984: 127; Okhotina-Lind and Møller 2001: 180; Zuev 2009: 72–73, 80.
[294] Gondatti 1897b: 170; Bogoras 1934: 79; Antropova 1957: 174.

on the Koryak. Since in this same year, 1730, Russian-Chukchi relations moved into a phase of sharp confrontation, A. Mel'nikov's expedition did not create any activity.

In general, at the beginning of the 1730s the Chukchi on the Anadyr' maintained unfriendly relations with the Russians. From the 1660s Russian authorities required tribute and amanaty from them. However, the Chukchi rejected this, which led to conflict. Only an insignificant part of the southern Chukchi, more interested in peaceful relations for the sake of hunting on the lower reaches of the Anadyr', agreed to pay a light tribute. But even this tribute was paid on a case by case basis. Intermediaries in relations between the Chukchi and the Russians were the Yukaghir, who knew the Chukchi language. Attacks of the Chukchi and Koryak on the tributary Yukaghir, which increased from the 1680s, required the Russian authorities to defend their subjects, which also led to armed conflicts. Attempts to establish peaceful trade relations, in which both conflicting sides (primarily the Chukchi) were interested, did not have decisive significance for the course of events, and relations changed in the direction of open confrontation.

CHAPTER 4

CHUKCHI WAR (1727 to 1778)

1. Beginning of Activity of the Anadyrsk Party: The Expeditions of Afanasii F. Shestakov and Dmitri I. Pavlutskii (1727 to 1732)

1.1. The Expedition of A. F. Shestakov (1727 to 1730)

In the late fall of 1724 the Yakutsk Cossack leader Afanasii Fedotovich Shestakov arrived again in St. Petersburg. He submitted to the cabinet (actually, the chancellery) of His Imperial Majesty the project of an expedition for pacification and accession to the empire of Siberian lands around Chukotka and the Sea of Okhotsk. Attached to the project was his creation of a map of eastern Siberia, only divergent copies of which have been preserved. A. F. Shestakov himself (1677–1730) came from a Cossack family, received no education, remaining illiterate; beginning service as a common Cossack, he received a noble title in 1721 and received the rank of Cossack leader; in Yakutsk he was occupied with economic and financial affairs, collection of tribute, and in doing this knew seven local languages (probably Yakut and dialects of Tungus and Yukaghir); based on his character Shestakov was a direct, persistent, bold and at the same time authoritarian commander. Battle associates—the Okhotsk Tungus (Even)—assessed the leader in this way soon after his death: ". . . and he, Sir Afanasei Fedotovich, was to us truly a father, and we did not hear a harsh word from him, and he was a real commander."[295]

[295] AAN. F. 21. Op. 4. No. 33. L. 259 ob.; the characteristics of A. F. Shestakov: Bushnev 1996: 64–65, 67–68.

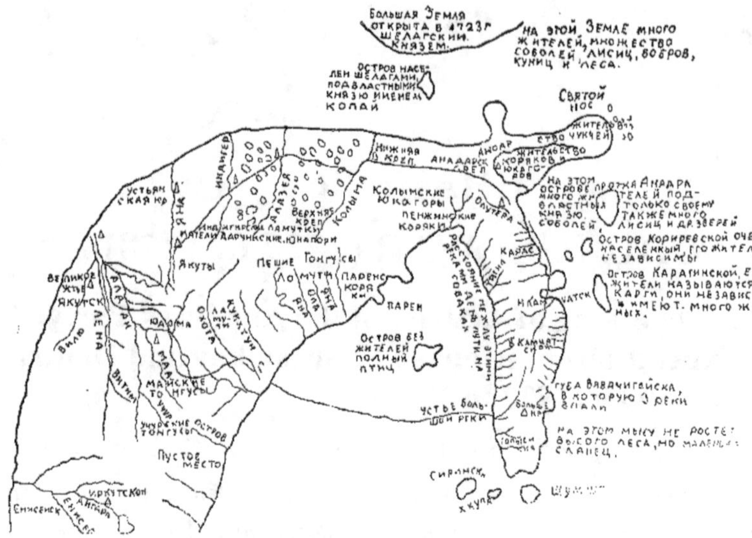

Figure 21. Part of A. F. Shestakov's (1725) map of eastern Siberia. For the complete map, see: Grekov 1960: 47, fig. 12. (Reproduced after Zubov and Badigin 1953: 11, fig. 1).

In St. Petersburg Shestakov enlisted the support of dignitaries; he was evidently presented to Prince A. D. Menshikov. Between October 1726 and March 1727, the leader's project was discussed in the Senate and Supreme Privy Council, where Shestakov himself reported as well, answering questions. The plans of the leader followed the general course of Russian policy directed toward the study and accession of the lands in the east: at this same time, the First Kamchatka Expedition of V. Bering (1724–1730) was being organized. On 23 March 1727, the empress signed an edict of the Senate "On calling into citizenship under Russian dominion unpeaceful foreigners, who are adjacent to Siberia, also traitors, who were formerly in allegiance, and about sending the Yakutsk Cossack leader Afonasii Shestakov for this calling and search for new lands." Described in the edict were the goals, composition, and organizational aspects of the expedition. Its primary task was to bring into obedience the Koryak of the Penzhina and Olyutora rivers who had lapsed, to build outposts there, to open up new lands in the east, to take into Russian citizenship the population, and to develop subsequent trade with Japan and Korea. It was recommended, as before, to call the "foreigners" into citizenship with "kindness." In Shestakov's party were to be 400 Siberian servicemen from the 1,500 Cossacks of the Yakutsk Regiment (almost a quarter of the personnel!), ten sailors, workmen for building ships, six small 4- to 5-pud (64- to 80-kg) cannons and two or four small mortars for equipping ships and outposts, ammunition, gifts for newly sought out "foreigners" (cloth, beads, needles, tobacco, and so on), and to recruit volunteers from the Yakut and other "foreigners" for the campaign, giving them goods instead of a salary, based on the established custom.

At the beginning of August 1727 Shestakov left St. Petersburg, arriving in September in Tobolsk, the capital of Siberia, where he was joined by a captain of the Siberian Dragoon Regiment, Dmitrii Ivanovich Pavlutskii (1692–1747). The members of the Supreme Privy Council, not completely trusting Shestakov, had ordered a field officer selected by the Siberian governor to be added to the expedition. It turned out to be Captain D. I. Pavlutskii. The captain himself was a decisive, forthright, and honorable soldier. He was supposed to lead the military command of the expedition, and both commanders were supposed to make joint decisions. Because of the unclear separation of powers, constant disputes arose, which came to fighting between supporters of the two commanders. According to the clarification of the Senate edict of 16 September 1728 by the Siberian governor Prince M. V. Dolgorukov, who carried out overall management of the expedition, the captain was supposed to be the "first commander" in the expedition, while Shestakov—the "second commander," though the latter referred to some imperial edict about his status, a copy of which the governor requested.[296] On 29 June 1728 the expedition arrived in Yakutsk where, based on the edict, 426 men (of them 385 Cossacks) were joined to it. Thus, the total in the party amounted to 591 men. In Yakutsk preparation for the campaign continued. It was not until 22 June 1729 that Shestakov set off to Okhotsk. In the fall Sotnik Vasilii Shipitsyn was sent from Yakutsk to Nizhnekolymsk where, together with promyshlennik I. Vilegin, he was supposed to occupy himself with the search for the Great Land, which was located, it was believed, opposite the mouth of the Kolyma. However, this task was not carried out. On 4 December 1729 D. I. Pavlutskii, with the primary forces,

[296] Pokrovskii 1941: 71–72.

moved from Yakutsk to Anadyrsk, where he arrived only on 3 September 1730.

Having reached Okhotsk in August 1729, Shestakov sent two ships from the port for investigation of the basin of the Sea of Okhotsk on 1 September: the *St. Gavriil* led by his twenty-four-year-old nephew, the noble Ivan Grigor'evich Shestakov, and the *Fortuna* led by his fourteen-year-old son Vasilii, who still had not been elevated to Cossack. In 1730 the first ship examined the Shantarskie Islands and the mouth of the Amur, and the second ship investigated five Kuril Islands, making 20 people each pay tribute in the form of a beaver pelt and receiving two amanaty from the Kuril people. In September 1729 the leader himself set off with 93 men on the boat *Vostochnyi Gavriil* to the land of the Koryak for future departure to the Anadyr'. On rounding Cape Alevina the ship encountered a storm and the party reached the Tauisk stockade on foot at the end of September. On 30 September the leader sent on the boat *Lev*, which arrived from Okhotsk, a detachment led by Captain Ivan Ostaf'ev (3 Cossacks and 30 Koryak, including the interpreter Chaibuga) to the Penzhina and Yama rivers for the collection of tribute and for bringing the Koryak into citizenship. Shestakov himself left Tauisk on 17 November 1729. The detachment consisted, based on information from Tungus eyewitnesses, of 18 servicemen, 33 Reindeer Lamut, 13 Okhotsk foot Tungus, 18 relatives of the Tungus Danila Babin, the guide from Tauisk, 10 Yakut, and 19 Koryak. The Tungus themselves were attached to the group voluntarily, receiving, besides local clothing and footwear, provisions and two rubles each in cash.

Shestakov acted cruelly in bringing the Settled Koryak into citizenship. The entrance to a house was blocked with a tree, after which it was proposed to the inhabitants to give tribute and provide amanaty. In the case of refusal, the house was burned. Thus, on the Tavatum River the detachment found seven homes of "unpeaceful" Koryak. At night they blocked the entrance and through the interpreter urged them to pay tribute. Since a refusal followed, three homes were burned along with the occupants, and the residents of the four remaining paid tribute of fox and ermine furs and provided one amanat. Terror worked: the Koryak, fearing such fate, began to provide tribute and hostages, fled, or even committed suicide. Shestakov left amanaty and tribute on the Ayakana River under the supervision of servicemen and several Tungus.

On 1 January 1730, on the Tumana River, Shestakov's expedition caught up with the detachment of I. Ostaf'ev. The latter, on the way from the Yama to the Penzhina, acting "gentle," managed to take tribute from 42 "old payers" and 168 "new payers" Koryak. The Reindeer Lamut of the Uyagan and Dolgan clans also joined the detachment of the leader. After unification the expeditionary forces numbered 23 Russians and 143 Lamut, Yakut, and Koryak.

The expedition, upon reaching the Gizhiga River, stopped in a Koryak settlement, the residents of which fled; there they seized a certain number of reindeer. In the course of six weeks the detachment waited for planned reinforcement from Anadyrsk, but it did not arrive and the detachment moved on. I. Ostaf'ev was sent to the sea for collection of tribute. At the Gizhiga a storehouse was built, in which were placed the various things unnecessary for the expedition at this moment. The ill Tungus interpreter D. Babin and a Koryak with a yaranga were left with the storehouse. D. Babin only pretended to be ill, deciding to flee from Shestakov, and he advised his relatives of the Shenyagir clan to do the same. Indeed, D. Babin worked some magic, and the results of the seance indicated an unfavorable future. He went back to Tauisk together with a Cossack.[297]

---

[297] About events of the beginning of the expedition, see: AAN. F. 3. Op. 10. No. 137. L. 81, 82–83 (report of F. Ch. Plenisner, 1763); F. 21. Op. 4. No. 33. L. 247–247 ob. (I. Ostaf'ev's story), 248 ob.–249 ob. (testimony of the foot Tungus in Okhotsk, 23 May 1727); Collection of the Imperial Russian Historical Society 1888: 44, 84–85, 101, 194–195, 249, 285–291 (documents of the Supreme Privy Council and Senate); about the expedition, see: Miller 1996: 63–65; Berkh 1819: 4–9 (briefly based on Miller); Sgibnev 1869: 1–14 (the most detailed presentation of events based on data of the Siberian Archives); Golder 1914: 153–157; Okun' 1935: 67–70; Efimov 1948: 156–159; Belov 1956: 258–260; Divin 1956: 13–18; Grekov 1960: 45–49; Magidovichi 1984: 95–96; Gol'denberg 1984: 8–72; 1985: 35–43; Fedorova 1984: 35–37, 59–61 (documents of the Senate and edicts of the Supreme Privy Council); Zuev 2002a: 53–75; 2009: 47–67 (dates of the expedition are given based on this monograph). Maps of A. F. Shestakov, see: Golder 1914: 111; Zubov and Badigin 1953: 11; Grekov 1960: 47, fig. 12; Efimov 1964: No. 60, 68; about the map and its copies: Gol'denberg 1980: 137–147.

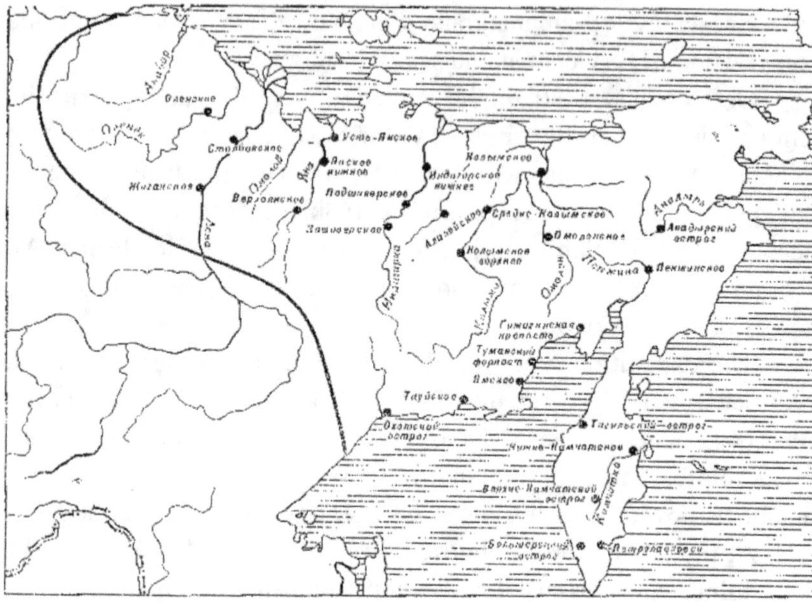

Figure 22. Russian stockades and winter quarters in northeastern Siberia. (Reproduced after Gurvich 1966: 267).

On 9 March the expedition crossed the Paren’ River, and on the third day on the Tylke (Tylkhoi) River news was received of an attack by the Chukchi against the Koryak. In testimonies of the participants of the battle, three foot Tungus, given in Okhotsk on 23 May 1730, we read: “And on the third day one non-peaceful Koryak came to meet us on a reindeer with sleds; on them the Koryak and we went on the campaign. And this Koryak told him, Mr. Shestakov: ‘Non-peaceful Chukchi came to us, and killed, and destroyed 4 yurts’.”[298]

In general, it turned out that the Chukchi carried out their raid on those non-tributary nomadic Koryak, to conquer whom servicemen also went. The Reindeer Koryak were caught between two fires and they clearly chose Russian citizenship; indeed, some negotiations had already been conducted with Shestakov. The Chukchi themselves, probably on the return route from a raid, subjected to destruction the nomad camp of the Koryak chief Yallakh, which consisted of four yarangas, where up to thirty people were killed—probably all or almost all the male population of the camp that accepted the fight.

[298] AAN. F. 21. Op. 4. No. 33. L. 253 ob.–254.

Having received news of the Chukchi raid, A. F. Shestkov set up camp, "constructed of sleds." The coffers and the baggage were left in this "Wagenburg" [sled fort] under the protection of three servicemen, led by Andrei Shitukhin, three Koryak "yurts," and the Reindeer Lamut of Bangachan, who did not want to advance farther.

On 11 March Shestakov sent an order to the Cossack Trifon Krupyshev who stayed in Tauisk with the servicemen to sail to the mouth of the Anadyr', to find opposite it the Great Land (that is, Alaska), and to bring into citizenship its residents. In addition, in this document was a description of the actual situation in the expedition: ". . . and from the Parena beside the sea to the Pyanzhina there is a large number of people and Reindeer Chauki Koryak, who prepare opposition to us, gathering an army. And we pursued them and made an treaty with them; and based on this treaty the Olansk (Aklansk.—*A. N.*) chief Yulta came and inflamed them. These Chauchi went from us with him and fled to Ogilansk (Aklansk.—*A. N.*); and . . . they wish to perpetrate . . . opposition to us; however, asking the All Merciful Lord's assistance in the deed we decided to go after these Chauchi."[299]

Though it seems that the ethnonym Chauki/Chauchi in the document is another name for the Chauchi—Reindeer Chukchi—Shestakov nevertheless knew who opposed him from intelligence and from the Koryak. Based on the document it turns out that the Reindeer Koryak are the "Chauk" (in reading of A. S. Slovtsov: the Chacheb—as the Settled Koryak called the Chavchuveny), who lived between the Paren' and the Penzhina, and who gathered forces against the detachment of Shestakov, probably moving away from the advancing Russians. The Russians approached the Koryak who evidently had already agreed to surrender tribute, having engaged in talks with them, but the Aklansk Koryak elder Yulta arrived and disrupted the talks.[300] He evidently had some agreement regarding a relationship with the Chukchi. The latter went

[299] AAN. F. 21. Op. 4. No. 33. L. 263 ob.; cf.: Sgibnev 1869: 14. About the Chacheb see: Dolgikh 1960: 562.

[300] Probably from these hostile actions of the Koryak the opinion developed that the Koryak had defeated A. F. Shestakov's detachment (Maidel' 1894: 543: in the opinion of the Yukagir, these were Koryak who called themselves Chukchi).

north from the Russians to the Oklan (Aklan) River and, in Shestakov's opinion, decided to take up battle with the servicemen.

On 13 March Shestakov moved after the Chukchi with a detachment of about 150 warriors, including 19 servicemen, 6 of whom were Anadyr' people. On this day the foot Tungus (Lamut) Baibachan with relatives, not wanting to go farther on the campaign, ran off before reaching camp—they probably remembered the predictions of D. Babin. In the evening, the detachment set up camp; scouts were sent out, who, upon returning, announced that the "Chyukchi stand in camps at . . . a clear place," the number of their "yurts" and people the scouts were not able to make out in the darkness, but there were many of them, "innumerable." At the war council the Russians and "foreigners" advised Shestakov to attack the Chukchi at night, after approaching closer to the camp of the latter. This same evening they went closer to the Chukchi. Again they set up camp. Again a war council was assembled and again it was proposed to attack at night. However, the tireless Shestakov, to the surprise of those around him, became overcome with fatigue, and in a kukhlyanka [kind of shirt] threw himself on the snow301 and slept through to lunch on 14 March. Waking up and going up a hill, the leader saw beyond the hills thick smoke from the Chukchi camp in which it was impossible even to make out a flying seagull, as well as another, separate Chukchi camp. At the next war council, they again debated between attacking by day, which Shestakov supported, or at night. In order that the separate camp did not unite with the Chukchi camp, the prince of the Reindeer Koryak Soblyak was sent there with 18 relatives and 9 Foot Okhotsk and Tauisk Tungus. Those sent killed five Chukchi, a sixth fled. One woman and a boy fell captive, whom the prince himself took; the Chukchi reindeer simply ran off.[302]

The Chukchi now noticed the Russians. Shestakov moved toward the enemy, the latter doing the same. Battle occurred on 14 March 1730 at a hill not far from the Egacha (later, Shestakovka) River, which falls into Penzhina Bay west of the Penzhina River.

---

[301] We note that in this case the influence of local customs is perceived. Thus, the Chukchi could sleep in their double warm *kukhlyanki* directly on the snow during a campaign (Korotaeva 2008: 8).

[302] The events, see: AAN. F. 21. Op. 4. No. 33. L. 243 ob. (official paper of D. I. Pavlutskii at the Yakutsk office of 31 May 1733—about the Vakhla River), 254–255 ob. (report of the Tungus in Okhotsk); Sgibnev 1869: 14–15; Zuev 2009: 67–68.

About two and a half months after the battle, on 23 May 1730, three foot Tungus (Lamut)—members of the fight—Ivan Largachanov, Nyagiyaku Galchigin, and Kudak Dedyutev gave in Okhotsk the following description of the fight:[303]

> And they came together; and he, Mr. Shestakov, ordered the Russians and foreigners to arm themselves and to dress up in battle dress, in armor (kuyaki) and helmet, and to join in arms and take up battle. And they armed themselves; and those Chukchi are also armed and put on their bone armors. And he placed us, the foot Tungus, with the Reindeer Tungus of the Dongan clan with Burgoda and with Tauisk on the right hand, and on the left hand he placed the Koryak of Prince Soblyak with relatives and the Tauisk Koryak, Etitka and relatives, and the relatives of Danilka Babin, and . . . placed the Russians and with them the Yakut. And he, Mr. Shestakov, at which time we were armed, and he dispersed us, and placed sleds with the interpreter Chaibuga. He made an outpost near us and he was in this stockade with the interpreter. And as we were armed, and there was no conversation between us and those Chyukchi; and he, Mr. Shestakov, ordered to give battle; and began to shoot at them, the Chyukchi, from guns and from bows, and they reciprocated; and it began to be battle; and in this battle people fell among them and among us; and they, the Chyukchi, began to bring down a crowd on us; and let the Russian people fire only two shots from guns, crushed them, since they were countless, so many of them fought us. And two or three sit at each fire. And the archers, not yet crushed, still fought; and on the right hand from us many archers began to be wounded. And the Koryak and relatives of Danilka on the left wing seeing that the Chyukchi were overcoming, all betrayed and fled to escape. And we didn't see this, since we were fighting with them, we look ahead, and back we don't look; and then before us there was a small number of Chyukchi, not so many fall, and the place is cleared of people. And we looked behind us: and the left wing was encircled by a raid; and the outpost, where Mr. Shestakov and the interpreter were, and he was still in the outpost; and the

[303] AAN. F. 21. Op. 4. No. 33. L. 255–257 (division into paragraphs and punctuation were added to facilitate reading). The fighting is described most detailed in an article by A. S. Sgibnev (1869: 14–17) based on descriptions of those same Tungus, and in my monograph (Nefëdkin 2003: 240–244), whereas in the works of A. S. Zuev (2002a; 2009: 46–69) accent is given to a general description of the expedition; also a brief description of the expedition: Berkh 1819: 4–9; Maidel' 1894: 545; Slyunin 1900: II: 40–41.

Koryak and Danilka's relatives were not there. And we in this battle thought that all had been killed (we learned about the betrayal and about the escape later); but even we looked back, but not often (when it was possible we looked); and we fought as much as our strength was able; for this, since we were in armor and helmets, we fought with them; and no matter how many we killed they had no losses. And those sitting by the fire, they didn't move from their place.

And Chaibuga yelled to the interpreter, our relative Poteshnoi, since he, Poteshnoi, was fighting behind us and close to the stockade: "Cry and yell to your relatives, that we lost Afanasii Feodotovich"; and this Poteshnoi cried in Tungus, and the cry in such a battle could hardly be heard. He roars and says that, you know, we lost Bol'shoi [the big one]. The Tungus Bol, hearing those speeches, began to call to us, that "we will be beaten when there is no Bol'shoi, we should run back if possible." And we took spear in hand, and went all in a group, and fled to escape; and they, the Chyukchi, let us go; and we fled past the stockade; and the view we saw of the killed Russians at this stockade, but the interpreter Chaibuga was not there. And where and how Mr. Shestakov was killed, we did not see, only Boris Zhertin and Vasilei Orefin stood, and they shot at the Chyukchi; and the same Chyukchi ran to the stockade with spears and pierced them; and we, fearful, did not run to the stockade, because there was a multitude of them, and they would also kill us, and we ran past.

And in this escape, going along the road, our relative, the Tungus Kulikan, who today is at the Tauisk stockade, told us, how the Chyukchi overcame Mr. Shestakov; he fought in the battle behind us and saw how the Chyukchi took the stockade in the raid; and Mr. Shestakov with the interpreter Chaibuga standing, and he sees that they took a gun from the Russians, and began also to bring down a crowd; and he looked at us, that many fell, and could not be patient, and popped out from the stockade to rescue his people, and shot from a gun, and killed several men, and soon an arrow hit him in his throat, and he pulled the arrow out and ran, and threw himself on a sled, and fled on reindeer back to the stockade; but the reindeer was of the Chyukchi and did not run to the stockade, rushing in the camp to them, the Chyukchi; and he was not able to manage it, and jumped from the sled, ran to his people; and they, the Chyukchi, not allowing him to his people, 4 men rushed and pierced him with spears; and the interpreter Chaibuga was with him on foot, and as they slashed him with spears, and Chaibuga

> fled to escape, and ran away, but he didn't see where he ran to; and all that was in the stockade of the leader Shestakov and clothing and kettles, and axes, and a banner of the Holy Virgin Hodegetria[304] and of servicepeople, and all the foreign belongings in the stockade those Chyukchi looted and divided among themselves.

In distinction from the paraphrase of the document by A. S. Sgibnev, this story contains a whole series of important details omitted by the historian, who reduced the material. The battle is recounted by the eyewitnesses, who from their point of view tell what happened to them in the battle and what they themselves saw and heard.

The layout, given at the order of Shestakov, was standard: in the center were the Cossacks and Yakut; on the right flank were the foot and Reindeer Lamut in three detachments; on the left were two Koryak detachments led by their princes and one Tauisk Lamut. The narrators themselves were on the right flank near the center. The battle formation of these Tungus-Lamut was not a compact line: some were behind and to the right, which is explained by the need for dispersal of warriors for convenience of the archers. And behind this battle line, fires had been built for every two or three men—a characteristic feature of Tungus warfare: lit fires on the battlefield for connection with the spirits.[305] In order to lead the battle, Shestakov stood not far behind the center of the battle line in a field fortification made of sleds, where the arsenal and a banner with the image of the Virgen were also located. For transmission of orders by voice, a Koryak translator was near Shestakov—the leader did not know this language. The Tungus do not mention the Chukchi formation. The narrators did not know the numbers or the names of the Chukchi commanders. However, it can be supposed that among the leaders were the "toion of the Eastern Sea" Keyu, who at that time seized his future wife Yuttu; and Khypayu, after whose defeat in the second battle (30 June 1731) D. I. Pavlutskii's servicemen found things belonging to Shestakov.[306]

---

[304] Hodegetria (Guide)—image of the virgin with Christ child.
[305] Miller 2009: 348.
[306] Zuev 2009: 209; about the booty of D. I. Pavlutskii see below.

Figure 23. A Chukchi in laminar armor with a left wing. An engraving based on a drawing by Luka Voronin, illustrator for the J. Billings Expedition (1785–1792). The only known image of a Chukchi warrior in armor of the 18th century. (Reproduced after Antropova 1957: 205, Fig. 22).

The battle lines of the enemy were formed within the range of a bow shot, that is, on the order of 100 m from each other. The fighting was on foot, the Reindeer Lamut dismounted and stood in line. The fight began on Shestakov's order. Evidently, the leader counted on the power of firearms, not knowing precisely the number of Chukchi. Based on the testimony of the Tungus, it is evident that the number of the enemy was not determined—it is clear only that there were more of them. Usually the detachments, that were sent on a raid on the Koryak in this period, numbered several hundred men. In the document it is specifically indicated that no negotiations with the Chukchi were conducted, though theoretically it was first necessary to bring them into citizenship, wherein they were to provide tribute and amanaty. The allies opened fire with guns and bows; the Chukchi answered them with shots from bows. The tactics of the Chukchi in the battle can be traced rather clearly: to not give the enemy the use of its advantage of long-range weaponry, both firearms and cold iron, and move directly into the melee with spears, with which the reindeer herder were masters, using with this their numerical superiority. First, they attacked the primary detachment, which consisted of Russians and Yakut. The Chukchi allowed the enemy only two times (not one time, as according to S. A. Sgibnev) to shoot from guns, after which they attacked the center of the battle order (that is, after about a minute). The most heated fighting also occurred there, about which clearly speak the losses: where about half of the Russians and most (6 of 10) Yakut were killed. Simultaneously, the Chukchi rounded Shestakov's left flank and began to encircle it. Evidently, the Chukchi moved part of the forces of the left flank to the right since the Tungus noted that the enemy before them became fewer. The Koryak and Tungus of the left wing fled, in order not to be surrounded. And also the Koryak, a part of whom had only recently been conquered, did not especially want to fight for the Russians. About the fact that the Koryak retreated almost immediately, attest their small losses in the battle—a total of one man. As a result, the Chukchi began to encircle the center. Seeing the losses of the servicemen, the leader rushed to their aid, shooting a few Chukchi. During the fight Shestakov was wounded in the throat, but pulled the arrow out. He wanted to save himself by fleeing and climbed onto the first of the available freight sleds, drawn by a single reindeer; however, his strength evidently ran out, and he was not able to manage the sled. The reindeer brought Shestakov to the Chukchi camp. Possibly the owner of the sled, together with three other warriors, stabbed the leader.

Until the death of the leader was known, the right flank kept fighting, probably conducting a long-range battle—the Tungus did not even have spears. The latter noted that the helmets and lamellar armor in fact protected them from the Chukchi's cold weapons. In general, the losses of the Tungus in battle were not so great—approximately a fifth part of them were lost. As is noted in the document, the Chukchi cleared the field of battle of their fallen comrades. The Tungus of the right flank learned of the death of Shestakov from the Koryak interpreter Chaibuga, who by a cry (evidently in Russian) made it known about this to the nearest Tungus, and this one relayed the information farther. The Tungus as a group turned and fled. The Chukchi did not hinder them, providing a "corridor." Running by the "sled fort," the Tungus saw that the battle still went on near it: probably retreating, Vasilii Orefin and the Yakut princeling nobleman Boris Zhertin carried on a firefight with the Chukchi. The former, with the attack on them by the Chukchi, were gripped in the melee, using spears, and were killed in an unequal battle. The Tungus were afraid to provide aid to the servicemen. The Chukchi did not pursue the broken enemy. They seized the booty in the "stockade" and divided it.

A total of 29 men in Shestakov's detachment were killed: the leader himself, the Yakut nobleman B. Zhertin, who evidently led the unwillingly recruited detachment of his compatriots, nine Cossacks, six Yakut, eleven Tungus (including five Okhotsk, including Poteshnoi), and one newly baptized Koryak; in the "fort" the Chukchi seized the banner, twelve muskets, three rifles, twelve hand grenades, and twelve sets of iron armor. Among the surviving servicemen there remained only four muskets, three bayonets, three spears, and a drum.[307] Part of this booty was recovered from the Chukchi in the following year during the first campaign of D. I. Pavlutskii in Chukotka.

In the example of this fight we can see some features of the tactics of both sides. The Russians were formed in the usual linear order: they themselves were located in the center, while their Siberian allies were on the flanks. In the battle, stake was made on the strength of the fire power and deflection of the enemy attack by it, since Shestakov's tactics had to be defensive because of the substantial numerical advantage of the enemy. The Chukchi used methods of battle that they had worked out during struggles with the Russians: first, surround the enemy; second, attack the main forces of the enemy in the center; third, do not allow the enemy to use the advantage of possessing guns in a long-range battle but rather move to hand-to-hand combat, where it is possible to use both their numerical and physical superiority; and finally, fourth, not to pursue the enemy, avoiding unnecessary losses. In general, for the Chukchi themselves this fight was peripheral, it was not imprinted on people's memory or in the folklore.

The Russians, Koryak, and Tungus who survived met after the battle in a camp guarded by the Cossacks of A. Shitukhin. I. Ostaf'ev, who

---

[307] Human losses and plunder: AAN. F. 21. Op. 4. No. 33. L. 262 (report of I. Batankuev). L. 265 (report of I. Ostaf'ev in Tauisk); Gol'denberg 1984: 72; Zuev 2002a: 63; 2009: 68; cf.: Sgibnev 1869: 16. The weapons remaining among the Cossacks: AAN. F. 21. Op. 4. No. 33. L. 264 ob.

had gone on reconnaissance to the sea and did not participate in the battle, arrived here with a guide. The defeated spent two days there, fearing attack by the victorious Chukchi. Shestakov's Koryak prisoners, one boy and three girls, were divided. Luka Zapletaev took one amanat and gave him for armor to some Lamut of the Uyagan clan; another Koryak amanat was taken by Chaibuga, who gave it back to the relatives. It was also noted that the foot Tungus cut bags containing state property and took spear heads. At a council of war the servicemen decided where to go next. At this time reinforcements arrived from Anadyrsk—ten servicemen. They convinced a large part of the Cossacks led by I. Ostaf'ev to go to the Anadyrsk stockade. The Cossacks went, taking the body of Shestakov, tribute in three trunks, and the remaining weapons. Three servicemen and Tungus were sent to the Tauisk stockade. In April 1730 the detachment arrived in Anadyrsk, where the leader was buried. The local allies of the Russians went home.[308]

1.2. First Campaign of D. I. Pavlutskii in Chukotka (1731)

Captain D. I. Pavlutskii received news of the defeat and death of A. F. Shestakov on 25 April 1730 while in Nizhnekolymsk. There he took command of an expedition, sending out orders to detachments of the expedition to move to Anadyrsk. The captain himself went on 5 July, leading a detachment that consisted of 150 servicemen and 55 Cossack children and promyshlenniki, for a total of 207 people, part of whom were newly acquired from the Zashiversk, Alazeisk, Indigirksk, and Nizhnekolymsk stockades. On 3 September the captain arrived in Anadyrsk, where at this time there remained only 18 Cossacks, who in addition had no guns.

On 30 November 1730, the tributary Alyutorsk Koryak appeared in Anadyrsk with the complaint that the Chukchi had attacked their relatives between the Paren' and Penzhina rivers. They requested protection by the Russians and offered their help. D. I. Pavlutskii, as a decisive and independent commander, would not await order from above but rather decided to go on the campaign, determining to go around the

308 AAN. F. 21. Op. 4. No. 33. L. 257 ob.–258 ob., 264–264 ob. (report of I. Ostaf'ev). Sgibnev 1869: 16–17.

whole Chuchi Peninsula (Anadyrskii Nos, as they said at that time) and return back to Anadyrsk. Indeed, the Tobol'sk provincial office confirmed the powers of the captain as commander of the expedition only by an order of 10 August 1731; a campaign against "non-peaceful foreigners" was not required at all, but this order was received in Anadyrsk only on 18 August 1732. In the spring of 1731 ten Cossacks and three grenadiers arrived in Anadyrsk from Kamchatka. They brought cannon balls, bullets, gunpowder, and wicks. Preparations for the expedition were thus completed, and the campaign against the Chukchi began on 12 March 1731.[309]

After 32 years, in 1763, veterans of the expedition were questioned about the events of the campaign at the direction of Lieutenant Colonel F. Plenisner in Anadyrsk. They were the sotniki of the Anadyrsk detachment Ivan Popov, Osip Nizhegorodov, and Stepan Pavlov, Pyatidesyatniki Aleksei Rusanov and Fedor Kurbatov, Cossack Ivan Isakov, and five additional servicemen. They told the following:

> While they were on the campaign in 731 in Chyukotskii land for eradication of the Chyukchi rebels, the whole detachment consisted of two hundred thirty six pyatidesyatniki and Cossacks, and two hundred and eighty Koryak and Yukaghir. March 2nd from the Anadyrsk fort they set off through swamps to the north, crossing directly to the Kovyma Sea, and went near it on the right hand to the east through the swamp, and mountainous and unforested places.
>
> And the 9th of May they arrived at the first Settled Chyukchi yurts, near this sea, in which they killed the Chyukchi who were there; on the whole a hundred reindeer were taken, and from this place they went near the shore of this sea to the first bay, which was not very large, and what name it had, they didn't know; they crossed on the ice from evening to sunrise, and, having crossed, they stopped and then saw from this place in the near distance in the sea a reef, on which was one yurt of those Settled, and we killed the Chyukchi who lived in it. And then they went near this sea shore and arrived at another bay, which was bigger than the previous, and what name it had, they didn't know, which they crossed in one day on the ice and followed near this sea and at that time met thirty people of the Reindeer Chyukchi. And when they found out about them, they

309 Sgibnev 1869: 22–28; Zuev 2009: 69–75.

returned back, and there was no battle with them then. And after these Chyukchi they went five days near this sea, where again they were met by those Chyukchi, who were now about seven hundred, with whom battle was conducted, and to four hundred fifty of those Chyukchi were killed, in all a hundred and a half male and female sex and young were taken, to five hundred reindeer, and then the remaining Chyukchi fled. On our side were killed: one pyatidesyatnik and another Cossack, and to seventy men wounded with light wounds.

And they arrived at a Chyukotsk stockade, which was constructed of their traveling cargo sleds and walrus hide and scattered with stone, tussocks, and sand, and was linked around with thongs, and in this outpost were to eight yurts, which they ravaged and burned.

From this stockade, turning to the right, they went across Chyukotskii Nos and went through this land about three weeks and did not see anyone, and no large rivers falling into the sea and there was no forest, and as they had arrived in the middle of this Chyukotskii Nos about the 29th of June, then the mentioned Chyukchi, on reindeer, first suddenly began to make an attack on them from behind, and then also in front, at which time there were up to a thousand of them, and there was a battle with them from morning to noon, and three hundred of them, for example, were killed, and in all no more than ten people were taken captive, while the remaining Chyukchi all fled. On our side no one was killed.

At that time, through the captured Chyukchi, they found out their reindeer herds, of which a total of forty thousand draft and non-draft large and small reindeer were taken, and after this they went through Chyukotskii Nos land to the named stone Serdtse, and before arriving at this stone, five hundred of those Chyukchi met them, and there was a battle with them, in which up to forty men of them were killed, for example, while the remaining all fled, up to thirty reindeer were taken, in which case one Cossack was killed on our side, while they do not remember how many were wounded. And from this place they went to the former Yukaghir fair on the 1st of August, where they spent the night and went near the bay that lies between the mouth of the Anadyr' and the sea, while the mentioned Kamen'-Sertse, as well as the Anadyr' mouth, remained on the left, and they returned the same way to Anadyrsk, which they had taken to go ahead. They arrived at Anadyrsk early in November.

> While they were on this Chyukotskii campaign, both going and also returning, they did not see any forest in all this land, and all the Chyukotskii land was swamp, hummocky, and wet, on this land also in rare places were small twigs rising from the ground no more than a *chetvert'* [17.8 cm], and in the absence of forests they cooked food for themselves necessarily taking these twigs, wherever they were able to find them, and building a fire, which they fanned with fans taken with them, and where they could not find them, then taking moss growing on hummocks, with this they cooked. During this campaign they went in summer on foot, and they carried with them a gun and the rest, the belongings that they had they put on pack reindeer; in this campaign no one of the detachment died from hunger, but they died from sudden illnesses. And how many of them died, and more than the above-written shows, because much time passed they don't remember.[310]

Since the servicemen were reminiscing about the campaign three decades later, they naturally forgot or confused some dates, numbers, and sequence of events. In October 1772 three of these veteran Cossacks told about this campaign at the order of the Gizhiga commander Captain Ya. M. Peresypkin, adding to their earlier account a number of significant details and alternative numbers. Since not three but eleven men were recalling about the campaign in 1763, it is clear that together they were better able to recall events. Especially because they were recalling nine years before, and among them were Cossack commanders, sotniki, and pyatidesyatniki who were better informed about the military plans. Therefore, if there is a difference between the depositions of the different years, one should rely on information obtained from the servicemen in 1763. In addition, J. I. Lindenau in 1742 composed in Okhotsk a small essay "Opisanie o chukotskoi zemli, gde onaya imeettsa" [Description of Chukchi Land, and Where It Is], which, properly speaking, is an annotation to his map of Northeastern Siberia. In this essay he gives valuable indicators on the route traversed by D. I. Pavlutskii's expedition. Also, G. F. Müller, in his work "Izvestiya o severnom morskom khode iz ust'ya Leny reki radi obreteniya vostochnykh stran" [News of the Northern Sea Route from the Mouth of the Lena River in Order to Attain Eastern Countries], compiled a brief description of the route and events of

[310] KPTs: 158–160; Berkh 1825: 172–176.

the campaign proper, based on oral accounts of Cossacks in Yakutsk in 1737. In his later (1758) essay "A Description of the Maritime Voyages along the Arctic and Pacific Oceans, Carried out by Russians" (see Müller 1996) information from written documents is added to this data, in particular from documents sent to the author by D. I. Pavlutskii. Also preserved were the text of a "Promemoriya" [Official Memo] of the captain to the Yakutsk office of 10 February 1732 and "Vedomost', uchinennaya ot Anadyrskoi partii" [Statement Made by the Anadyrsk Party] of 20 November 1738, also compiled on the request of G. F. Müller. In the last, the text of the "Promemoriya" is reproduced with a few minor details, where very briefly and in general features the overall course of the expedition can be traced. The brief statements of the three veteran Cossacks, taken from servicemen on the Indigirka by Lieutenant D. Ya. Laptev in October 1739, add several useful comments to the block of information about the campaign.[311]

In general, for addition and correction of the Cossacks' statements of 1763 it is necessary to use documents that A. S. Zuev examined in a special article in 2001. This article, partially reproduced in the monograph of 2009, is the most detailed work on the theme, though the author himself was basically limited to simple retelling of the sources. It must be noted that this campaign is one of the most studied topics among the subjects of the Russian-Chukchi wars. The campaign itself was studied as early as 1819, when the work of the maritime historian V. N. Berkh appeared, written based on information gathered by G. F. Müller; then, in 1852 an article by the scholar of Siberian studies N. Shchukin followed, printed twice in different editions, and finally in 1869, based on materials from the Siberian archives, the campaign was briefly described by the Kamchatka historian A. S. Sgibnev.[312]

---

[311] Lindenau 1983: 162–164; Miller 1996: 65–66; Efimov 1948: 219–228; Okhotina-Lind and Møller 2013: 91–92 ("Izvestie o severnom morskom khode. . ." G. F. Miller); about the map of J. Lindenau: Vdovin 1943: 52–53; the account of the Cossacks of 1772: RGADA. F. 199. No. 528. Ch. 2. D. 3. L. 9–11; No. 539. Ch. 2. D. 6. L. 65–68; transcription of this information: Zuev 2001: 3–38; reports of D. I. Pavlutskii, cf.: AAN. F. 21. Op. 4. No. 33. L. 241–243; Zuev 2003: 134–138; statements of 1739: Okhotina-Lind and Møller 2015: 368–369.

[312] Berkh 1819: 13–15; Shchukin 1852: 8–11; 1854: 409–437; Sgibnev 1869: 28–31; Pokrovskii 1941: 328 (letter of D. Ya. Laptev of 30 October 1740); Zuev 2001: 3–38

Thus, on 12 March (not the 2nd, as in the document) 1731 the detachment of D. I. Pavlutskii moved along the route on reindeer sleds supplied by the Koryak and Yukaghir. The party consisted of 215 service people (Cossacks and soldiers), including Cossack children, and promyshlenniki, 160 Koryak, and 60 Yukaghir; they hauled one cannon with them. Based on local custom they drove a herd for supplying the expedition with provisions. All in all, the Koryak of the Anadyrsk department provided 641 draft reindeer and 3,756 karaginnye (herd) reindeer. While still in Anadyrsk they prepared wood (?) charcoal, packed in bags, for campfires; and along the way they acquired willows and moss. Participating in the campaign were tributaries of the Anadyrsk department, whom D. I. Pavlutskii appointed based on the general registry of tribute payers. As the servicemen indicated in 1772, the Yukaghir performed as guides for the detachment; according to Captain D. Ya. Laptev, Koryak knew the locality better, as did the Cossacks Ivan Modin and Yakov Osipov, who were also interpreters. From March to the beginning of May the expedition traveled on sleds about 10 km per 24 hours. This was not so little for a large expedition—in October–November 1791 the detachment of J. Billings, returning from Chukotka on reindeer sleds, traveled 6 to 22 verst per 24 hours.[313] In general, for the 52 days of the trip along the coast from the beginning of May (tentatively, the 2nd) to 30 June (excepting an eight-day rest after the first battle) the detachment of D. I. Pavlutskii could have traveled 500 km, not counting days when camped. Moving along the coast, to the second battle (30 June) they traveled over the ice, and then overland. Five or six servicemen brought a fur tent for sleeping at night. On 15 May it became warm, and when the snow disappeared, the servicemen went on foot, carrying their guns, while baggage weighing 1.5 to 2.0 pudy (24 to 32 kg) was carried by pack reindeer.

According to G. F. Müller, the detachment passed along the Uboinaya (Ubienka) River, across the tops of the Belaya and Chernaya

---

(does not examine in any detail the route of return to Anadyrsk, simply retelling the information); 2009: 75–80 (an abbreviated text of the same article). Also briefly about the expedition, cf.: Golder 1914: 163; Belov 1956: 260–261; Grekov 1960: 53–54; Vdovin 1965: 117–118; Magidovichi 1984: 96–98; Gol'denberg 1984: 77–90.

[313] Merck 1978: 147–148.

(Tanyurer) rivers and out onto the coast of the Chukchi Sea, reaching on 2 May the mouth of a large river. According to the depositions of 1739, the party, crossing through the mountains, moved north to the ocean along a large river for 14 days (that is, went about 140 km) and, not reaching the shore, turned east. M. I. Belov, on the basis of J. Lindenau's map—where three rivers that fall into the North Sea between Cape Shelagskii and Chukotskii Nos are indicated—supposes that these rivers were the Ekiatap (Ekvyvatap), Amguema, and Vankarem. Based on the diagram of J. Lindenau, it turns out that the detachment went to the sea in the region of the first river. Based on another supposition, the party crossed directly from the upper reaches of the Tanyurer to the nearest upper reaches of the Amguema.[314]

On 9 May the detachment stumbled upon a dwelling of the Settled Chukchi on the shore. The captain did not make an appeal to turn in tribute (which emphasizes the punitive nature of the expedition), but quickly attacked. Six men, who probably tried to resist, were killed, while the women stabbed to death themselves and the children. A hundred reindeer were seized there. It is difficult to determine to whom the reindeer belonged. The Settled Chukchi hardly had such a number of animals. In the 1772 reminiscences the Cossacks recounted that two "yurts" of Reindeer Chukchi were encountered; 30 men were killed, the remaining fled; the servicemen seized up to 2,000 reindeer and divided the property captured from the enemy. Contamination of the two events probably occurred in both reminiscences. In the beginning of May, the Reindeer Chukchi could migrate to the shore to spend the summer; the reindeer probably belonged to them. Indeed, J. Lindenau's description indicates that from the yurts (not yurt) to the place of the first battle two weeks' route passed, which gives us a date of encounter of 24 May. It is possible that this evidence also hints at the second encounter of the Russians with the Chukchi, this time with a nomadic camp of two yurts.

Later, the servicemen, crossing the ice of some bay, noticed a small island in the sea, on which there was a winter semisubterranean dwelling of settled residents. Those sent by D. I. Pavlutskii killed six men there, with a total, together with women and children, of 20 people. Again

---

314 Belov 1956: 261; pro: Gol'denberg 1984: 91; Amguema: Magidovichi 1984: 96; Zuev 2001: 3–7.

no negotiations were conducted—the servicemen did not mention them. Then, crossing the ice of another large bay, the detachment continued its movement to the east along the shore of the sea, when it met up with 30 Reindeer Chukchi, with whom D. I. Pavlutskii tried through a translator to enter in negotiations. The Chukchi not only did not endorse the captain's proposal to accept Russian citizenship, but sneeringly stated that they would win over the Russians with lassos alone, after which they retired. This group was most probably reconnoitering or going to join up with the primary forces of the Chukchi. The expedition encountered the latter five days later, on 7 June.

Figure 24. A warrior in Koryak armor with two wings. A mannequin in American Museum of Natural History (New York). The photo by P. Yakushin, 2017.

This battle is described in detail in the depositions of the Cossacks in 1772. Having arrived at the mouth of some large river and because of a flood not seeing the possibility to cross it, D. I. Pavlutskii ordered the detachment to cross on the ice of the Chukchi Sea in front of the mouth of the river—on the sea the ice was firm, whereas on the mouth of the river it had been thawing. Approaching the other bank of the river, the Russians noticed an army of Chukchi numbering 700 men (based on the testimonies 1739 and 1772—1,000), which, as was clarified, was led by the toion of the "Northeastern Sea" Naikhnyu. Initially the Chukchi, based on the words of captives, were tactically gathered to attack the Russians as soon as they reached the bank, but then decided to give battle on the bank; indeed, it was clear to the Chukchi that their number exceeded that of the expeditionary force. The captain, in spite of the unfavorable position (near the bank the thaw water stood waist-deep), ordered to go onto the bank. The convoy was left on the ice of the sea under the protection of Cossacks. The Chukchi rejected the offer of the captain, given through the interpreter, to provide tribute and to give amanaty. Using their numerical superiority, they began to deploy and surround those coming onto the bank. D. I. Pavlutskii, who was in command of the right flank, lined up the Russians, Koryak, and Yukaghir in a solid line (in a "front"), whereas on the left flank the Cossack Sotnik V. Shipitsyn, who knew the local conditions better, stretched out a line of Cossacks, placing them next to one another at a distance of three meters in order to avoid encirclement. The Russians went on the attack. The allies of the servicemen on the right flank wavered and began to retreat back to the sea, but three Cossacks from the rear guard, leaving the convoy, rushed to the rescue and helped those retreating resist. The left flank of the servicemen successfully advanced, pushing the Chukchi beyond eight "yurts" (evidently part of the camp with families). The battle lasted many hours: from morning to sunset (on the polar day on 18 June 2014 at the latitude of Vilyuisk sunset was at 11:19 pm). The field of battle remained with the Russians, which at that time was viewed as a victory. The Chukchi remaining alive, led by Naikhnyu, fled. As A. S. Zuev correctly noted, a battle of such length was a long-range one, which is also indicated by the insignificant losses to D. I. Pavlutskii's detachment: a pyatidesyatnik, a Cossack, and five Koryak were killed; about 70 men were wounded. The losses of the vanquished were substantially more: 70 (based on testimonies of 1739) or 450 or even 700 killed, and 150 men fell captive, while part of the Chukchi women managed to kill their children. In fact, it is possible that among the Chukchi 70 warriors fell, whereas the remaining who died were family members, while the rest were taken prisoner. In such case, a large part of the warriors survived. As trophies went 500 (or 4,000) reindeer and the property, which was divided among the detachment. It is clear that the Chukchi bore substantial losses from gunfire; however, they probably managed to drive off a significant part of the reindeer: for 700 warriors 1,400 draft animals were required.

After stopping for eight days in one place for rest and treatment of the wounded, D. I. Pavlutskii's detachment moved farther east. They arrived at some fortified nomad camp consisting of eight yarangas. These fortifications can be imagined as large cargo sleds covered with walrus hide, to which were attached thongs forming a type of corral, and then all this was reinforced with stones, sod, and sand. This was possibly some nomad camp of Reindeer Chukchi who decided to give battle, unified because of the danger. Or perhaps this was simply a camp fortified as a consequence of the warlike times, so the men could leave in summer for the pasture. The servicemen did not even mention the fate of the residents; the residents were probably killed or taken captive and the fortification burned.

Further, based on the reminiscences of the servicemen in 1763, they went to the right, into the depths of the Chukchi Peninsula, whereas G. Müller notes that the party reached Chukotskii Nos several days after the second battle; according to the reports of D. I. Pavlutskii himself, after the second battle the party moved from the Kolyma to the Anadyr' Sea. According to J. Lindenau's map, the second battle occurred at the base of Chukotskii Nos, which is going sharply away to the north from this place. If the third river on this map is the Vankarem, then the battle occurred east of it. According to G. Müller, the battle occurred at the second of two rivers that fall into the Arctic Ocean at a distance of a day's travel from each other (10 to 30 km). To the east of the Vankarem there are the Vel'mai and Linatkhyrvuvaam rivers at a distance of about 16 km from each other. If it is accepted that the battle occurred beyond the second river, then it is possible to imagine that the party had already begun to move to the southeast along Kolyuchinskaya Bay, which was also fixed in the memory of the Cossacks.

Thus, on 30 June, during the march in the tundra, D. I. Pavlutskii's marching column was suddenly attacked at some river, first in back, and then in front. Now the enemy had 1,000 warriors (in the statements of 1739—1,500, while in the reminiscences of 1772—3,000). These were

Reindeer Chukchi led by the same Naikhnyu and the toion of the "Eastern Sea" Khypayu who arrived to help him. Also in this militia were residents of Uelen; indeed, according to Uelen legends, in almost every family there was a warrior who went to fight against the Russians.[315] Probably one of the toiony attacked the rear guard, while the other attacked the advance guard. According to D. I. Pavlutskii's order, a field fortification for protection of the sick and wounded was made from the cargo sleds. The battle itself again lasted several hours—from morning to noon (according to the reminiscences of 1772, from the first hour of the day to 9 o'clock in the evening—about eight hours). The servicemen probably managed to deploy in a combat formation and open fire, keeping the enemy at a distance. Indeed, at the end of the battle no one in the detachment had been killed, and only 20 wounded. Among the Chukchi 300 were killed (according to the depositions of 1739—210), only fewer than ten people fell captive to the Russians, and an insignificant number of reindeer were also seized. It seems that the overall losses of the Chukchi in this battle were fewer than in the preceding—less than a third. Those Chukchi remaining alive withdrew, evidently on sleds. The pursuers killed several more people. Apparently the Chukchi camp itself was not seized; it was probably in a secluded place, not in the direction of the flight, which as a result saved the lives of the families of the warriors. Therefore, the number of captives and seized reindeer, probably draft animals, was insignificant. Among the trophies found were items belonging to Shestakov's detachment and to the leader personally (armor, firearms), which indicates that some of the Chukchi warriors had fought against the Russians at Egache.

After the battle D. I. Pavlutskii's detachment stopped at a place for three days to rest. It was learned from the prisoners where the primary Chukchi herds were, and as a result up to 40,000 reindeer were seized. This was evidently the united herd of a large number of Chauchu, perhaps warriors of one of the toiony. A scouting party sent by the captain reached the (Chukchi) sea, where it found two yarangas of Reindeer Chukchi who had arrived there for summer, which were burned, since the residents themselves simply fled.

---

[315] Rytkheu 2004: 93.

Figure 25. A Chukchi yaranga in St. Lawrence Bay. Photograph from the boundary of the 19th and 20th centuries.

Later, according to J. Lindenau, D. I. Pavlutskii intended to go to Chukotskii Nos in order to destroy the "Chukotskii Stone Fort," but, for lack of reindeer feed on the way, the expedition moved in a southern direction—across the mountains to the Anadyr' Sea (modern Anadyr' Bay). On J. Lindenau's map this "Stone Fort" is shown in the mountains in the center of Chukotskii Nos; if its location is placed on a modern map, it is apparently in the Genkanai Range on the Daurkin Peninsula. The bulk of the seized reindeer required food, which became the deciding factor in altering the course of the expedition. The reindeer livestock was broken up into 12 herds of 1,000 to 2,000 head each, which was optimal for caring for them. Consequently, the total number of reindeer reached 24,000 head. They placed the herds under Koryak and Yukaghir shepherds. However, the first did not care about the Russian trophies, while the second were simply not skilled at managing such large herds. As a result, part of the reindeer simply ran off and part the Chukchi possibly drove back. It is apparent that the Chukchi took their herds back by force since five Koryak, according to the "Promemoriya" of D. I. Pavlutskii, perished not only in battles but also in the pasture. Perhaps five Koryak died there; however, about this D. I. Pavlutskii wrote that five Koryak (1732) were killed, then ten (1738)—the contradiction can be removed by assuming that the first number represents only the losses in battle, while in the second all the losses are included—in battle, from illness, and in the march.

On 14 July the last, third, fight of the campaign occurred somewhere in the central part of the Chukchi Peninsula, based on D. I. Pavlutskii's "Promemoriya": the Chukchi, ". . . not allowing us to the Anadyr' Sea," gave battle. According to the more detailed reminiscences of the Cossacks in 1772, the events evolved in this way: 13 July the toion of the "Eastern Sea" Keyu agreed to pay tribute. At noon of the following day D. I. Pavlutskii, with half the detachment, moved to the camp of the Chukchi and demanded the promised tribute, to which a refusal followed. The Chukchi at this time numbered 500 warriors from both seas (the Kolyma and Anadyr'), that is, the number of the enemies was approximately even. Among the warriors were also Eskimos from the Diomede Islands, which they themselves told M. S. Gvozdev in 1732. G. F. Müller also particularly noted that "among those killed in the last battle was a man who had two large holes in the upper lip,[316] in which teeth carved from walrus teeth are set, and they declare that this is revered for great beauty among some Chukotskie clans living on the islands."[317] The fight lasted from noon to evening: at some stage the Chukchi attempted to attack with spears, depriving the enemy of the possibility of using its superiority with firearms. Nevertheless, Keyu fled with his people. Probably belonging to this battle is a brief remark by the Cossacks in 1739 about the fact that the Chukchi "saw one discharge from our guns and fled," this clearly attests to the fact that this militia (or, at least, its significant part) did not participate in the previous battles and was encountering gunfire for the first time. Therefore, losses of those killed, judging by the depositions of 1763, were insignificant—less than a tenth part of the existing forces: up to 40 killed (incidentally, based on the reminiscences of 1772, it was 200 killed and as many wounded); several captives and 30 reindeer were seized, probably draft animals belonging to the deceased. The Russians lost one Cossack killed and 30 wounded.

Judging by length, the battle was primarily a long-range one. The number of Chukchi, even based on the enemy's information, did not exceed the number of D. I. Pavlutskii's detachment. Only half of the detachment (probably the servicemen—the most battle-ready part of the

---

316 In the text of G. F. Müller there is clearly an error: labrets are set in the lower lip.
317 Efimov 1948: 225; cf.: Zuev 2009: 115; statement of the Eskimos to M. S. Gvozdev: Berg 1946: 100.

detachment) participated in the battle, the other half (probably primarily allies) was occupied protecting the camp and herds. In spite of the summer period and lack of possibility of moving on sleds, the Chukchi had time to pull together forces from different parts of the peninsula and even the Islands of Gvozdev [Diomedes]. Though toion Keyu led the troops his army probably also contained the remnants of the forces of toiony Naikhnyu and Khypayu. During the battle, the Chukchi tried to attack with spears but were repelled by gunfire. If one judges based on the reminiscences of 1763, the losses to the Chukchi in this battle were not so great—less than a tenth part of the fighters, a large part of them escaping. The Russians captured the Chukchi camp, where they took captives and thirty reindeer that were probably being kept for food, since reindeer were not ridden in summer.

Figure 26. Armor of a Chukchi warrior of the 18th century: iron lamellar helmet; armor of plates made of reindeer antler; lamellar arm guard for the left arm and greave of antler and seal hide. Russian Ethnographic Museum, No. 4859: 14 (armor), 15 (arm guard), 16 (greave). (Reproduced after Zuev 2009: Fig. 18).

After spending the day in the place of battle, D. I. Pavlutskii moved across the mountains (the Chukchi Highlands) to the south coast of the Chukchi Peninsula. Five days later (20 July) the expedition, according to J. Lindenau, reached a nomad camp of "Chingagiskie yurts," that is, the nomad camp of toion Chimkaigan, brother of Keyu. The residents had evacuated to the mountains. Chimkaigan and the elder Kopeikin came to D. I. Pavlutskii and offered themselves as hostages so that their relatives would give tribute for them. Evidently doing so, they wanted to avoid the annihilation of their nomad camp. The clan of Kopeikin had probably already paid tribute earlier; at least, Prince Kopenko had paid tribute of walrus tusks to the Anadyrsk collectors in 1688. However, the relatives of the leaders fled the locality in terror, not paying anything. As a result, Kopeikin killed himself and Chimkaigan was killed at his request.

According to G. F. Müller, the expedition reached the south coast of the Chukchi Peninsula after 10 days (25 July); the detachment moved farther along the coast, which extends to the southeast, that is, the southern shore of the peninsula going from capes Bering and Achchena to Provideniya Bay. A large part of the expedition (evidently with booty) led by the captain went along the shore, while the other part sailed along the coast in baidary. Large traveling baidary were evidently seized from the coastal residents. It was probably at the time of this return that some nomad camp on Masko Island was looted, which D. I. Pavlutskii probably calls Maiskoe in the "Promemoriya"—this is evidently Maskyn Spit at the entrance into Kresta Bay. They seized there 136 furs of red foxes and 5 of *sivodushka* foxes, 65 draft reindeer, as well as a musket, and the banner, found by the Koryak, that belonged to Shestakov's detachment. It can be supposed that the Reindeer Chukchi hid themselves with their draft animals on the spit, having taken with them their most valuable property. Therefore, they found a substantial number of fox furs and foreign trophies there.

Moving along the coast two weeks after the battle (3 August), according to J. Lindenau, the expedition reached the mouth of some river (the Chautakan?); approximately 10 verst from the river was the cape with Serdtse-Kamen' Mountain, from where, according to G. Müller, they continued on foot. After a week and a half (13–14 August) they reached the Bol'shaya [Large] River, which falls into St. Kresta Bay. The Yukaghir fair, where the Cossacks spent the night, judging by J. Lindenau's map, was south of the Bol'shaya River. This river was most probably the

Etel'kuyum River, which falls into the bay of the same name (Etel'kuyum—Yukagirskaya Bay) on the north of Kresta Bay.[318] The expedition then turned west. Following farther along this same way, across the upper reaches of the Chernaya (Black) and Belaya (White) rivers, and along the Ubienka river, the detachment returned to Anadyrsk on 21 October 1731.

In this campaign three Cossacks, ten Koryak, and one Yukaghir perished; one Cossack went missing in the tundra and eight Russians died of illness, primarily of inflammatory diseases. Freed from captivity were 42 Koryak, of whom 11 men were taxed with tribute for 1731, while the women and children were given back to the Koryak; also freed were two Russians, while one serviceman escaped from captivity himself. Less than 500 reindeer (of over 40,600 or 24,000 head!) were driven to Anadyrsk, the remaining animals possibly either escaped, were taken back by force by the Chukchi, killed for food, or given to the servicemen with the reduction of two rubles per head deducted from their salary. The captain himself indicated in reports that the booty taken from the island at the end of the expedition was only 65 draft (not herd) reindeer, as can be understood. The remaining reindeer were sent to the nomad camp at the Olyutorskii stockade. Part of the trophies taken by the Chukchi from Shestakov's detachment were returned: all 12 suits of iron armor, the banner, muskets, and rifles. They inflicted both material and physical harm on the Chukchi. The servicemen captured 160 people (men, women, and children), part of them being returned home from Nizhnekolymsk (women), while others died along the way to Yakutsk and Irkutsk.[319] If the losses of the Chukchi given in the stories of the Cossacks in 1763 are summarized, it turns out that more than 800 Chukchi perished, while in 1772 the Cossacks counted 1,452 enemy warriors killed. D. I. Pavlutskii himself notes in the "Promemoriya" that "no small number of them, the Chyukoch, died." Evidently the captain himself did not know the precise number of the enemy losses; by contrast, he provides absolutely precise

[318] Efimov 1964: No. 122; cf.: Antropova 1957: 174; Vdovin 1965: 77; Leont'ev and Novikova 1989: 438.
[319] Sauer 1802: 95; L. 1901. No. 48: 3. We note that the return of an unredeemed prisoner was the usual practice of the Russians (Bagrin 2013: 50).

data when reporting on his losses and the trophies seized. In any case, the losses for the population of Chukotka were substantial.

The first campaign of D. I. Pavlutskii was a purely punitive expedition. As early as 1763 the merchant N. P. Shalaurov noted in his report to the office of the Senate: “And when in past times parties left the Anadyrsk stockade, in the 731st year it was Major Pavlutskii; then they went to the north coasts to those Settled Chukch, and as I heard from some Anadyrsk Cossacks who were in this party, those Settled Chyukchi had no opposition to this party, but because those clans of the Chyukchi migrating near the Anadyrsk stockade and inflicting offence on the loyal Koryak and Yukaghir and taking away the reindeer, therefore for this reason alone, that these were Chyukotskii people, by order of said Pavlutskii they were killed and ravaged. . . .”

The campaign itself added little to the then existing geographic knowledge.[320] The captain himself had not yet received an order to compile a map or carry out geographic research. After the campaign of 10 February 1732, D. I. Pavlutskii wrote only brief reports to the Tobol’sk provincial and Yakutsk district offices. Later, academician G. F. Müller, in order to acquire information about the campaign, specially questioned members of the expedition in Yakutsk. The unsatisfactory level of the ethnographic knowledge about the Chukchi after the expedition is represented by the answer of the manager of the Anadyrsk tribute house Ivan Batankuev to the question of the Tobol’sk office about the peoples of the Anadyrsk Region (8 April 1733). Briefly telling about the Yukaghir (Chuvan and Khodyn peoples) and the Reindeer and Settled Koryak, he writes: “On the north side close to the sea and near the Arctic Ocean to the Anadyr mouth in ~~to~~ the cape with that name were the non-peaceful Chukchi who do not pay tribute; they are many and have a large number of reindeer among them, and they also go to war on the Reindeer and Settled Koryak, and they kill many, and they take captive their wives and children; and it is unknown what they eat and are satisfied with; and what faith they have and to whom they bring sacrifice are also unknown; and

[320] The opinion of N. P. Shalaurov: Belov 1954: 181; Vdovin 1965: 123, note 79. Incidentally, there is the opinion of a large contribution of the campaign to the geography of Chukotka, for example, see Grekov 1960: 54; Gol’denberg 1984: 84, 221; cf.: Vdovin 1943: 54; Orlova 1957: 129.

the Settled Chyukchi pay tribute of a small number of red foxes, but another year they don't pay."[321]

As we see, the Anadyrsk official who, based on duty, was supposed to interact with the Chukchi, in fact knew little about them. The ethnographic knowledge of I. Batankuev was limited by external factors: by raids and tribute. The raids of the Chukchi were conducted against both the Reindeer Koryak and the Settled Koryak, there was no difference for the Anadyr people in the direction of the Chukchi raids. It also pays to note that in the mid-1730s some Settled Chukchi paid tribute with fox furs, but they did this irregularly, probably when collectors went to them. At this time, the total number of Yukaghir tribute payers in the Anadyrsk statement, together with Koryak and Chukchi, numbered 682, delivering 638 furs of red fox, 3 sivodushchatye, and 38 sable. On the whole, after D. I. Pavlutskii's campaign it became clear that "to bring the Chukoch into citizenship was impossible, because there was nothing sacred among them: the children abandon the fathers, the fathers abandon the children."[322]

1.3. The Voyage of the Boat *St. Gavriil* (1732)

For organization of the maritime part of A. F. Shestakov's expedition, the Admrialty College sent from St. Petersburg the navigator Yakov Gens, assistant navigator Ivan Fedorov, geodesist Mikhail Gvozdev, and ten sailors. They arrived in Okhotsk in October 1729. Receiving D. I. Pavlutskii's order of 26 April 1730, navigator Ya. Gens sailed in September of this years with the boats *Vostochnyi Gavriil* and *St. Gavriil* from Okhotsk to Kamchatka, from where it was planned to sail farther to Anadyrsk and then to carry out a secret maritime part of the military expedition against the Chukchi. In October 1730 the boat *Vostochnyi Gavriil* was wrecked near Kamchatka, while an uprising of the Itel'men in 1731 and participation in its suppression by the servicemen from the boat, equipped with the ship's guns, detained the crew there for the winter, until the summer of 1732. Because of an eye disease of the navigator Ya. Gens, the assistant navigator I. Fedorov was supposed to

---

[321] AAN. F. 3. Op. 10. No. 137. L. 14 ob.–15, 92; cf.: Vdovin 1965: 118; Zuev 2009: 80.

[322] Sgibnev 1869: 31; tribute: Bakai 1895: 92–93.

command the *St. Gavriil* which remained in working condition, but he also was unwell, suffering an ailment of the right leg, and by order of Pavlutskii the boat fell under the command of geodesist M. S. Gvozdev. On 23 July 1732 the boat, having on board 4 sailors, 32 servicemen, an interpreter (newly baptized Egor Buslaev), and a seafarer who had sailed with V. Bering, Kondratii Moshkov, left the mouth of the Kamchatka River and set off to the north. According to Pavlutskii's order of 11 February 1732, the crew was set the task of bringing into Russian citizenship the residents of the islands opposite Anadyrskii Nos (the Chukchi Peninsula), that is, in fact to collect tribute from them.[323]

On 3 August, the crew reached the mouth of the Anadyr', where they met no Russians, and two days later, 5 August, the boat dropped anchor at the Chukchi shore south, it is believed, of the modern Cape Dezhneva. A dinghy with seven men led by Gvozdev was sent to replenish the supply of fresh water. On the shore the mariners noted a reindeer herd numbering more than 150 head and two people (evidently herders), who, seeing the new arrivals, hid in the mountains. The Cossack Efim Permyakov shot two reindeer to replenish food supplies. On the following day two baidary moved toward the boat. Having watched the ship from afar (at a distance of a verst), they sailed away, not engaging even in dialogue.

On 13 August the mariners dropped anchor again, at a village on the coast that consisted of six semisubterranean dwellings. With the approach of the first dinghy the residents evacuated in three baidary.

---

[323] Documents: Polonskii 1850: 394–402 (paraphrase with the addition of unique information from an unpreserved or lost report of M. S. Gvozdev to the Okhotsk office of 13 April 1741 from the archive of Irkutsk (Gel'mersen 1874: 456, note; Gol'denberg 1984: 176); quotations, see: Polonskii 1850: 399, 400, 402 [I. F. Skurikhin]); Sokolov 1851: 88–107; Andreev 1948: 101–103 (deposition of I. Skurikhin); Efimov 1948: 236–249 (quotation from the reports of M. S. Gvozdev, see: p. 246 [meeting of 13 August], p. 240 [of 17 August; Ratmanova Island]); Fedorova 1984: 90–94, 97–110; Divin 1979: 150–155; description of the voyage: Sgibnev 1869: 23–26; Golder 1914: 158–163; Berg 1946: 99–103; Efimov 1948: 159–174; Divin 1956: 18–28; Grekov 1960: 49–53; Gol'denberg 1984: 123–138; 1985: 56–80; Zuev 2009: 69–72, 83–86. Data of I. Kobelev of 1779: Efimov 1948: 229–230. The name "rotastye" ["mouthed"]: Bogoras-Tan 1934: 9; resettlement of the Eskimos to Naukan: Galkin 1929: 11, 73; cf.: Nefëdkin 2003: 212, note 3. Wild reindeer on Ratmanova Island: AAN. F. 21. Op. 4. No. 33. L. 134. A preliminary version of this part, see: Nefëdkin 2015a.

Gvozdev sent a second party of 20 servicemen to the village. In the report of 1 September 1743, Gvozdev thus describes the encounter:

> . . . and the Chukchi departed from the yurts in baidary and arrived at a rock opposite the yurts, and there seemed to be just six people on the rock; and I sent a serviceman with an interpreter to them to bring them into the citizenship of H.I.H. [Her (Anna Ioannovna) Imperial Highness] autocratic hand for tribute payment, and they the Chukchi said: "We don't know and don't pay tribute, and don't hunt"; and three times I sent the interpreter to bring them into tribute payment, and they told the interpreter that "a steward should come to us, and we will pay tribute in parkas"; and I sent the serviceman Permyakov with the interpreter, and this Permyakov went and, having arrived, said that they asked him for a knife, and they gave him a parka of marten and said to the interpreter: "What tribute are you asking for from us? I have the greatest power among my relatives. What tribute do you want? And we had a fight with the captain, and you caught us at the wrong time"; and he swung his spear at the interpreter.

A. S. Polonskii adds: "For such expression Gvozdev ordered to be burned, to instill fear, two summer seal-skin yurts and, having taken from the others whale oil for food for the servicemen, and returned to the boat."

Consequently, the settled residents of the west coast of Chukotka chose to evacuate the village, while the six men led by an elder who stayed refused to pay tribute, not understanding why they, properly speaking, should pay tribute to foreigners. Calling the chief, the residents were going to give him a parka. Since Gvozdev had sent a Cossack, the latter received the parka as a gift, in exchange for which he had to give a knife—iron weapons were especially valuable. The elder of the settlement noted that the arrival of the boat had caught them by surprise, that they were not prepared for a landing from the sea, and that they viewed the Russians as enemies; indeed, they had fought against Captain Pavlutskii with the other Chukchi.

On 17 August, seeing on the Chukotka shore a village (Naukan?), Gvozdev with ten servicemen and one sailor set off in a dinghy to the settlement; however, the residents went to a nearby cliff. Then the dinghy went to a cliff on which "the Chukchi sing together and one of them leaps and cavorts" (probably a cult activity, some kind of ritual). Because of the substantial distance between the sides, no negotiations were conducted.

Then the Eskimos got into two traveling baidary, which accommodated 20 people each, and sailed toward the boat, remaining at a proper distance from it. To the interpreter's questioning "they said they themselves were the Toothed Chukchi, and about the land they said, where they had yurts, was Chukotskoi Nos." The words of the Eskimos, transmitted by the translator who spoke in the language of the Chukchi, indicated that they called themselves in Chukchi *Iykyrgalyt* ("Mouthed"). In fact, according to Uelen legends, Alaskan Eskimo had resettled in Naukan.

Then they sailed to the northern part of Big Diomede (Ratmanova Island), where they lowered the dinghy with ten servicemen, sailor L. Petrov, and Gvozdev himself. When the dinghy approached the shore, the islanders began to shoot arrows from a hill. The Russians answered by firing three rifles. Then, to the interpreter's question about the population of the island, they answered that they, "the Chukchi, that their relatives went with the Reindeer Chukchi to fight against the captain and there they all were killed." Gvozdev landed and inspected two pit houses, the residents of which had fled. In A. S. Polonskii's retelling of the actions of the mariners we read: "The yurt was burned, while the hide baidara and a light hide boat they took with them." In general, the hostility of the Eskimos was explained by the fact that they were allies of the Chukchi in the war with the Russians.

Then they sailed to the southern extremity of Ratmanova Island, where they discovered a large settlement (Imaklik) of approximately 20 dwellings. In the dinghy Gvozdev, with sailor L. Petrov and 21 servicemen, approached the settlement, but the Eskimos forbade them to land. Then the interpreter landed from the baidara and began to conduct negotiations with those sent from the settlement. On the appeal of the interpreter to pay tribute, the residents answered with a rejection. Gvozdev turned from forced action since the population was numerous.

On 20 August, they went to Little Diomede (Kruzenshterna Island). The Cossack E. Permyakov, with nine servicemen and the interpreter, approached the island in the baidara, but were shot at from bows. Therefore, no landing or negotiations occurred.

On the next day, the boat arrived at Cape Prince of Wales in Alaska, and on 22 August, because of leaks in the ship and inadequate provision, Gvozdev turned back. On the return trip he again dropped anchor at one of the Diomede Islands (Ratmanova Island?) and went to land in the dinghy with servicemen. According to the account of Cossack Il'ya Skurikhin who sailed in the boat, Gvozdev obtained up to 20 reindeer on shore, took away reindeer hides and walrus tusk to 100 pudy (1.64 tons—sic!) from three local dwellings, for which he went ashore three times. When the dinghy returned to the boat, more than a hundred residents ran onto the shore and began to shoot at the Russians with bows, whom, however, because of the distance they were not able to hit. They also fired guns from the dinghy. The residents, seeing the fire, ran off. It should be noted that the reindeer, which pastured on the shore by the dwellings, must have been domestic. However, at this time they were domestic only in Chukotka, but not on the islands of Bering Strait or in Alaska. It is possible that after the passage of nine years I. F. Skurikhin was confused in his description of the events. In addition, it is possible that there was some exaggeration of the number of residents who armed themselves against the Russians. According to I. Kobelev, in 1779 203 men and 195 women lived in two settlements on Ratmanova Island, and on Kruzenshterna Island—85 men and 79 women. If this information is extrapolated back a half century, then all the battle-ready men of the two settlements of the big island combined came out in battle with those who had sailed up. On 27 September 1732 the expedition returned to Kamchatka.

In general, the secret expedition of Gvozdev did not carry out one of Pavlutskii's primary tasks: it did not acquire tribute from the residents, and therefore, did not annex these territories to Russia. Indeed, the coastal residents of Chukotka and the Eskimos of the Diomede Islands (at least Ratmanova Island, located closer to Asia) stood as allies of the Chukchi in their struggle with the Russians. They participated in the war against Pavlutskii in 1731. Thus, a cautious and hostile attitude arose toward those who sailed up and the reluctance to provide tribute.

1.4. The Expedition of D. I. Pavlutskii down the Anadyr' (1732)

From February to April 1732 D. I. Pavlutskii went with a detachment of 225 servicemen and Yukaghir and Koryak allies on a campaign against the rebellious and still non-tributary Paren' and Aklansk Koryak and subjugated them. Returning to Anadyrsk, Pavlutskii began to prepare for another campaign to the mouth of the Anadyr'. Indeed, as early as 11 February 1732 he ordered the boat *St. Gavriil* brought there for joint action against the Chukchi. In the "Promemoriya" of 31 March 1733, to the Yakutsk provincial office, Pavlutskii himself described the campaign, as usual, very briefly: "And on May 27th with the opening of the spring thaw water a trip conceived on ships down the Anadyr' River to examine the Anadyr' mouth and finding the non-tributary and non-paying and non-peaceful Settled Chukchi." However, they did not find the Chukchi at the mouth of the River, where they arrived on 14 July, and, after staying there at a place with old Chukchi dwellings, on 25 July they set off on the return trip to Anadyrsk. Pavlutskii did not have maritime ships, only old river boats, thus he did not dare to embark on them on a voyage on the sea. Pavlutskii was possibly waiting not just for the Chukchi but also the approach of Gvozdev's boat, which arrived there on 3 August, after the departure of the captain. Pavlutskii left three servicemen for a reindeer hunt in the Rodionovo tract, on the left bank of the Anadyr', where the Belaya River falls into it. The Chukchi evidently knew about the campaign and thus did not go there to hunt wild reindeer. Having learned of the approach of the Russian, the Settled Chukchi who lived, as J. Lindenau (1742) reports, on the Nerpich'ya (Kanchalan) River and at the Krasnaya River (near Krasnoe Lake) evacuated. Incidentally, they nevertheless came in baidary to the Anadyr', probably for the fall reindeer hunt, and suddenly captured all three service men.[324]

On 18 August 1732, having received an edict of the Tobol'sk office of 3 September 1731, Pavlutskii busied himself repairing the Anadyrsk stockade (August to October 1732). Now the stockade was a square wooden (of cottonwood) fort with sides of about 40 m and walls 3.2 m high. At each of the four corners of the stockade stood a tower, with a fifth tower standing above the gate. On the island outside the stockade were a chapel and 80 households.[325]

---

324 AAN. F. 31. Op. 4. No. 33. L. 245; cf.: Zuev 2001: 36–37; 2009: 83; Nefëdkin 2003: 26; Okhotina-Lind and Møller 2015: 369 (with the dating of the campaign to 1731). About D. I. Pavlutskii's campaign against the Koryak in 1732: Zuev 2002: 60–80; 2009: 82. Information about the Settled Chukchi on the lower Anadyr': Lindenau 1983: 162.

325 AAN. F. 21. Op. 4. No. 33. L. 246; cf.: Zuev 2003: 138; 2009: 86; AAN. F. 21. Op. 5. No. 103. L. 7 ob. (Ya. Lindenau's description, 1742); AAN. F. 3. Op. 10. No. 137. L. 82 (according to F. Plenisner's report, in 1763 only one wall of this stockade

Figure 27. Yakutsk stockade in the 19th century. (Reproduced after Middendorf 1878. Pt. 2, sec. 6: 763).

D. I. Pavlutskii himself, following an order from the Siberian office, went from Anadyrsk to Yaktsk on 5 November 1732 to investigate extortion by the local tribute collectors and servicemen. On 6 March 1733, by Senate edict, Pavlutskii was promoted to major for "service and for his long term on the far edge of existence." The Kamchatka part of the party was disbanded in the second half of 1733: the servicemen were assigned to various stockades, and to Okhotsk and Yakutsk.[326]

On the whole, the expedition of A. F. Shestakov-D. I. Pavlutskii (referred to from the 1730s as the Anadyrsk Party), in distinction, for example, from the V. Bering expeditions, bore a purely military character. Its primary tasks were conquest of the rebellious Koryak and bringing the Chukchi and other peoples in the newly discovered lands into citizenship, that is, the same tasks that were set before the servicemen on the campaigns in the 17th century. The expedition did not carry out most of the tasks set for it: the Koryak were not conquered, nor the Chukchi, and no new lands were discovered or the population of these lands brought into citizenship. Also, geographic knowledge did not receive substantial

---

remained, all the rest had been washed away in floods). The traditional date of reconstruction of the Anadyr' stockade—winter 1730/1731: Sgibnev 1869: 27; Golder 1914: 163; Gol'denberg 1984: 83; cf.: Vdovin 1960: 42 (based on the edict of 10 August 1731).

[326] Departure of the captain from Anadyrsk: AAN. F. 21. Op. 4. No. 33. L. 246 ob.; Senate edict: Andreev 1948: 79–80; disbanding of the detachment: Gol'denberg 1984: 155.

augmentation and no maps were compiled, which was clear to G. Müller as early as 1737;[327] only later (in 1742) was the map of J. Lindenau with a very inaccurate depiction of Chukotka revealed. The Kuril and Shantar Islands were not described at all. Gvozdev's reports to the Okhotsk office (22 June 1733) and to Pavlutskii (19 December 1732) were not sent to higher authorities and did not become known, and the geodesist himself never brought anyone into Russian citizenship. Only in 1741 did the Okhotsk commander A. Devier assume that Gvozdev's boat visited the shores of America—the Great Land. The Koryak conquered by A. F. Shestakov, his former allies, immediately abandoned submission, and in 1732 Pavlutskii had to reconquer them. Only in 1742 was chief Etitka, who rebelled in 1730, "pyatiren" ["five-folded"] (his extremities and head were chopped off) and his relatives were beaten with whips.[328]

2. Warfare in the 1730s to the Mid-1750s

2.1. Chukchi Raids in the 1730s to the Early 1740s

D. I. Pavlutskii considered attempts to conquer the Chukchi futile. And the Chukchi themselves, because of losses born during the campaign of 1731, had to take revenge on the Russians for their killed relatives. And this revenge was not long in coming. The Chukchi, not wishing to engage the Russians in open battle, which had caused the significant losses, switched to guerilla warfare, to surprise attacks on small detachments of Russians in the region of the Anadyr'. In February 1733, 600 Chukchi attacked the reindeer herd of the garrison near Anadyrsk. Based on a report of Yukaghir of the Anadyrsk Department, "in 733, when Major Pavlutskii was not there, . . . those non-peaceful Chukchi approached the Anadyrsk stockade and took some reindeer that were by the stockade and killed the shepherds, and took captive their wives and children and killed several tributary Koryak and twelve servicepeople."[329] As the Koryak recalled in 1739, the fires of the Chukchi camp were seen from the stockade, that is,

[327] Efimov 1948: 225; Grekov 1960: 54; Okhotina-Lind and Møller 2013: 91; cf.: Vdovin 1965: 123, note 79.

[328] Senate archive Vol. VI. 1893: 202–203.

[329] Pokrovskii 1941: 129; Okhotina-Lind and Møller 2015: 365, 369–370; cf.: Sgibnev 2005: 52; Vdovin 1965: 119; Zuev 2009: 86; cf.: RGADA. F. 199. No. 539. Ch. 2. D. 6. L. 26 (the Chukchi killed six tributary Koryak at Nakipnaya).

at least one night the Chukchi camped near Anadyrsk, not especially fearing attack. It is also known that on 10 February 1733, in the vicinity of Nakipnaya not far from the Anadyrsk stockade, the Chukchi unexpectedly attacked the detachment of R. Yakimov, who was bringing back tribute obtained from the Koryak. The less numerous servicemen managed to construct a "Wagenburg" from slaughtered reindeer and sleds. Yakimov, going out for negotiations with the enemy, was killed. The Chukchi attached shields to sleds and over the course of five days repeatedly tried to approach the fort under such cover, but were repelled by gunfire from the servicemen and retreated. After the departure of the Chukchi, the servicemen, led by Ivan Modin, arrived at the stockade.[330]

In 1737, a group of Chukchi approached the Nizhnekamchatsk stockade in the center of eastern Kamchatka, killing six servicemen and several local residents. This raid was probably conducted against the Koryak, though the Chukchi, in search of booty, simply moved on. In March 1738, the Chukchi came close to Anadyrsk, killing 8 Cossacks and 20 tributary Koryak. In 1739 news was received in Anadyrsk from the Koryak that the Chukchi threated to lay siege to Anadyrsk.[331] Chukchi raids on the Koryak followed in December 1738, in March 1739, and in March 1741. In answer to the last raid in July, the Koryak prepared a campaign to the mouth of the Anadyr', calling on the Russians from Anadyrsk to help them. The commander of the stockade, Sotnik Vasilii Shipitsyn, answered this request by sailing on 6 August in 10 shitiki with 70 servicemen (almost the whole garrison of the stockade) down the Anadyr' for a meeting with his tribute collectors, and possibly also for a reindeer hunt (though this is not mentioned in the documents). On this same day 14 Koryak captives who had been left by the Chukchi with their herds ran to the sotnik. They reported that the Chukchi were gathering in summer for an animal hunt on the Anadyr', and in the fall they would go to the stockade. According to the report of tribute collectors who arrived from the mouth of the Anadyr' on 17 August, of 69 Chukchi who were supposed to pay tribute for 1740, only nine people paid (and one more who was new), the others refused to pay; meanwhile, the "toyon of the

---

[330] RGADA. F. 199. No. 539. Ch. 2. D. 6. L. 26; Zuev 2007a: 288; 2009: 87, 294.

[331] Pokrovskii 1941: 129; Sgibnev 2005: 63; Vdovin 1965: 119; Gurvich 1966: 105; 1982: 203; Zuev 2009: 101.

Northeastern Sea" Naikhnyu boasted that he was going to burn Anadyrsk and annihilate its population. And a captive Koryak woman had told the collectors that the Chukchi planned to attack the Russians at Chekaevo. The Russian detachment arrived in the Chekaevo area at the confluence of the Anadyr' and the Tanyurer rivers, where there was a reindeer crossing. V. Shipitsyn himself in his "Izvestie" ["News"] to the Okhotsk office of 24 October 1741 described further events this way:

> And against the above-mentioned of the Koryak report and the questioning of the Koryak captives, and the tribute collectors of Popov and his comrades performed on August 22nd 1741 for the disobedience of nonpayment of tribute and for ravaging the Koryak who pay tribute, the best Chukchi toyony, precisely: Kanlan, Pilit, Epancha Echakov, Apalko Nankha, nephew of Pilitov, Nankhan, an other Chukchi Pukhaech, and Tiemga, brother of Tintionov, arrived at the Chakaevo area in baidary in a multitude with spears and bows, but to them, and to Shipitsyn, no damage was done at this time. And in the morning the mentioned Chukchi toyony cunningly approached in three baidary, and those toyony, praise be to all-gracious God, bringing them to the bank, we killed, and in four baidary the Chukchi were killed and several men wounded. And we were afraid to pursue those Chukchi from Chekaevo to the mouth of the Anadyr' River. And with him, Shipitsyn, were only seventy servicepeople, while they, the Chukchi, along the bank of the Anadyr' River to the sea went a large number of baidary, one hundred and more Nose and River, and other sea Chukchi.[332]

Obviously, V. Shipitsyn tells too briefly of the murder of the Chukchi elders on 22 August 1741, giving, so to speak, the official version of the events, explaining the murder of the Chukchi as the result of cunning plans by the latter, which became known to the sotnik from the statements of the Koryak and tribute collectors. According to the documents, it turns out that the Chukchi who planned the murder of Russians at Chekaevo are blamed for everything, while they themselves want to secede from Russian citizenship and cease paying tribute. However, this explanation suffers from obvious subjectivity and understatement.

[332] AAN. F. 21. Op. 5. No. 142. L. 31–33 ob.; quotation: L. 33.

Three decades later, in 1772 in Gizhiga, three Cossacks who participated in the events recalled them in much greater detail: V. Shipitsyn

> reached Chekaevo in three days. Here, upon arrival, was the Chukchi toyon Yultuv'ya in ten baidary, which had sailed down to Buyan Island, which from Chekaevo is three verst distant where Shipitsyn made camp and set up a defense around it. On the next day, from this Buyan arrived the Chukchi toyony Keyu, Tegruv'e, Yat'girgyn, Apavno, and of the Northeastern Sea Naikhnyu, and Yatygirgyn in forty baidary, in which were up to two hundred troops, who brought the Cossacks of Ankudinov and three comrades, previously sent to them, with collected tribute of foxes and walrus bone (and what number they do not know). And about the Cossack Ustinov, they explained: "He will be brought with the other Chukchi who are still afar." These toyony in the evening of this day went in their baidary to the aforesaid Buyan Island. . . . Then on the tenth day, from Chekaevo were seen toward Buyan Island those Chukchi baidary that arrived from below along the Anadyr' River and in them a multitude of people. From there on this day the toyon Apavko and relatives arrived to Chekaevo in twenty baidary; in which the Cossack Ustinov was brought. Apavko, Southeastern Sea Naikhnyu, and other best people had asked Sotnik Shipitsyn to give them, allegedly for collection from them in the coming winter of tribute, the Cossacks Petr Modin, Nikita Kobelev, Aleksei Shipunov, Efrem Kotel'nikov, Evsei Shchukin, Vasilii Berezhnoi, and also Kurkutskii and Rusakov, and Chertovskii. However, for that which was promised to be given, those toyony came the next day and in the Cossacks' reception, based on Russian custom, they signed in books; then they went to Buyan Island. At which time Sotnik Shipitsyn, agreeing with the detachment, that those Chukchi (as the Koryak released from captivity declared, also based on the praise of Naikhnev) because of their multitude would not murder him and the detachment at Chekaevo, and then they would not be able to devastate the Anadyrsk stockade and inflict with death the people in it, when the next day the Chukchi will be there, having seated the elite in a circle, allegedly for receipts, and presumably pleased with what had been collected from the detachment, whatever who could find, *mokhovoi* [Chinese small tobacco] tobacco and *korol'ki* [pieces of coral], and then to kill the leaders. And the next day the Chukchi arrived in several baidary, in which were up to three hundred people (and still others are in baidary from Buyan and

> other places below haven't come yet), and stopped on the bank near the water thirty *sazhen* [1 sazhen = 7 feet or 2.13 m] from our camp, from which the toyony of the Northeastern Sea Naikhnyu, the Eastern Sea Naikhylpyn, Apavko, An`yatyrgin, Yultuv'e (and other names they do not remember), twelve men, came to our camp itself (while others remained with the baidary). Sotnik Shipitsyn sat in a circle with them, and with him the Cossack Vasilii Popov (who was the scribe). First, they gave two or three *zolotniki* [~1.5 to 2.0 ounces] of tobacco and several *korol'ki*. Meanwhile, this Shipitsyn ordered the detachment to be readily and operable to fight, and to the other best Cossacks to stand by the Chukchi sitting in a circle, one man each, without a gun, but secretly having knives on themselves, and to await his, Shipitsyn's, order, who after sitting a while and stood up, said "Kolite" ["Stab"]; and all the Cossacks with knives readied on them began to slash, and immediately all those Chukchi who were in the circle were killed. And then he went with the whole detachment to the enemy force that was at the baidary and began to shoot with guns. The Chukchi, seeing the killing of their main toyony, threw themselves into their baidary and paddled down the Anadyr' River, and immediately after Sotnik Shipitsyn with the detachment pursued them giving chase in the ships. Several of their baidary were smashed and the people killed (how many they do not remember), and others, due to the lightness of their baidary fled from the Russian ships to their homes through the marshes and shallow places in the Anadyr' River, whom, in our vessels because of their heaviness, were not possible to chase. From there we returned to Chekaevo, where upon arrival we gathered up belongings, and granted satisfactory weather, raised sails and went to Anadyrsk, where we arrived in the last days of August.[333]

On the whole, the events developed in the following way: at the mouth of the Anadyr' came Reindeer and Settled Chukchi and even island residents, who were led by the former enemies of D. I. Pavlutskii, the toyony of the Northeastern Sea—Naikhnyu and of Chukotskii Nos—Keyu. For his part, having arrived on the lower Anadyr', Shipitsyn set up a fortified camp in the Chekaevo area; the Chukchi made camp 3 km down the Anadyr' on Buyan Island. As can be understood from the document, Shipitsyn decided in a military council to attack the Chukchi based on

[333] RGADA. F. 199. No. 528. Ch. 2. D. 3. L. 18 ob.–19 ob.; cf.: Zuev 2009: 102–107.

information obtained about hostile intentions of the latter. Incidentally, these intentions of the Chukchi could have been, as in other cases, simply some kind of braggadocio. This same Naikhnyu, who chanted about his future campaign against Anadyrsk, peacefully arrived to negotiations with Shipitsyn. Since the war went on and attitude toward the enemy was by no means friendly, the sotnik decided not to act peacefully, as was prescribed by the primary direction of tsarist policy, but by force. He decided to act cunningly and deprive the Chukchi of their leaders, which would bring about demoralization of the enemy and its flight. Indeed, the forces of the sides were unequal: 70 servicemen against more than 2,000 Chukchi—the document mentions more than a hundred baidary, and in each such large traveling baidary were more than twenty people (in this number it is also necessary to include families).[334] The Chukchi elders, not suspecting anything, arrived at the negotiations, not having at this moment hostile intentions, but on the contrary, wishing to agree to further pay tribute. The Chukchi simply asked that they be sent specific servicemen for collection, men who evidently best understood them and were more acceptable in their behavior. On the morning of 22 August the Chukchi toyony were invited to sign a document and on the next day, 23 August, during negotiations, at Shipitsyn's predetermined sign, 12 elders were slaughtered by the Cossacks. Then the servicemen opened fire on the remaining 300 warriors who had accompanied the toyony, located 60 m from them. The Chukchi began to flee in their light baidary along the Anadyr'. Because of shoals and marshes the Russian shitiki were not able to pursue them. Only four baidary with the people were destroyed. Most of the Chukchi retired in their baidary to the mouth of the river. Shipitsyn returned with the detachment to Anadyrsk. This event, by its deceit, induced fear for a long time in the Chukchi, who were afraid of its repetition.[335]

Based on a decision by the Admiralty College, the commander of the Eastern Lena Detachment of the Great Northern (Second Kamchatka)

---

[334] Based on information of the Anadyrsk serviceman Antip Karmalin, 170 baidary arrived with 20 or more people in each (AAN. F. 21. Op. 5. No. 142. L. 33 ob.), that is, the total number of Chukchi turned out to be 3,400 people.

[335] AII. F. 36. Op. 1. No. 643. L. 583 ob.; RGADA. F. 199. No. 528. Ch. 2. D. 3. L. 34 ob.; No. 539. Ch. 2. D. 6. L. 36–36 ob., 39 ob., 83 ob., 84 ob.

Expedition, Captain Dmitrii Yakovlevich Laptev (1701–1771), had to compose a description of the seacoast from the Kolyma to the mouth of the Anadyr', not going onto the shore in the territory of the enemy Chukchi, but at the same time gathering information on them. Leaving Nizhnekolymsk in 1740 and 1741, and trying to sail around the Chukchi Peninsula, but not succeeding because of the ice, Laptev went in October 1741 to Anadyrsk on dogsleds to carry out his task. Since the threat of attack by the Chukchi existed, the captain ordered that they move in "military order" on the Labaznaya River (a tributary of the Bol'shoi (Great) Anyui). Having arrived on 18 November in Anadyrsk, the captain built two rowing ships, to which he added four that were previously at the stockade. On the ships, equipped with cannons, were placed people from Laptev's detachment and a substantial part of the Anadyrsk servicemen led by V. Shipitsyn, for a total of 90 men. From 9 June to 3 August 1742 the detachment was occupied with describing the course of the Anadyr' River, while the servicemen led by the sotnik burned ten Chukchi "yurts" discovered at the river's mouth. In October Laptev returned to Nizhnekolymsk, leaving three cast-iron cannons of two-pound caliber at Anadyrsk.[336]

2.2. The Campaigns of D. I. Pavlovskii in Chukotka (1744 to 1747)

2.2.1. The Campaign of 1744

Chukchi raids of the on the Koryak—Russian subjects—whom the empire had to defend, presented the Russian central authorities with a significant problem. On 21 May 1740, the Secretary of the Senate M. I. Danilov, based on materials of the Siberian decree (based on a report of the Yaktusk governor), reported on the situation in the region. Cabinet Minister Prince A. M. Cherkasskii, Siberian governor in 1719–1724, proposed defending against the Chukchi not passively, but rather to go on the offensive, going against them in a campaign, to destroy and bring them into citizenship. M. I. Danilov went to Vice Chancellor A. I. Osterman with this proposal and received his support. On 4 June 1740 the Cabinet

[336] AAN. F. 3. Op. 10. No. 137. L. 105; Sokolov 1851a: 127–129; 1852a: 166; Yanikov 1949: 115–129; Belov 1956: 322–324; Divin 1979: 206; Zuev 2009: 118, 296; cannons: Shashkov 1864: 67.

of Ministers and the Senate directed to the Irkutsk vice governor L. Lang, the Selenga commandant Major General I. D. Bukhgol'ts, and officers of the Kamchatka expedition orders obliging them—having clarified the forces of the Chukchi and gathered servicemen—"to go to those Chukchi with a military hand and by every effort strive not only to return and avenge the offense to the Koryak loyal to H.I.H., but also to finally devastate them, the Chukchi themselves, and bring them into the citizenship of H.I.H." However, having examined information from the Siberian authorities about the remoteness of Chukchi territory and about the small number of the Russian garrison there, the Senate delivered on 25 November 1740 an "opinion" to the Cabinet of Ministers:

> inasmuch as . . . the non-peaceful Chukchi live from Yakutsk and from Anadyrsk at a great distance, it is 10 weeks travel to them, and from Yakutsk more than 30 weeks . . . in this regard the Senate declared its opinion, that based on the inconvenience of this route and the remoteness for ravaging them, the non-peaceful Chukchi, detachments should not be sent, since on such a distant and difficult route, because of the food supply that would be necessary to bring with them, it is recognized as an inconvenience; it is needless to lose and ruin people from hunger, and in order that henceforth the Koryak and the Anadyrsk stockade not be inadvertently attacked and ravaged by those Chukchi, to add for precaution to the Anadyrsk stockade regular and irregular light troops.

On 14 January 1741 the Cabinet of Ministers ordered an increase in the number of the Anadyrsk garrison at the descretion of L. Lang.[337]

Having received Shipitsyn's news about the new raid on the Reindeer Koryak in March 1741, the Senate on 3 February 1742, based on Lang's recommendation, issued an order regarding extermination and deportation of the Chukchi: ". . . on these non-peaceful Chukchi place an armed military hand and eradicate them altogether, those of them who will go into citizenship of Her Imperial Highness, take those, as well as their wives and children, captive and bring them from their dwellings, and for further security distribute them in the Yakutsk Department on various stockades."[338] In general, Russian leadership decided to return to the path

---

[337] Quotations: the first: Senate Archive. Vol. II. 1889: 35–37; the second: KPTs: 161.
[338] KPTs: 162–163; cf.: AAN. F. 3. Op. 10. No. 137. L. 99–100 (Imperial order of 1 June 1742); Senate Archive. Vol. V. 1892: 326; Gol'denberg 1984: 177.

of a forceful solution to the problem, believing that the Chukchi could not be conquered by peaceful methods.

On 3 February 1742, by order of the Senate, based on the recommendation of the head of the Okhotsk port A. Devier, Major D. I. Pavlutskii was again assigned as commander of the Anadyrsk Party. Now the major was ordered to gather information about Chukotka and its population, maintain a travel log, and make a map of the locality and route of the expedition. The Irkutsk provisional office, in an order of 22 June 1742, placed the Anadyrsk Party under its direct supervision, and ordered D. I. Pavlutskii

> to send to the highest (best) Chukchi elders, in order that they come to him, Pavlutskii, and when they have come to him, then to explain to them the order, that they, for their many infidelities and disobediences as well as for robbery of others loyal to Her Imperial Highness, are commanded to be placed under arrest and sent to Yakutsk, and then for him to go with the troops available to him to their dwellings and if he himself does not see any of their opposition or hostile activity, then to deal with them the same way as with the elders. If they show themselves as enemies, then to deal with them in a military way.[339]

Thus, it was planned first to summon the Chukchi elders to negotiations and immediately arrest them, thereby decapitating the Chukchi clans. And further the major was supposed to carry out a campaign to Chukotka and deport or annihilate upon any resistance the local population. In general, full deportation or annihilation of the Chukchi were planned.

On 7 November 1743 D. I. Pavlutskii arrived in Anadyrsk with reinforcements of 234 men. Carrying out orders from Irkutsk, the major outlined making several campaigns to Chukotka. On 2 February he moved to the next campaign. This campaign is described in detail in the major's report (with detailed events) and in the collected memoirs of eleven Cossacks, written in 1763. In the report the route of the expedition is presented geographically: the hydrography and toponymy (more than 30 names) that the servicemen learned from captive Chukchi. Part of these geographic names can be identified based on maps of 1744 and 1746

---

[339] Quotation: AAN. F. 3. Op. 10. No. 137. L. 101–102; cf.: AAN. F. 3. Op. 10. No. 137. L. 104; Zuev 2009: 117–118.

compiled by a member of the expedition Ensign Timofei Perevalov by order of the major, where the routes of the campaigns are depicted.[340]

D. I. Pavlutskii's detachment, bearing one banner and carrying one iron cannon, consisted of 40 soldiers, 367 Cossacks, 170 Koryak, and 67 Yukaghir, for a total of 644 men. The servicemen traveled on draft reindeer collected from the Koryak and Yukaghir—the Koryak alone provided the expedition with 5,000 head; in the general convoy were 20 "carts" of rye bread and other provisions intended for supplementary issuance to the servicemen.

In 1773, Captain Ya. M. Peresypkin retold in this way the course of events of the campaign in D. I. Pavlutskii's report:

> . . . And being in this locality on March 2, on the Nerpich'ya River (which falls into the Anadyr' River), those sent driving along both sides of this locality saw the small sled trail of the enemy Chukchi, which this same day Sotnik Krivogornitsyn was ordered to follow, and with him one Cossack sergeant, one pyatidesyatnik, thirty Cossacks, and thirty Yukaghir and Koryak people. When the enemy was reached and called first into citizenship of Her Imperial Highness; others did not go; and of them this Krivogornitsyn killed four men at the outset, took captive one man, two women, and twenty draft reindeer with sleds and outfits.
>
> The mentioned captives explained through translation that the enemy Chukchi had three yurts near Sertse-Kamen', and with them three herds of reindeer, to which was six days' travel without baggage from this Nerpech'ya River, and other Chukchi were located farther than the mentioned Sertse-Kamen', where previously this Pavlutskii (as he was captain) had a third battle with those Chukchi.
>
> When this information from the enemy was received, he followed with the detachment to Sertse-Kamen' without

---

[340] Statements of 1763: KPTs: 163–165; Berkh 1825: 176–182; Anadyrsk documents: AAN. F. 3. Op. 10. No. 137. L. 106; RGADA. F. 199. No. 528. Ch. 2. D. 3. L. 14–16, 17–18; the map and about it, see: Vdovin 1943: 53–54; Orlova 1957: 127; Grekov 1960: 206, Fig. 38; Efimov 1964: No. 123 (map of 1746, erroneously inserted in the Atlas instead of the described map of 1744—p. 83; also No. 124: map of T. Perevalov 1754; No. 126–127; maps of F. Plenisner; No. 130: map of N. I. Daurkin, after 1771); Gol'denberg 1984: 93; about the campaign, see: Shchukin 1852: 10–11; Okun' 1935: 75–76; Zuev 2004a: 212–215; 2009: 119–122 (without analysis of the details of the campaign).

baggage, and the Cossack Sergeant Mokrushubov was left with the convoy, with whom were fifty soldiers. On the morning of the 15th they saw the enemy, which consisted of three yurts, who went on reindeer through the tundra over the hills. And as they began to come into view in the vicinity of those yurts, to the reindeer herd ran some twenty people of the enemy, eighteen of whom the major himself and the lightly armed people killed, while two fled; after whom they pursued, but seeing seven enemy yurts in which was the chief Chukchi toyon Tention with eighty-eight people, to whom the major arrived with fifteen men; however, several men who were lagging behind also arrived; then, calling on almighty God for help, he started a fight where this toyon Tention with eighty-eight people were killed; and on our side ten military people were wounded.

On this same day, based on news from captives, Sotnik Krivogornitsyn was sent with thirty men and twenty Koryak and Yukaghir. They found three enemy yurts, in which was the toyon Tegruv'ya, and killed five men during a battle, and the rest fled. Five captive Koryak men were recovered there, as well as eleven women who were given to their relatives, the Koryak; eleven Chukchi people were taken captive. As booty in the first and second battle three thousand reindeer were obtained, from which the detachment got one riding reindeer each and another non-riding kargina, and one kargina each to all the old Koryak and Yukaghir for training and possession, while all the remaining were used by the detachment as food.

On April 1st, those left behind with the convoy went in advance without the commander's knowledge to where the previous battle had occurred and two Yukaghir people were killed by the enemy Chukchi. On the 14th Sotnik Ivan Butusin, along with one hundred men, was sent to the sea for surveillance of the enemy. On the 15th a report was received from the indicated Sotnik Butusin: "Four yurts of the enemy Chukchi were found beside the sea, in which twenty-five people were killed and four women taken captive, while on our side five men were wounded." On the 24th the enemy Chukchi villainously approached our reindeer herd in the night, in which they speared twenty reindeer; after which, though they were chased, they made it to their homes on their reindeer. On the morning of 7 May six enemy people came toward our camp, but seeing the circumstance of our military discipline, turned away.

The interpreter Yakov Osipov, and sixty Cossacks and sixty Koryak and Yukaghir people with him, was sent to the sea to find the enemy's reindeer herds. And on this same day twenty enemy people arrived to him from the bay of the sea on the left, having gone around in a detour, and with whom this Osipov had negotiations, and he himself caught one Chukchi and arrived in camp safely. This Chukchi reported that allegedly all the Chukchi, gathered and went from Chukotskii land to the Nos, and if they find out about the approach to them of the Russian troops, they intend to go to the Northeastern Sea. On the 12th, encountering the rear of the convoy of the mentioned enemy, one sled with two reindeer was captured. On the 14th Sotnik Ivan Butusin was sent with a hundred and sixty men and fifty Koryak and Yukaghir to scout out the enemy sled route. On the 16th Sotnik Butusin reported: "Among the enemy a thousand reindeer were found, three women were taken captive, and several men fled." On the 19th, of the Koryak and Yukaghir sent from the camp to scout out the enemy route, the enemy killed three men. On the 20th, Sotnik Kotkovskoi was sent near the sea with seventy men and thirty Koryak and Yukaghir, while Cossack Dmitrei Shadrin searched from an available rock on the left side of our camp for enemy herds, and reported upon arrival: "Kotkovskoi, finding Foot Chukchi near the sea, killed fourteen of them, while the others paddled into the sea in two skin baidary, four people were taken captive."

On the 22nd he went to Chukotskii Nos. Pyatidesyatnik Yakov Skrebykin was sent ahead; with him were sent ninety Koryak and forty Yukaghir men to scout near a bay of the sea for Foot Chukchi; he reported upon arrival: "The enemy was found, among whom four people were killed, while the rest ran off; while on our side one man was wounded." On the 24th Sotnik Butusin was sent, and with him a hundred and six soldiers and Cossacks and fifty Koryak and Yukaghir for search of the enemy trails in the rocks on the left of our camp, while Pavlutskii himself went to Chukotskii Nos. On the 29th, interpreter Petr Modin and forty men were sent to the rocks on the right side of Chukotskii Nos, and in addition to this number were sent one pyatidesyatnik and forty-five Cossacks and infidels. On the 29th the report of Sotnik Butusin was received, who reported: "At the Northeastern Sea he killed thirty people of the Foot Chukchi, and ten young people, and took thirteen young men and one woman captive. Upon appearance of these Chukchi, I went near the sea and met with the interpreter Modin, and the whole Foot Chukchi stockade, built of stones, was taken, in which a hundred and thirty enemy people were killed, and twenty-seven men and women were taken captive, while the rest fled through the rocks from the mentioned stockade near the sea; while on our side twenty-seven men were wounded and a hundred reindeer were acquired as booty."

On the 31st of May, based on a statement of the captives, the enemy went to escape with reindeer herds near the Northeastern Sea; that is why, on this date, Sotnik Krivogornitsyn was sent in pursuit, and with him was a detachment of a hundred and fifty soldiers and Cossacks, fifty Koryak and Yukaghir, amounting to two hundred people. On the 3rd of June, a report was received from Pyatidesyatnik Skrebykin, sent on assignment May 29th along the right of the Chukotskii Nos, that of the enemy Chukchi ten men had been killed, two women taken captive, and two hundred reindeer acquired as booty; on our side one Cossack and one newly baptized Yukaghir were killed. And on the above-written 3rd day the major himself went from the very Chukotskii Nos to the Northeastern Sea, where Sotnik Krivogornitsyn had been previously sent. On the 16th Pyatidesyatnik Dyagilev was sent with fifty men and twenty Koryak and Yukaghir to the sea to acquire food of whale and walrus fat for the detachment, but more so to search for Sotnik Krivogornitsyn. On the 18th a report was received from Pyatidesyatnik Dyagilev: "While being at the sea one Chukchi was taken, who states that Sotnik Krivogornitsyn is still ahead, while the two Chukchi herds are near the sea at the mouth of the Chafontyveim River. He also went there with troops." On the 20th day Pyatidesyatnik Dyagilev arrived in camp and reported, based on the statement of a Chukchi that he captured, the herds that the Chukchi said were at the mouth of the Chafontyveim River could not be found, and only two Chukchi were caught, who were brought back alive. On the 7th of July Sotnik Krivogornitsyn arrived with fourteen men; he had killed the Chukchi toyon Kiniyam and taken seven people of the male sex captive, and three hundred reindeer had been acquired as booty.

From there Major Pavlutskii went with the whole detachment back to Anadyrsk with great difficulty, for all the food supplies without remains had been eaten, including the draft reindeer on which the detachment traveled they had killed for food, and in the absence of all that, they had only grass and roots for food.

On the 13th of August Cossack Petr Modin was sent along with ten men and two chiefs of the Yukaghir to Anadyrsk to bring the ships along the Anadyr' River. On the 14th a detachment of two hundred and eighty people was sent with Sotnik Butusin by water in baidary made of reindeer hides along the Chernaya River, which flows into the Anadyr' River. On the 25th two Yukaghir chiefs, previously sent with Cossack Modin, arrived from the Anadyr' River from the Rodionov tract. They announced that the aforesaid Modin, on his arrival at Rodionovo, having taken their four *vetki* [small boats], left there by Modin's relatives for a reindeer hunt, in which he set off to the Anadyrsk stockade while six people remained at Rodionovo to make a baidara. On the same night that Modin departed to Anadyrsk they heard that the enemy Chukchi had attacked Modin's remaining six men; and they didn't know whether they were killed or not, for they, the Yukaghir, were out pasturing their reindeer fifty sazhen from them in the tundra. On the 30th Major Pavlutskii arrived in the Rodionovo tract, where he found the five men, who had been sent with Cossack Modin, that were killed. And the Cossack Modin, who arrived from Anadyrsk, told the story, that he had left six men at the Rodionovo tract to make a baidara, of whom one Cossack Vorontsov soon arrived in Anadyrsk after him and stated that five of his comrades had been killed by the enemy Chukchi, while he, Vorontsov, had broken free and fled up the Anadyr' River and, as he went along the river there were four Chukchi baidary and ten vetki, which he cut up with a knife, and departed in one vetka; and in those baidary he found a rifle and took it with him; this was the gun of the Anadyrsk townsman Vasilii Fedotov who lived by the Anadyr' River with his wife and two young children and a newly baptized Yakut Nikolai in the Yukaghir tract below Rodionovo, whom the enemy killed and took captive his wife and two children. And from the mentioned Cossacks killed at Rodionovo, the enemy took three rifles, one Turkish flintlock [Oriental type rifle], six bows with arrows, and six spears.

The indicated Major Pavlutskii arrived in Anadyrsk on September 22, and Sotnik Butusin arrived safely on the 26th on the ships gathered by Cossack Modin.

And in the above-designated campaign against the enemy three hundred and sixty people were killed, ninety-four people of the male and female sex were taken captive, and as booty four thousand six hundred and twenty reindeer were acquired, all of which the detachment used for food with no

> remains. In his campaign, Major Pavlutskii himself had an extreme shortage of food.[341]

D. I. Pavlutskii's report poorly traces the route of the campaign, but it is possible to establish it based on the reminiscences of the veterans in 1763 and the route marked on the maps of T. Perevalov and later F. Plenisner, as well as using data from the map of N. I. Daurkin (1775). Going from the stockade to the south from the Anadyr', on the sixth day the detachment crossed the Main River on the ice, then going on farther through the modern Alganskii Ridge, went to the Anadyr', which it crossed at the mouth of the Krivyi (Kymil'neiskii) Channel, a little above the Chekaevo tract (east of the Chernaya River); then, cutting through the middle course of the Chernaya (Tanyurer) River, it arrived at the Nerpech'ya (Kanchalan) River, which it crossed on the upper course. There, on 2 March, traces of Chukchi Reindeer sleds were discovered, which S. Krivogornitsyn was sent to follow. He, as ordered, tried to invite the Chukchi to come into citizenship, to which he received a rejection, which was followed by the destruction of a yaranga. Receiving information from captives that the closest Chukchi camp was near the mountain Serdtse-Kamen' (Mt. Priskal'naya), the major, leaving behind the convoy with a guard of 50 people, went ahead with the primary bulk of the army (594 men). After six days he reached Nochen (Kresta) Bay, having been moving along the Poperechnaya (Vaamchergyrgyn?) River, which flows from the west into the bay. The detachment crossed the bay on the ice approximately in the middle (about 45 km) at night, probably so the Chukchi would not notice the appearance of the enemy. In the morning of 15 March, north of Serdtse-Kamen' D. I. Pavlutskii saw three reindeer herds and as many herders' yarangas. He killed 18 of 20 herders. Pursuing those remaining alive, the major, with fifteen servicemen, reached the primary camp of the main Chukchi toyon Tention, which consisted of seven yarangas. The details of the subsequent battle are unclear, but the Chukchi, probably seeing the small number of the enemy, began to fight, coming out in front of the yarangas. In the battle 88 Chukchi were killed, including the toyon, whereas among the Russians 10 men were wounded. On the 15th of March, after this battle, having learned that not far away was another camp of three yarangas of the toyon Tegruv'e, the major sent

[341] RGADA. F. 199. No. 528. Ch. 2. D. 3. L. 14–16.

Sotnik S. Krivogornitsyn with fifty soldiers, who attacked this camp. A fight occurred in which five Chukchi died, while others hid themselves (among the latter was Tegruv'e himself). Sixteen captive Koryak were freed, 11 Chukchi seized, and 3,000 reindeer were brought in from the two camps. Among the Chukchi of those seized on this day, there was the family of the ten-year-old boy Tangitan, whose father, the Koryak Okhmat, fled, while his mother, a relative of Tention, died under interrogation during torture by fire. Tangitan became the major's prisoner and later, at baptism, received the name Nikolai Daurkin.[342]

Having spent two weeks resting and waiting for the convoy, which arrived on 1 April, the major went farther into the mountainous interior part of the southern Chukchi Peninsula, sending the detachment of I. Butusin to the coast to search for settlements of the enemy. The major, having crossed the middle course of the Eponcha (Chautakan?), Chimgan (Erguveem?), and Kurupka (Kurupkan) rivers, gave the detachment a three-day rest on the south shore of Yurupka Bay, which is named Urupka on T. Perevalov's map, and is opposite an island (probably in Senyavin Strait).[343] On 7 May six scouts were seen there and Yakov Osipov's detachment was sent out. While going around the bay it clashed with a scouting party of 20 Chukchi, one being seized for interrogation. The last reported that the Chukchi were gathered at the Nos, and if the Russians follow them, they will be moving to the Northeastern (Chukchi) Sea. It turned out that the Chukchi had assembled an army, but, remembering the campaign of 1731, did not dare to engage in battle, and left the place, taking the herds, counting on wearing the enemy down. Probably to this

---

342 RGADA. F. 199. No. 539. Ch. 2. D. 6. L. 71 ob., note; Divin 1979: 323.

343 Establishing a clearer route for the 1744 campaign will be possible only after satisfactory publication of the above-mentioned maps of T. Perevalov, F. Plenisner, and N. I. Daurkin, where the geographic names can be read on a map itself. In A. V. Efimov's *Atlas* (1964), for example, names are given in a separate list; determining which toponyms on the maps correspond to the names in the list is often impossible. On N. I. Daurkin's map the Eponcha River is shown flowing into the gulf between Imylin and Imtunu (Efimov 1964: No. 130). If these toponyms match Enmelen and Imtuk, then this river is located more to the east. However, an island is shown in front of Imylin, which in such case might correspond to the spit of Achchen Lagoon. But then the question arises, how did the Russian troops on the return trip go so far to the east and cross the upper reaches of the Eponcha River. On T. Perevalov's maps the river is separated from Nochen Bay only by two ranges parallel to it.

stage of the war belong the Chukchi traditions, according to which the Chukchi, in fear of Yakunin, were going to flee to America, and hid themselves with their herds on an island, probably on Arakamchechen.[344]

Figure 28. Chukchi sleds. (Photograph from the turn of the 19th–20th centuries).

Crossing Yurupka Bay on the ice at night and passing along the Gizyang (Getlyangen?) River to Silega Bay (Getlyangen Lagoon?), the major moved along the coast to Micheng (Michegmenskaya) Bay and farther through "kamen' Mecheng" (the Teniyanskii Range?). On 12 May the detachment set off in pursuit of the Chukchi, but the latter managed to escape, losing only one riding sled.

On 22 May the major's detachment went in the direction of Chukotskii Nos. On Perevalov's map the western part of the peninsula is called by the last name, though on Daurkin's map this toponym corresponds to Cape Dezhneva. On the first map the latter cape is not distinguished from the others; this is evidently explained by the fact that the detachment did not go to this cape, passing along the interior part of the region.

On 24 May, in the search for traces of the enemy, the major sent I. Butusin's detachment of 156 men to the north, into the mountains by the coast. The major, crossing the ice in Tenyakh (Lavrentiya) Bay, reached Khumatam Bay (Uelen Lagoon?), where he made camp, dispatching detachments against the Chukchi. This bay is clearly different from the

344 Retreat to a safe place: Bogoras 1900: 336; flight to America: Bogoras 1902a: 71, 84; refuge on an island: Sarychev 1952: 244.

eponymous bay not far from Kolyuchinskaya Bay on Perevalov's map, which the major approached later.

I. Butusin's detachment destroyed a large coastal village, probably west of Cape Dezhneva (Uelen?), where 40 residents died. On 29 May the Pyatidesyatnik Ya. Skrebykin, sent to the south, engaged in battle with Reindeer Chukchi he discovered, seizing 200 animals, but having lost two men during this. On 29 May the interpreter P. Modin was sent with 86 warriors in a southern direction, into the mountains. On this same day two detachments met by the sea, coming first, Butusin—from the northeast, and second, Modin—from the south, they approached a stone stockade of maritime residents, whom the Reindeer Chukchi had joined. A great battle occurred there: the defenders came out and began to fight, but were defeated, having lost 130 men killed and 27 taken captive, whereas among the servicemen 27 were wounded. Probably the Naukan historical tradition "Nashestvie" ["Invasion"] has preserved the memory of this storming: the residents of Sanluk (6 to 7 km east of Uelen) built a fort on the mountain; since the enemy at first did not notice this fortification, the defenders made a sortie. Led by the elder Anaruk, the Eskimos actively defended, shooting from bows and from slings. However, the enemy stormed the fort and killed Anaruk. Some of those found in the fortification were able to save themselves by fleeing. Perhaps a folkloric echo of these events is contained in the Chukchi story from the Yakunin cycle about the fact that the Reindeer Chukchi, fleeing from Yakunin, went from the depths of the country to the coast, constructed a fortification in a gorge between cliffs, which the enemy destroyed, rolling boulders down from above; after this a second fortification was erected by the survivors under a plunging mountain in the land of Neten (a village west of modern Enurmino), which Yakunin was not able to destroy.[345]

D. I. Pavlutskii, learning that the Chukchi, as planned, went with the reindeer herds to the north along the seacoast, sent S. Krivogornitsyn with 200 fighters in pursuit on 31 May. The major himself, remaining in camp until 3 June, moved from Chukotskii Nos following the sotnik to the northeast. For more than a month, to 7 July, Krivogornitsyn tried to find

---

[345] The "Nashestvie" tradition: Menovshchikov 1987: 168–170; 1988: 314–317; Nefëdkin 2003: 307–312; about Yakunin: Bogoras 1900: 332–333; Bogoraz-Tan 1934: 169–170.

the Chukchi and their herds, but was able to kill only one toyon with 14 people, take 7 people captive, and drive in 300 reindeer. By this time the detachment began to experience inadequate provisions: on 16 June Pyatidesyatnik I. Dyagilev with 70 fighters was sent not only in search of the missing Krivogornitsyn, but also for a search of the coast for whale and walrus fat. After joining up with Krivogornitsyn, the detachment began to move back to Anadyrsk.

The major moved along the northern interior part of Daurkin's peninsula. After reaching the Nochan (Inchounveem?) River, the detachment went west along the river's upper course, then crossed the Pokoechi (Chegitun?)[346] and Ittei (Varevarvaam?) rivers and went farther north to the Stavy (Netteveem?) River. Then, having crossed three rivers—Itynochan (Ymylinkuul?), Nachelty (Inluveem?), and Avandy (Tenynvaam?), which flow into Khumatam Bay (the western bay of modern Neskenpil'nyn Lagoon); the detachment then went to the eastern shore of Anakhyi (Kolyuchinskaya Bay). Reaching this bay, the detachment began to experience extreme inadequacy of food; since the herd of the reindeer it had taken was exhausted, there was no new confiscation, and hunting birds and wild deer in July did not provided the necessary supplies. Passing the bay from southeast and south, crossing the rivers that flow into the bay, the detachment moved west toward the Amguema. According to the map of Perevalov's expedition, the detachment reached the Omgayana (Amguema), but this is a simple generalized route—the major's route, depicted in the servicemen's account of 1763, looks more complex: the detachment crossed the courses of three right tributaries of the Omgayana, separated by ranges: the rivers Netten (in the memories of the Cossacks—the Nutten), the Chachaum (Chagnum), and Echa (Yecha), then making a loop, turned southeast,

[346] Chukotskii Nos (Cape Dezhneva) is represented on N. I. Daurkin's map between the Popotchi River (Popoechi according to T. Perevalov) on the south and Chevantu (Chavaktu) on the north (Efimov 1964: No. 130). In this case the major's camp, located at Khamatan Bay, could have been situated at Puoten (?) Bay; in this case it is necessary to reexamine the attributions of hydronyms down to Kolyuchinskaya Bay mentioned in the depositions of the Cossacks. However, on T. Perevalov's map the Diomede Islands are shown opposite Tenyakhi Bay, south of the Nochan River, which speaks of views different from those of N. I. Daurkin's. Clarification of the campaign route will be aided by the work with published high quality 18th century maps.

where it crossed the upper reaches of the Eponcha River, which it had crossed earlier. Moving farther west, the detachment reached the Kukuna (Kenynyn?) River, which flows from the north into Nochen Bay (into the northeastern Kangynin Bay, Kresta Inlet),[347] and up along the course of this river (to the north), crossing it on the upper course. This route can be explained by natural obstacles, search for summer pasture for the remaining reindeer, or looking for the enemy, for which, incidentally, the major usually sent individual detachments. Only then they went approximately parallel to the upper reaches of the modern Amguema and, going around the upper reaches of the Nerpich'ya River, reached the upper reaches of the Chernaya (Tanyurer) River and moved south along it to the Yukaghir region. The Chukchi did not pursue them. There half of the detachment began to fish with nets, while 280 men were sent on 14 August, led by Sotnik I. Butusin, in makeshift baidary of reindeer (not of walrus, as usual) hide for reindeer hunting in the Chekaevo tract where the Chernaya River enters the Anadyr', where the sotnik found the carcasses of reindeer slaughtered by the Chukchi. Subsequently, the Chukchi, in spite of the campaign against them, sailed to Chekaevo, where they killed reindeer, but the appearance of the Russians did not permit them to take away their prey. On 13 August the major sent P. Modin with 12 men on foot to Anadyrsk, from where they were supposed to bring shitiki and to deliver supplies to the Rodionovo tract (4 km from the confluence of the Belaya and Anadyr's rivers). On 30 August the major arrived through swamps and mountains (the Pekul'nei Range) to Rodionovo, where P. Modin's five ships arrived on the next day, bringing provisions. On 14 September, E. Kotel'nikov was sent to Anadyrsk for additional supplies, bringing back ten bags of state provisions. Having sent I. Butusin part of the ships and awaited news from him, the major sailed in two ships to Anadyrsk, where he arrived on 22 September, and on 26 September Butusin's detachment returned on the remaining three shitiki. On the return trip the servicemen caught fish and hunted for reindeer and birds to supplement their provisions.

During the campaign 360 Chukchi were killed, 94 people taken captive, a large part of whom died on the return to Anadyrsk, and 4,620

---

[347] This loop is visible on T. Perevalov's map (1746), but is absent on the later maps of F. Plenisner (Efimov 1964: No. 123, 126–127).

reindeer seized, which incidentally were all used as food, as were the reindeer of the Koryak. D. I. Pavlutskii's detachment lost 6 Cossacks killed and 6 Koryak and Yukaghir. On the return trip, when the detachment was coming back on foot, the insufficiency of food was felt especially strongly. Roots were dug-up for food, and the servicemen hunted geese, ducks, and wild reindeer. Those weak from hunger were carried on makeshift stretchers of spears, of whom 11 soldiers and 39 Cossacks died. A draft reindeer was worth 10 rubles—approximately the annual salary of a Cossack! The major himself, suffering from a leg ailment, kept for himself only one riding reindeer, the rest of his reindeer, 49 head, he gave to the detachment for food.

The campaign of 1744 was the largest military expedition against the Chukchi. This was the second campaign of D. I. Pavlutskii, the second and last to the territory of the Chukchi Peninsula proper. The campaign lasted almost eight months (against seven in the campaign of 1731); 644 men participated in it (as opposed to 516 in 1731), but fewer of the enemy were killed—360 men (instead of at a minimum 800 in 1731), 94 Chukchi were taken captive (in 1731—160), no captured reindeer were brought to Anadyrsk (in 1731 they brought 500 head), and the losses of the detachment amounted to 62 men (as opposed to 23 in 1731). As we see, the successes of the expedition were much more modest than in the preceding campaign of 1731. Such modest successes are explained by the strategies of the Chukchi. Though the major always sent reconnaissance detachments from the main force, the major's scouting was worse than the enemy's: neither the number of Chukchi troops or the names of their leaders were known. Among the Chukchi, reconnaissance went well, even the Russians sometimes noticed the enemy scouts. Seeing a significant number of the major's troops, the Chukchi preferred to move in front of him, not engaging in battle, that is, used essentially the "Scythian retreat" strategy and lured the enemy deep into their territory with surprise raids on the herds and the enemy's detached soldiers. Knowing their locality, the Chukchi simply moved away from the enemy. Judging by everything, they assembled an army with which they moved their reindeer herds. If at the beginning of the campaign the herds were in mountain pastures, later they were eventually driven to summer pasture on the north coast of Chukotka. Knowing that with time the large number of soldiers of

Pavlutskii's army would experience food shortages, the Chukchi strove to keep the herds away and protect them. However, since there was no common leadership among them, they were not able to evacuate the whole population; the settlements of the coastal residents, who did not actively participate in the raids, especially suffered. The nomads mainly suffered at the beginning, in March, when the Russian army unexpectedly appeared at Kamen'-Serdtse mountain. However, with this, the Chukchi were not even deterred from their usual economic cycle: in summer they arrived on the Anadyr' for the reindeer hunt!

2.2.2. The River Expedition in the Summer of 1745

After the campaign of 1744 there were 479 servicemen in Anadyrsk, including 40 soldiers and 437 Cossacks. Meanwhile, provisions and monetary allowance were still in route from Yakutsk and had not arrived, while reindeer hunting on the lower reaches of the Anadyr' was skipped because of the late arrival from the campaign at the end of September 1744. As a consequence of this, inadequate provisions were experienced in the stockade. Therefore, immediately upon the river opening up in 1745, D. I. Pavlutskii was forced to go for procurement of food instead of a new campaign against the Chukchi.

Pavlutskii himself described the expedition in the greatest detail in a report to the Irkutsk Provencial Chancellery on 10 October 1745:

> The past spring of this year, upon departure from the Anadyrsk stockade, the tributary Koryak took their reindeer to their homes, and moreover for tributary hunting, the detachments of my soldiers and servicemen began to live in great need and hunger, which was now pitiful to see. Because of this, I determined to build twelve ships, which, with great need from starvation, were made: larch are sewn with rods and caulked with moss; and for food, fishing and reindeer at crossings, I went down along the Anadyr' River from the Anadyrsk stockade upon the opening up of the Anadyr' River on June 4th. And on June 8th arrived in the Rodionovo tract, set up camp for hunting with soldiers and with servicemen for fishing and hunting reindeer at crossing the river; and being that fish and reindeer were few in number, again with zeal for the service of Her Imperial Highness, I went to the sea on July 12 in search of the enemy Chukchi, since in former years they have been at the mouth of the Anadyr' River for fishing and reindeer crossing the river. And I arrived at the sea on July 22nd from the mouth

of the Anadyr' River, and to the left I sent a detachment with Sotnik Semen Krivogornitsyn and with him 126 soldiers and servicemen, and told him to go through the land to the sea and to look for enemy Chukchi and signs of them.

And on July 25th, the designated Sotnik Krivogornitsyn, arriving back entirely safe, declared in a report: Going to the sea through the mountains, on the 23rd of July we saw two enemy Chukchi men coming our way to observe us, who, seeing us, ran back to escape, and our lightly armed servicepeople ran after them, caught one, who was resistant, and this one they killed, while the other ran to their army; we followed after the one who ran and saw a large number of yurts of our enemy Chukchi standing by the sea on a spit; and the enemy, seeing our army, took apart the yurts, put them in baidary, and departed to sea, for example, thirty baidary; of these baidary, but in one baidara they rowed to ours and stood at sea, as far as the voice could possibly be heard, and had negotiations. In this baidara were eight people, and one of them was called the main toyon Keyu's, brother of Tevryun. And the aforementioned Sotnik Krivogornitsyn spoke through the interpreter, that they should come into citizenship under Her Imperial Highness's autocratic hand and pay tribute, and give from themselves amanaty; to which they, the enemy Chukchi, said: To give amanaty is a sin. Thus, Sotnik Krivogornitsyn called them, the enemy Chukchi, to me, and they did not come, merely said that you went in our land, and now we will be your guests; and did not engage in any other conversation and went to sea.

Thus, I went from the sea at the mouth of the Anadyr' River back up the Anadyr' River on July 30th safely with the detachment.

On August 3rd I arrived with the detachment at the Chekaevo tract and made camp in a secretive location at the Krivaya channel, waiting according to the aforesaid that they will secretly follow us, I ordered to prepare—to remove the tops from four ships for lightness and to have four or five oars on each side and two light ships with covers.

On the 4th I sent Sergeant Aleksei Mokroshubov and with him sixty soldiers and servicemen below the camp to an island three verst to a former Chukchi dwelling to guard against the enemy Chukchi.

On 13 August, from the aforementioned Sergeant Mokroshubov I received a verbal report: The enemy Chukchi

> came villainously close in baidary and in skin boats to the island, on which this Mokroshubov was set with a guard, and those enemy Chukchi wanted to kill the guard and, seeing the vigilance and strength of the guard, got in their baidary and skin boats, and sailed away in flight.
>
> On this same day, taking soldiers and servicepeople from my detachment, on the above-noted light ships I immediately followed after the mentioned enemy in pursuit down the Anadyr' River (and in the camp left 60 soldiers and servicemen to guard the other ships and things).
>
> Following after them, we saw the enemy in the evening, going from us by the Krasnaya River with great haste to escape; we pursued those Chukchi in our four light ships two nights and a day, all paddling without rest, and seeing our approach after them, that night the enemy abandoned its skin boats, dishes, and hides.
>
> On the 14th we overtook those, who came out against us in battle, and with the help of Almighty God and the good fortune of Her Imperial Highness, we killed no small number of them, the enemy Chukchi, both on the river and on the banks; in this battle 17 enemy baidary were taken and of the female sex and small children 22 souls were taken captive and distributed to the servicepeople, and with this we took the rifle of a serviceman of the party killed earlier, who had been killed by the Chukchi in the Rodionovo tract in the past 1744.
>
> In the aforementioned battle, on our side one sotnik, one soldier, and three servicemen were wounded.
>
> On the 17th everyone arrived safely in camp at the river place in the Chekaevo tract.
>
> At noon on the 24th I went from Chekaevo on ships back to the Anadyrsk stockade with the whole detachment, and arrived on September 4th all safely.[348]

In general, D. I. Pavlutskii described the expedition in rather good detail, not indicating, however, how many servicemen were in the 12 paddled shitiki, but this was obviously a significant part of almost the half-thousand personnel composition of servicemen of the stockade. On the other hand, the Chukchi had 30 baidary (approximately 300 people), and based on Ya. M. Peresypkin's retelling of the events, even 100 baidary.

---

[348] Quotation: KPTs: 165–166; about the campaign, see: RGADA. F. 199. No. 528. Ch. 2. D. 3. L. 16 ob.–17; Okun' 1935: 76–77; Zuev 2004a: 216–217; 2009: 123–124.

More than a month, from 8 June to 12 July, the detachment was occupied with fishing and reindeer hunting at the Rodionovo tract, though not especially successfully, since the spring river-crossing of the reindeer had already passed and the fall one not started. After the hunts, the major sailed to the mouth of the Anadyr' on 22 July. From there he sent the reconnaissance detachment of Sotnik S. Krivogornitsyn (126 soldiers and Cossacks) on foot to the north shore of the estuary. The latter discovered on the sea spit a summer camp of the Chukchi (probably awaiting the fall river-crossing). Stacking the yarangas into baidary, they had left. On 30 July the major sailed back to Chekaevo, where he sat in ambush, leaving a guard led by Sergeant A. Mokroshubov on Buyan Island, 3 km below on the course of the Anadyr'. The latter was situated in a place where in the preceding years there had been a Chukchi camp. On 13 August, learning of an attempted attack by the Chukchi on the detachment of A. Mokroshubov, Pavlutskii sailed in pursuit in four lightened shitiki. On the following day they overtook the Chukchi on the Anadyr' beyond its confluence with the Krasnaya River, where a battle in the form of a shootout occurred. The Russians shot at both those sailing and those escaping on the shore. After the battle, 17 baidary and 22 women and children were seized, who were distributed to the servicemen as prisoners. Having arrived back at Chekaevo on 17 August, the detachment spent a week there, putting itself in order and hunting.

The captured Chukchi women told the major that the Chukchi wanted, having assembled, to sail in spring in baidary for a surprise attack on the Russians in retaliation for last year's (1744) campaign of D. I. Pavlutskii to Chukotka. However, since the ice at the mouth of the Anadyr' remained until the beginning of June (OS), such campaign did not succeed in taking place, moreover that the Anadyr' detachment was already at Rodionovo. In addition, probably at this time all the Chukchi had still not gathered, and those staying in the camp on the spit with toyon Tegruv'e were awaiting others. Also clearly attesting to intentions to wage war are the words of the toyon to Sotnik Krivogornitsyn, who had destroyed the toyon's camp in March 1744, and implementation of the plan was clearly seen in the attempt to attack the Russian detachment on Buyan Island. In addition, the prisoners told Pavlutskii about the impending campaign of the Chukchi from the mouth of the Anadyr' with the first snow—the place of the assembling of the troops—against the stockade or against the Koryak. Viewing these threats as entirely real, the major on the eve of freeze-up on the river sent storekeeper P. Troshin on 7 October 1745 to the mouth of the Anadyr’ with 60 soldiers and Cossacks for observation of possible gathering there of the enemy.[349]

On 17 January 1746 Captain M. P. Spanberg, having returned from the Second Kamchatka Expedition to St. Petersburg, gave a report to the Admiralty College in which he proposed sending one ship into Bering Strait for aid to D. I. Pavlutskii; the ship was to cruise in the strait and not permit aid to the Chukchi from the islanders and Alaskans. However, this proposal was not adopted because of lack of development of the project.[350]

2.2.3. The Summer Expedition of D. I. Pavlutskii (1746)

In June of the following year, 1746, immediately upon the return from the Chaun campaign, D. I. Pavlutskii carried out his next voyage on the lower reaches of the Anadyr’ for fishing and reindeer hunting and for “observation of the enemy,” though details about this expedition are lacking in the documents.[351]

Since such expeditions down the Anadyr’ became common for the residents of Anadyrsk, it pays to dwell on the economic cycle of the Anadyrsk residents, whose number of Russians in the population in the mid-18th century has been determined to be 650 to 700 people.[352]

The reason for these expeditions was the lack of food, since official delivery of which from Yakutsk was periodically delayed. Later, having arrived in Anadyrsk and becoming acquainted with the situation, Lieutenant Colonel F. Kh. Plenisner described in detail in his report (1763) the hunting cycle: on about 20 May (OS) the reindeer herds went from south to north and swam across the Anadyr, in particular at the Bani tract approximately 300 verst (about 320 km) below Anadyrsk. At this time, when the ice on the river had already broken up, the residents of the stockade sailed there in shitiki for the hunt. They lived in tents and did not

[349] KPTs: 166–167; Zuev 2009: 124; ice at the mouth of the Anadyr until June 5th (OS): Olsuf’ev 1896: 9; freeze-up on the river—about 20 September (OS): Merk 1978: 143.
[350] Belov 1956: 324.
[351] Zuev 2009: 125.
[352] Gurvich 1966: 124.

even kindle campfires, so that the animals would not smell the smoke. The reindeer swam across the river in different places over the course of three days. The hunters in light cottonwood single-log dugout boats, waylaying the animals swimming across the river and therefore rather helpless, killed reindeer with special long spears. They killed, as F. Plenisner noted, 1,000 to 1,500 head. The hides were removed from the carcasses, the bones were removed, and the meat was dried there. On about 1 June, after the end of the crossing, the hunters returned back and splitting up at this into three groups. One group having sailed to the mouth of the Main River, dispersed among lakes, caught fish, and made *yukola* [dried fish]; the second returned to Anadyrsk and only then sailed to the Main River; and the third group immediately sailed to the Main for fishing. Then the two last groups (up to 200 men) sailed to the Chekaevo tract (on the left bank of the Anadyr' at its confluence with the Tanyurer River) for fall reindeer hunting: on 1 September the reindeer moved south. As C. Merck noted: ". . . also at this time the Settled Chukchi arrived in baidary, both for hunting and for trade." The armed detachment guaranteed protection for the promyshlenniki. In September the Anadyrsk people returned from Chekaevo. The first group made yukola and rendered fat until the beginning of August, and later proceeded to the Anadyr to catch red fish [salmon]; then, from the middle of September, they fished for herring. Around 1 October, when the river was covered with ice, they all returned to the stockade. Still before November they fished for white fish, and in winter, until March, for burbot. In September the women went out from the stockade and made supplies of edible roots, which they mixed with salmon roe for food. From the beginning of October the Anadyrsk people went out for firewood and to hunt partridges and hares. Everything that was prepared went into personal supplies, and not for the sale. Trade was conducted by visiting Yakutsk merchants.[353]

2.2.4. The Last Campaign of D. I. Pavlutskii (1747)

On 12 March 1747 about 500 Chukchi attacked the tributary Koryak at the mouth of the Orlova River, a left tributary of the Main,

---

[353] Report of F. Plenisner of 31 October 1763: AAN. F. 3. Op. 10. No. 137. L. 138–141; quotation: Merck 1978: 144–145; also see: AII. F. 36. Op. 1. No. 643. L. 584 (deposition of G. Sheikin); Berkh 1823: 198–200.

which enters the Anadyr', took eight women captive, and drove off seven herds of draft reindeer and reindeer for slaughter, including riding animals of the Anadyrsk party. Receiving news of this about midnight, Major D. I. Pavlutskii, not delaying, immediately on the night of the 12th to 13th of March organized a pursuit. He took with him 97 soldiers and Cossacks led by two sotniki, who were joined by 35 Koryak. On reindeer and dog teams collected from the latter the detachment moved lightly (but with a cannon) in pursuit. According to the traditions of the Anadyr' Chuvan people, told to G. Maydell, the servicemen went on freight sleds harnessed to one reindeer, while the major, being a tall man, went on a sled harnessed to four animals. In addition, they drove the reindeer that were to be slaughtered. The primary detachment of the servicemen, 202 soldiers and Cossacks under the command of Sotnik A. Kotkovskii, had to go on skis, carrying their weapons and pulling sleds. Only 53 men were left for protection of Anadyrsk. Before dawn on 14 March the major caught sight of the Chukchi camp on a hill located at the mouth of the same Orlova.

The fight that ensued was described rather in detail in the sources of both sides. Depositions of the Koryak and Cossacks about the battle rather objectively present the events from the Russian side. Information of those who worked with Siberian documents, P. A. Slovtsov, A. S. Sgibnev, and G. L. Maydell, adds to the reports. Details about the death of the major are contained in the historic traditions of the Kolyma, Markov, and Anadyr' people. The Chukchi point of view on the events can be seen in a series of folkloric stories about Yakunin, who usually correlates with D. I. Pavlutskii; though Chukchi folklore tells about the death of this character differently.[354]

Thus, before dawn the major gathered a council. The veteran Esaul (Captain) Petr Modin and the Koryak princes believed that it was

---

[354] Documents: RGADA. F. 199. No. 528. Ch. 2. D. 3. L. 19 ob.–20; No. 539. Ch. 2. D. 6. L. 76–76 ob.; AAN. F. 3. Op. 10. No. 137. L. 27 ob.–28, 113–114; KPTs: 169–174; Slovtsov 1886. Bk. 1: 253–254 (he used a handwritten Siberian chronicle from the library of the Tobol'sk Seminary, which describes the events from the campaign of Yermak before 1760); Maidel' 1894: 560; traditions of the Anadyr' Chuvan and Kolyma people: Maidel' 1925: 23–24; Olyksandrovich 1884: 295; D'yachkov 1893: 37–39; Chukchi folklore: Bogoras 1900: 90–91, 330–334; 389–390; Menovshchikov 1958: 156–158; Takakava 1974: 103–105; about the battle on the Orlova River, see: Sgibnev 2005: 64; Nefědkin 2003: 244–249; Zuev 2004a: 219–223; 2009: 126–127.

necessary to set up camp, make a fence of the sleds, and wait for the rest of the detachment, whereas Sotnik S. Krivogornitsyn—actually the major's adjutant—advised attacking the Chukchi immediately while they were concentrated in one place and not dispersed in the surrounding areas. Pavlutskii, following the second advice, at dawn gave the order to begin the attack. (It should be noted for a better idea that on 25 March 2014 sunrise in the region of the city of Anadyr' was at 5:55 am). The major's detachment went on the attack against the Chukchi, who were located on a hill, with an unfavorable position at the bottom.

In their oral depositions of 17 March, participants of the battle Pyatidesyatnik Efrem Kotel'nikov and Cossack Vasilii Antipin gave the most adequate description of the fight, which in the retelling of Captain Ya. M. Peresypkin sounds like this:

> . . . the fight was in an inconvenient place, the enemy on a high hill and our troops under the hill. The enemy, rolling from this hill and attacking with its multitude of people, who were up to five hundred, with just spears. Against them from our side, they were first hammered with the cannon, and then with rifles, from which we managed only one shot, since the enemy did not permit reloading the guns another time, and, they becoming stronger, right away our troops were led to extreme confusion; against which our troops made a strong battle with spears. However, the enemy increased, which they, Kotel'nikov and Antipin, upon the enemy taking the banner and drum, where coordinated, not aware of each other, and for a short time connecting with gunfire, then few men began to move back from the enemy retreating while fighting. And finally, it came to the fact that the enemy was taking guns and spears from our hands, and ours were pulling bows and spears from the enemy with their hands and defended themselves with knives, where the enemy took sixteen guns; and on our side thirteen were wounded, fifteen Koryak. Those who [Chukchi.—A. N.] pursued them (but not all, the rest remained in the place of battle to gather up guns and other things) to the Koryak stronghold, which was made of reindeer sleds. Where fifty men of our troops arrived, seeing whom, the enemy turned back to the place of battle; and though several of the enemy troops were killed and wounded in the fighting, in such confusion the number is not known.[355]

---

[355] RGADA. F. 199. No. 528. Ch. 2. D. 3. L. 20.

The eyewitnesses to the event, the servicemen, told only about what occurred with them and around them, though the most furious fighting was played out in the place where D. I. Pavlutskii was. The major himself fought bravely. According to the traditions of the Anadyr' people, he fought holding a sword in his right hand and the barrel of a gun in his left. With the threat of being surrounded (according to P. A. Slovtsov, the Chukchi, by going around the hill, came upon the Russians from behind) and the retreat of the Koryak and servicemen, the major also began to retreat with the few who remained with him.

> A Chukchi, who fell captive to the Chuvan people, described the death of the major in the following way. After the escape of the traitors, the Chukchi could not kill Pavlutskii for a long time because he was wearing armor. The Chukchi for a long time shot at him from bows and stabbed him with spears, but nevertheless were not able to wound him; finally, surrounding him like wolves around a deer, they entangled him with thongs, dropping him to the ground, and found a place to stab him under the chin.

G. Maydell gives some more details, which the Chukchi also told to the Chuvan people: the major attacked the enemy three times, but the Chukchi were not able to wound him, since he was in mail and helmet. Then they toppled him with lassos and began to choke him, "there he himself opened the iron breastplate and a blow of a spear ended the affair." The legend of the Kolyma people also tells about this, according to which the major loosened the collar of his chain mail. Based on Anadyr' tradition transmitted by G. Maydell: "When they [the Cossacks.—A. N.] reached the place of the fight, they found the snow trampled in three places and littered with bodies of the Chukchi, in the fourth lay Pavlutskii himself, now without chainmail and helmet, and all around a huge number of dead Chukchi." According to Chukchi stories, the major was at first wounded in the eye with an arrow and then finished off, or after wounding him, they captured and tortured him in retaliation for the suffering caused to the Chukchi. These descriptions arose under the influence of the folk subject, which is based on the custom of such severe treatment for a captured commander. There was even a tradition that the Chukchi, cutting off the major's head, kept it as a relic, and the residents of Nizhnekolymsk believed that the Chukchi cut the body of the major into pieces, dried them, and preserved them as memorabilia.[356]

The Russians and Koryak remaining alive went to the "Wagenburg" of freight sleds; some managed to flee on reindeer. Only the vanguard of 50 servicemen of A. Kotkovskii's detachment saved the situation—the Chukchi retreated from the camp to the battlefield. In the afternoon of 14 March Kotkovskii arrived at the Koryak camp with his detachment and received information about the major's death. According to legends, when the primary detachment of Russians, going to help the major, encountered those fleeing from the place of battle, it was decided that the major had been killed, and therefore the servicemen did not hurry.[357] Though Kotkovskii wanted to pursue the Chukchi, who had departed on reindeer, there was nothing to go on: the Koryak reindeer had been driven off during a raid on 12 March and after the defeat. They arrived to the field of battle (according to the Tobol'sk chronicler P. A. Slovtsov, at five verst) the next day, when the Chukchi had already left, and and discovered there the body of the major without helmet and armor. The last had been taken by the Chukchi as trophies. Only in 1870 the Chukchi *erem* Amvraurgin gave the Kolyma district constable G. Maydell the captured chainmail of the major, which the erem had inherited from his grandfather. Upon examination of the major's body three wounds were discovered: on the right side—from a spear, and in the neck and on the brow—from arrows, which in some degree correlates with data of the Chukchi folklore.[358] In the battle died the major himself, two sotniki

---

356 Story quotation of a captive Chukchi: D'yachkov 1893: 39; quotation: Maidel' 1925: 23–24; about the Chukchi going to the back: Slovtsov 1886. Bk. 1: 254; Kolyma legend: Olyksandrovich 1884: 295; wound in the eye in a Chukchi story: Bogoras 1900: 332; cf.: Menovshchikov 1958: 157; torture: Bogoras 1900: 333–334; 390; the major's head: Sgibnev 1869: 34; cutting the body into pieces: Olyksandrovich 1884: 295.

357 Maidel' 1925: 23–24; D'yachkov 1992: 185–187. According to Nizhnekolymsk legend, the Kolyma people hurried to help the major, but the Cossacks they met (Berezkin and Krigovornitsyn) coming from D. I. Pavlutskii told them not to hurry, and thus aid was delayed (Olyksandrovich 1884: 295; Bogoras 1902: 148; 1902a: 71). In Chuvan tradition, instead of Russian Kolyma people, it was the Chuvan people themselves who did not manage to help (D'yachkov 1893: 37–38).

358 Maidel' 1894: 264; Neiman 1871. Vol. I: 18; Olykandrovich 1884: 295; wounds on the body: RGADA. F. 199. No. 528. Ch. 2. D. 3. L. 20.

(including S. Krivogornitsyn), one pyatidesyatnik, one corporal, one soldier, thirty-three Cossacks (including P. Modin), eleven Koryak, thirteen servicemen and fifteen Koryak were wounded during the retreat, and one Cossack fell captive. The Chukchi captured the banner, the drum, the iron field cannon, forty guns with charges, and fifty-one spears. The Chukchi losses were unknown to the Russians. The bodies of the dead Russians were carried from the field of battle and on 20 March were delivered to Anadyrsk, where they were buried. The body of the major, wrapped in paper, was also brought to the stockade and, covered with wax, was placed in a coffin. From 19 March to 8 November it was preserved in an icebox-cellar under the storehouse in Anadyrsk, and then was sent to Yakutsk where his family lived. In March 1748 the body arrived in Yakutsk and buried at the church of the Yakutsk Spasskii Monastery.[359]

It is not explained in the sources why the Chukchi did not leave the mouth of the Orlova during more than 24 hours, which passed between their raid on the Koryak on 12 March and the appearance of the major on the morning of 14 March—they usually returned back with the booty. Perhaps they simply guarded the stolen herds or planned a continuation of the raid? The Chukchi occupied the dominant position on a hill and could not have feared the sudden appearance of the enemy. D. I. Pavlutskii, even seeing the numerical superiority of the enemy, rejected the sensible advice to wait for reinforcements, but, accustomed to defeating the Chukchi, ordered an attack at dawn, not allowing the servicemen even to rest after the travel. The primary sources do not report on the formation of the major's detachment—the disposition was probably the traditional: servicemen in the center, the Koryak on one or both flanks. The Chukchi obviously noticed the appearance of the servicemen and their small numbers in comparison with their own and managed to prepare for battle and adopt a plan of action. Instead of having the troops shoot at the Chukchi from a distance, and thereby use the advantage of firearms, Pavlutskii ordered them to move toward the Chukchi. After the first discharge of the cannon and rifles the Chukchi threw themselves with spears on the enemy from the hill, thereby increasing the effectiveness of their attack. The Russians and Koryak, with the fourfold superiority of the

[359] Losses: RGADA. F. 199. No. 528. Ch. 2. D. 3. L. 20; about the major's body: RGADA. F. 199. No. 528. Ch. 2. D. 3. L. 20 ob.; Bushnev 1995: 17–19.

enemy that was bypassing them on the flanks, began to fall back. Conducting a retreating battle and firing from rifles, the Cossacks rallied with Pyatidesyatnik E. Kotel'nikov and fell back to the Koryak "Wagenburg." During this retreat, which gave moral support to the Chukchi, the Russians and Koryak bore substantial losses, while the Chukchi seized their trophies. Part of the fleeing managed to escape on reindeer sleds, the rest began to defend themselves, settling in the fortification. It is unknown how long the servicemen could have defended themselves, but the Chukchi, seeing after midday the approaching vanguard detachment of A. Kotkovskii, retired, driving the reindeer from the Koryak camp. The Chukchi managed to collect on the field of battle the guns not taken by those retreating and left.

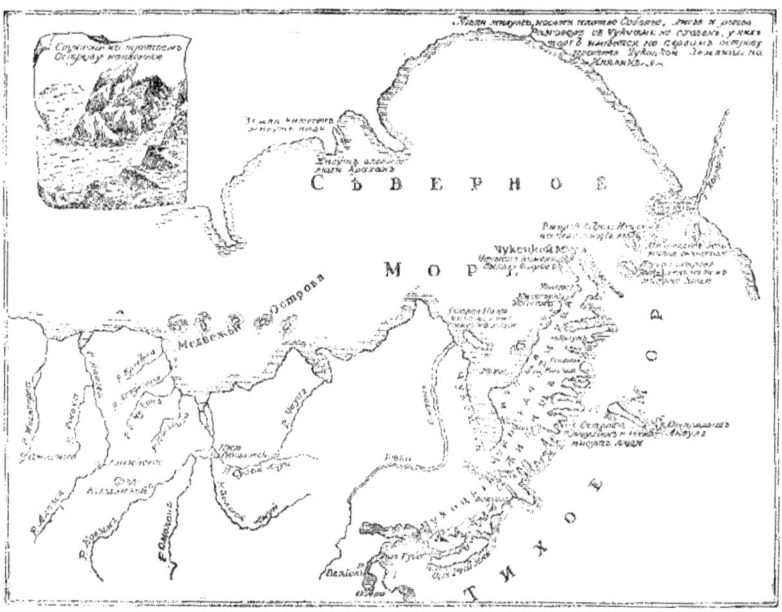

Figure 29. F. Ch. Plenisner's "Map of Chukotskii Nos," published in 1780. (Reproduced after Zubov and Badigin 1953: 21, Fig. 2).

The defeat of the Russians was complete: almost half the detachment perished—this was the most crushing defeat of the Russians in open battle with the Chukchi—even in the battle at Ergach fewer than a fifth part of the detachment died with A. F. Shestakov! This defeat was caused, on the one hand, by Pavlutskii's self-confidence, who was accustomed to defeating the Chukchi and who threw himself into the attack with the substantial superiority of the enemy, which also maintained the favorable position, and on the other, by the tactics of the Chukchi themselves, who learned how to confront the servicemen.

The battle occurred between the Anadyr' and Main rivers, in a place called Maiorskaya sopka [Major's Hill] and Maiorskoe ozero [Major's Lake] in the middle of the 19th century.[360] Even at the beginning of the 20th century bones were periodically found around this place of still unburied (Chukchi?) warriors, and a memorial cross was set on the hill in 1900.[361]

The period of active offensive action against the Chukchi ended with the death of D. I. Pavlutskii. Only Pavlutskii had twice (in 1731 and 1744) dared to invade Chukotka and abolish the enemy in its territory, trying to apply to the Chukchi maximal material damage and human loss. However, the results of the major's campaigns were insignificant: the Chukchi, as before, refused to pay tribute or to provide hostages, that is, to adopt Russian citizenship; they continued their hostile actions in relation to the Russians, Koryak, and Yukaghir. And there was no special increase in ethnographic knowledge—as before, there was little information about the Chukchi, though geographic knowledge was replenished—two maps of Chukotka by members of T. Perevalov's campaigns (1744 and 1746) appeared. In 1763 F. Plenisner noted, on the basis of a study of Anadyrsk documents that, concerning investigations of the region, "after him [Pavlutskii.—A. N.], the attempt to lead the Chukchi into citizenship was continued, but there was no attempt to know the

[360] It can be noted that, according to a Markovo tradition, the fighting men of D. I. Pavlutskii gathered at Maiorskaya sopka, while the Chukchi, approaching on sleds, fired at the enemy from afar and thus killed everyone (information of V. P. Kevkei, Anadyr', September 2013), that is, the historical events are represented as if reversed. A similar story was told to the American Richard J. Bush by the Anadyr' people in the mid-1860s: ". . .a good many years ago… the Tchuctchus advanced on Crepast in very large numbers, and put the Russians to fight, driving them to a hill two days journey distant, on the direct route to Penjina, called 'Ruskii khrebet' [Russian Ridge]. Here the Tchuctchus again attacked them, killing a great many, and thoroughly defeating the remainder" (Bush 1872: 384). What is more, based on the same Markovo tradition, the Koryak killed D. I. Pavlutskii, and not the Chukchi (Shentalinskaya 2003: 61; 2008: 267).

[361] Sgibnev 1869: 59; D'yachkov 1893: 38; Koltun 1904: 35; about the cross: Shentalinskaya 2008: 272.

position of the place and land on which the Chukchi live."[362] Because of the difficulty in supplying reinforcements, ammunition, and food, the Russian authorities there began to prefer defensive action, sending maritime expeditions down the Anadyr' to the mouth in summer. There were reindeer crossings there, where the Chukchi came, as well as overland parties that had to defend the Koryak and Yukaghir camps from Chukchi raids.

2.3. Warfare in the Late 1740s to Mid-1750s

On 3 April 1747 Sotnik Aleksei Kotkovskii, who became the person senior in rank in Anadyrsk, sent a report to the Irkutsk Chancellery with a description of D. I. Pavlutskii's death composed based on depositions of Koryak and servicemen; copies of the stories of the latter were appended to the report. In the stockade remained 309 servicemen, and the sotnik requested, together with a new commander, that one hundred Cossacks, seventeen soldiers, and four grenadiers be sent to the stockade.[363] In the absence of a suitable officer in the rank of captain, command of the stockade was assigned first to First Lieutenant Petr Kovalev, who arrived in Anadyrsk in December 1747 with fourteen servicemen. Then in December 1748 a new commander—Lieutenant Semen Kekerov—arrived at the stockade with a reinforcement of fifty-one soldiers and Cossacks. As a result, the Anadyrsk party consisted of 349 men in 1750.[364]

The Anadyrsk party continued activity in two directions—to the south, against uprising Koryak, and to the north, against the Chukchi. In January–April 1749 S. Kekerov went on a campaign against the Aklansk and Penzhina Koryak, and returning to the stockade, sailed down the Anadyr' to the Bani and Rodionovo tracts for hunting reindeer and fishing, as well as to search for Chukchi, who continued their raids on the Koryak.

Here is how Lieutenant Kekerov describes the expedition in a report to the Irkutsk Chancellery, which, in turn, sent this document to the Senate: "He, the lieutenant, with the detachment, in search of the enemy, set off on May 30th 1749 from the Anadyrsk stockade on ships, and were

[362] AAN. F. 3. Op. 10. No. 137. L. 83.
[363] KPTs: 172–174.
[364] RGADA. F. 199. No. 528. Ch. 2. D. 3. L. 20 ob.; Zuev 2009: 138, 428.

in the different places and there no signs were seen, other than a stockade made by the enemy in the Krasnyi Yar tract from an alluvial forest, and then he arrived safely back in Anadyrsk on 25 July." On 22 August the esaul (captain) interpreter N. Kurkin reported to Kekerov that at the tract of the Vakarina channel, near Anadyrsk, two Chukchi baidary had appeared. "And this same day, he, the lieutenant, with the detachment followed those enemies in ships, but because of opposing winds pursuing them was not possible." On 3 September the lieutenant came to Ust'-Maina for security. The Koryak woman Telna, who had fled Chukchi captivity, reported that the Chukchi

> had among themselves a great dispute in July 749 about a campaign against the Anadyrsk stockade; of whom one of the top men, Keyu and his relatives, went in three baidary in this hope: although the Russian people attack them, it was their intention, in the form of guile, to allegedly come into citizenship and pay tribute, and having paid be in the previous evil state; meanwhile the other Chukchi, nineteen baidary, remained on the Krasnaya River. And in the fall of this 749 of course the enemy, being gathered, goes on a campaign to the Koryak land.

For defense of the Reindeer Koryak, on 30 January 1749 the lieutenant set off with 270 soldiers and Cossacks on a campaign, leaving in Anadyrsk for security seventy-nine servicemen.[365]

In general, it turns out that Kekerov set off to the hunting expedition that has become standard for the Anadyr people, simultaneously thinking of attacking the Chukchi, who also came to the lower Anadyr' for the reindeer hunt. However, the opponents missed each other. The lieutenant supposed that the Chukchi would come to the spring reindeer crossing. However, they came to the fall crossing when Kekerov had left to the stockade. The reindeer hunt by the Russians was not successful, so they began fishing. Arriving on the Krasnaya River in July, most of the Chukchi rejected the proposition of Russia's old enemy, Keyu, to go to Anadyrsk, probably for robbery, and left on a campaign to Khatyrka against the Kerek. Keyu thought that if he were caught by the Russian servicemen, he would give them the tribute and they would release him. Three baidary (on the order of sixty men) of Keyu's "clan"

---

[365] Quotations: Senate Archive. Vol. VIII. 1897: 301; events: Zuev 2009: 140–141.

sailed toward Anadyrsk. As you can see, two of these baidary were noted at the channel in the tract of the Vakarin channel. But having learned about five of Kekerov's ships sailing up, the Chukchi retreated. In the fall, Keyu joined the remaining group, planning to carry out a campaign on the Reindeer Koryak. In general, there was no armed clash between the Chukchi and the Russians during the summer expedition of 1749 on the Anadyr'.

On 19 March 1750 Kekerov set off from Anadyrsk with a detachment on reindeer sleds collected from the Koryak and Yukaghir to the Talovka River against the Chukchi, who had carried out a raid on the Reindeer Koryak, but the Chukchi managed to get away. The lieutenant himself, with a hundred servicemen, went against the Settled Koryak of Kamchatka, who were rebelling, and returned to Anadyrsk on 26 April.[366]

In the summer of this same year Kekerov went down the river on another expedition, which was reported by Ya. M. Peresypkin in the following way:

> In the 1750th year, August 7th, Lieutenant Semen Kekerov went from Anadyrsk with a hundred and twenty men of regular military and non-regular ranks down the Anadyr' River in shitiki. . . . And on the 13th, upon arrival at the Chekaevo tract, the enemy Chukchi of the best toyon Keyu and his relatives, who floundered along (rushed) in baidary paddled down along the Anadyr' River, after whom the indicated Lieutenant Kekerov followed with the detachment in shitiki, and overtaking them, began a fight that lasted from morning to noon with our firearms, and from the enemy side with bows, after which we went before midnight to the mouth of the Krasnaya River (which from Chekaevo is no small distance), to which the military servicemen from the fight, (and even more so in the offensive storm), and from rowing had become very tired (and in addition, it was becoming dark). To chase after the enemy now following the mentioned fierce storm was not possible. And I was forced to return from the Krasnaya River while there was favorable good weather. Raising the sails, I went to the Chekaevo tract and, not reaching it, found standing on the shore several baidary and vetki, from which we took: one little brown fox, two worn-out marten parkas, five of red fox, and our troops found no few dead Chukchi.

---

366 RGADA. F. 199. No. 528. Ch. 2. D. 3. L. 21 ob.

> On the 14th, leaving Ensign Kovalev at Chekaevo with the detachment, Lieutenant Kekerov himself went in three ships with a small detachment in search of the mentioned fleeing Chukchi to the Krasnyi Yar tract, where in the morning of the 16th two men and one woman were taken, who through translation stated: "In his, Kekerov's, first fight with them, four Chukchi toyony were killed, who were taken in baidary to their land by their custom for cremation, and also other Foot Chukchi, not small number (and they don't know how many), were killed and wounded." And those captives were left behind because of their wounds. And all the enemy then on the Anadyr' River noticed by him, Kekerov, at the indicated Chekaevo in their various summer dwellings were up to a thousand people. From there on the 16th he returned to Chekaevo. And in the aforementioned fight three men on our side were wounded. And on the evening of the 17th, taking ensign Kovalev with a detachment, he went to Anadyrsk, where he arrived safely on the 5th of September.[367]

Although the goal of the campaign was not clearly indicated in the document, it can be suggested that it, as earlier, was of two purposes: hunting and warfare. Thus, Kekerov arrived at the Chekaevo tract not for the spring hunt, but for the fall one. Here on 13 August he clashed with the same toyon Keyu, who was attacked without any negotiations about accepting him into citizenship. The Chukchi were forced to engage in battle, which lasted several hours until midday. After midday the Chukchi began to depart to the mouth of the Krasnaya River. Pursuit by the Russians stopped because of fatigue of the crews of the shitiki, bad weather, and the onset of darkness (twilight began in the region of the city of Anadyr' on 24 August 2014 at 7:59 p.m. and ended at 8:59 p.m.). In this fight only three men of the Russian detachment were wounded, while among the Chukchi four toyony and a number of Settled "Chukchi" were killed; they evidently made up the primary bulk of those who had arrived, the number of whom the lieutenant himself estimated at a thousand men. The primary group of the Chukchi sailed away, which became clear upon reconnaissance of the stream on 14–16 August.

On 1 December 1750 Captain Vasilii Shatilov accompanied by 148 servicemen arrived at Anadyrsk to replace the wounded old S.

---

[367] RGADA. F. 199. No. 528. Ch. 2. D. 3. L. 21 ob.–22; No. 539. Ch. 2. D. 6. L. 76 ob.–77.

Kekerov. Instead of granting Shatilov's request for resignation due to ill health (because of scurvy and illness of the eyes and head), the Selenga commandant V. V. Jacobi assigned him as commander of the stockade. The captain had instruction, issued by the Irkutsk Chancellery, to continue warfare: ". . . henceforth with the traitors taken prisoner and in the search for non-peaceful foreigners to act toward them with the force of those instructions and orders given to the deceased Major Pavlutskii and Lieutenant Kekerov invariably in all." The captain acted on the instructions. In April–May 1751 the detachment of V. Shatilov suppressed the Taigonos Koryak. Returning on 22 May to the stockade, the captain already on 3 June had set off down the Anadyr' to the Rodionovo and Bani localities for fishing and reindeer hunting, and "the more so that the enemy Chukchi by water communication could not perpetrate any surprise attack and mortal murder, or damage to the most high Her Imperial Highness's interest, and ruin to the Anadyrsk fort." However, the hunt was unsuccessful, and the Chukchi in general were not met. In addition, in V. Shatilov's absence from Anadyrsk the Koryak amanaty arose and fled, driving off the government herd, which consisted of 2,500 captured Koryak and Chukchi reindeer. In this same year of 1751, the Anadyr' Reindeer Koryak also "betrayed," after which they stopped paying tribute in Anadyrsk.[368]

Having lost the herd and not receiving provisions from Yakutsk, in August Shatilov was again forced to set off down the Anadyr' to the fall reindeer hunt. In the most detail, based on his campaign journal, Shatilov tells about what occurred in the report to the Irkutsk Chancellery of 22 October 1752, which differs in the details and dates (sometimes substantially) from an earlier report to this same chancellery—of 11 March of the same year:

> Last August, on the 9th day of this year of 1752, I went from the Anadyrsk fort with a detachment down along the

[368] Assignment of V. Shatilov: Shashkov 1864: 78; quotation from the edict of the Irkutsk Chancellery: Senate Archive. Vol. VIII. 1897: 207; spring campaign against the Koryak: KPTs: 115–120; June expedition of 1751: KPTs: 120 (quotation); uprising of the Koryak amanaty: KPTs: 121–122; RGADA. F. 199. No. 539. Ch. 2. D. 6. L. 28 ob.–29; betrayal of the Anadyr' Koryak: RGADA. F. 199. No. 539. Ch. 2. D. 6. L. 29; AAN. F. 3. Op. 10. No. 137. L. 68 ob., 143, 178.

course of the Anadyr' River on eleven ships in search of the enemy Chukchi; in the party were 180 men of different ranks.

On the 12th I arrived at the mouth of the Main River and left the Pyatidesyatnik Verkhoturov with 18 men and one ship, and on this same day I left Pyatidesyatnik Petr Shatilov at the Pankov tract and with him 10 ordinary and one disabled (a total of two ships) to hunt reindeer.

At dawn on the 14th we saw the enemy in two baidary, above the Chekaevo tract, of which, one, with the help of Almighty God, I attacked with a military hand and killed the Chukchi in it, except five people who escaped from this baidara into a thicket on the shore, including one wounded. And they heard about this fight in the Chekaevo tract, where there was a Chukchi village, which the enemy Chukchi left with all kind of haste to escape; and the Chukchi who were on Buyan Island and at the mouth of the Krasnaya River and at Krasnyi Yar, those who fled to escape from Chekaevo, made known to each other by the firing of guns about my action upon them with a military hand. And, based on this sign mentioned above, those Chukchi ran with haste to escape from the indicated localities, after whom I pursued with great zeal, in particular those people, the Chukchi, who have among them baidary, will be accosted, and those people will be taken in other baidary; only three baidary, as I began to comprehend them, then they seeing that they can't go faster with the current, they rowed through the tidal flat toward the shore (that is, the shoal) and went onto the bank, leaving their baggage in those baidary, and they went in the river in quite inaccessible and overgrown with thickets. And I, having run around this tidal flat by sail, and came to the shore below, because I could not pass through this tidal flat by ship where they, the Chukchi, had gone in baidary. However, I sent soldiers and Cossacks after them, who, with the help of Almighty God, though they were not able to reach them, the Chukchi themselves. However, they took captive of the male sex one boy, six large and small of the female sex, and the indicated three baidary. And during this time many other ships began to gather, ship after ship, and after a continuation they, the enemy Chukchi, went hastily to escape in thirty baidary, because their baidary were light, as opposed to our ships, and progressed through the shoals, and though I had warily intended to pursue them, it was now not far from there to Prokop'evska, Onemenska, and Nerpich'ya bays, through which I would have to follow them, the Chukchi; however, for a great distance through these bays I did not follow them, the Chukchi, farther, since our ships did not come together soon; in addition, the military people were quite tired, because one of our ships had pursued them, the Chukchi, a distance of more than 80 verst, and the other ships remained behind me because they were heavy and only one of them could be seen behind me.

And upon return from there, one pyatidesyatnik and with him twenty ordinaries (?) privates were sent to Krasnyi Yar for reconnaissance of the enemy, who had again taken one woman with a small girl captive, and upon arrival at Krasnyi Yar one woman with a boy was taken captive. In this locality the military service people stopped for a rest and during the night all the ships withdrew, and stood at anchor across the whole river, since, as I first pursued them, the Chukchi, by ships from the Chekaevo tract, then from fear of battle a few baidary went to the mouth of the Krasnaya River, and one vetka went to the shore on the side of the river opposite the mouth of this Krasnaya River.

At dawn on the morning of the 15th I arrived with ships to the shore, and there was one woman who stated that because she was old her comrades left her and she could not go with them and then she came to see if the Russian people had gone or not, and if there were any Chukchi baidara or vetka left on which she could reach her comrades. Right away two uryadniki (sergeants) and with them twenty-four privates (ordinaries) were sent on both sides of the Krasnyi Yar to a search for her indicated companions. The sergeants, having arrived back, reported to me that no enemy could be found.

And the produced Chukchi woman, sitting on the ship, suddenly seizing a spear, stabbing herself, threw herself into the water, whom they pulled out, but she was now hardly breathing, and for this I commanded him to stab and cover with stones. And in this campaign, of the male sex two boys were taken captive, ten large and small of the female sex, twenty-six red foxes . . . and four cross foxes . . . three worn-out marten parkas and one iron armor (kuyak), 15 baidary, and 40 vetki. And on this same day, giving thanks to God, set off in order to reach in good time the Anadyrsk party and not miss a convenient autumn of low water and high-water fishing [weir.—A. N.], so that the military service people would not suffer extreme poverty and hunger in this coming winter through non-delivery of food and without fish to eat . I went to the Chekaevo tract and upon arrival at the mouth of the Krasnaya River again sent two sergeants and with them twenty-four privates to search for the enemy Chukchi, who from fear of the military had gone to the mouth of the Krasnaya River; and having arrived back, the mentioned uryadniki (sergeants) reported to me that the enemy Chukchi were impossible to find and they didn't see any signs of them.

. . . On the 16th, having gotten up in the morning, I went to the Chekaevo tract and, on the way, I left Ensign Kovalev on Buyan Island with five ships (this island is located at the short distance of about ten verst from Chekaevo) for hunting reindeer, and I arrived on the same day at Chekaevo with the other four ships and stopped to hunt reindeer, and continued to the 18th; in particular in these localities, both in Chekaevo and also on Buyan Island, there was no reindeer hunting, and for this reason Ensign Kovalev was ordered to come with all the five ships with him to the Chekaevo tract, who arrived on this same day safely. And upon his arrival I went with the whole detachment to the Pankov Plës [*plës*, a broad deep part of a river] tract, where I left one ship and one pyatidesyatnik with ten privates; where I arrived on the evening of the 22nd and remained until the 26th to hunt reindeer and from there went to the locality at the mouth of the Main River, where there was a pyatidesyatnik and with him eighteen privates with one ship. And on this same 26th, I arrived at the mouth of the Main, and on the 27th I remained in this locality to divide the reindeer hunt of the designated pyatidesyatnik to the soldiers and Cossacks, and I ordered them to load the ship. And at dawn on the morning of the 28th I set off, went to Anadyrsk, and arrived on September 2, by the grace of Almighty God, safely.

On the 4th of September I sent on four ships one Cossack sotnik with 50 men, including uryadniki (sergeants), corporals, and privates, from the Anadyrsk fort to the indicated mouth of the Main River as a precaution that the known enemy Chukchi would not then villainously approach the Anadyrsk fort by water and would not be able to perpetrate any sudden damage to the loyal subject peoples of Her Imperial Highness through attack and murderous death. The sotnik was ordered to be there on the 17th; he arrived back to the Anadyrsk fort from this mission on September 22 and reported to me in writing that neither the enemy Chukchi were seen nor any villainous, criminal signs of them.

. . . And the Chukchi prisoners produced, women and children, were sold at public auction, and the money from the sale, as well as the baidary and vetki without being sold, were given to the soldiers and Cossacks as a reward for the many

> difficulties endured by them in the designated campaign and for the faithful and diligent service to Her Imperial Highness.[369]

In general, this expedition was the usual hunting-fighting campaign of the Anadyrsk detachment, which sent 180 soldiers and Cossacks for the fall reindeer hunt. According to the report of 11 March, having set up camp in the Krasnyi Yar tract, from where the Chukchi had fled, V. Shatilov sent in pursuit of them one shitik led by Sotnik Kurkin, who was supposed first to call the Chukchi to tribute payment by peace, and after refusal, to act with force—the standard practice. The Chukchi, as usual, rejected Russian citizenship. Obviously, the Chukchi had already hunted and were going to leave. They set up camps at the reindeer crossings: on Buyan Island, at the mouth of the Krasnaya River, and at Krasnyi Yar. On 13 August Shatilov, being near the Krasnaya River, ordered the two other crossings at Chekaevo and Buyan to be occupied, in this way displacing the Chukchi. The latter were not going to fight, despite the fact that in the 30 transport baidary there should have been many times more of them than the Russians, on the order of 600 people, including women.

The primary battle occurred on 14 August. In the morning, upstream from the Chekaevo tract, the servicemen attacked two baidary, having overtaken one of them and killing a large part of the crew, the second baidara was able to escape. With firearms (indicators of Russian presence) the Chukchi gave a signal of alarm, after which the flight of the Chukchi became rampant. A shitik with twenty servicemen led by Shatilov began to pursue three baidary (and four vetki), the crews of which got out on the shore and hid. Mooring, the servicemen seized these baidary and captives—one boy and six women and girls. In the second half of the day the other shitiki arrived to Shatilov and began to pursue the primary part of the Chukchi, who were traveling in 30 baidary. To increase speed, the rowers abandoned half of the Chukchi baidary and their crews transferred to the others. More than 80 verst (85 km), to the Kopeikina rocky spit, the servicemen pursued the enemy, but because of bad weather, the tiring of the crews, and approach of darkness they were forced to stop the chase.

---

[369] KPTs: 176–178; cf.: RGADA. F. 199. No. 528. Ch. 2. D. 3. L. 24–24 ob.; No. 539. Ch. 2. D. 6. L. 77–78 (excerpt from a report of 22 October 1752); KPTs: 174–176 (abstract of the report of 11 March 1752).

The Chukchi, obviously did not actively fight, with only the servicemen shooting, thus loses of the latter are not mentioned. After the Chukchi left, the servicemen, in essence, became occupied with hunting reindeer. On 4 September, upon arrival of the detachment in Anadyrsk, V. Shatilov sent a sotnik with fifty men of the detachment to the mouth of the Main for defense against possible Chukchi attack. This post was removed on 17 September when the river, now covered with ice,[370] became unnavigable and there was no longer danger of the enemy approaching the stockade by a water route.

In August 1753 and in May–June 1754 Shatilov again went down the Anadyr to hunt reindeer and search for the Chukchi, but did not encounter the latter.[371]

In March 1754 a Chukchi group of 500 people led by Tegruv'e and Mego appeared in the region of Anadyrsk. On 21 March, in the Shtinnoe tract, the soldier D. Zykov was killed. On the following day, 30 verst from Anadyrsk, the Chukchi attacked a group of sixteen Anadyr' people traveling on dog sleds, who were returning to the stockade from hunting and fishing; three of them were killed, and a fourth (Cossack Boris Kuznetskii) was taken captive, the rest managed to escape. Then on the Nalucha River (35 verst from Anadyrsk) the Chukchi destroyed a Yukaghir camp that consisted of twenty-five households, killing with this seventeen men and three women, taking twenty-five boys and fifty-six women captive, carrying off property, and driving away 912 reindeer for a total sum of 8,113 rubles. According to the thinking of some Chukchi, if they did not get booty from the Yukaghir, they would have subjected Anadyrsk itself to attack. For defense of the Yukaghir, on 23 March Lieutenant M. Bel'skii, leading a detachment of 200 men on dog sleds, was sent in pursuit of the Chukchi, but they were not able to catch the enemy and returned to the stockade on 28 March.[372]

---

[370] Merck 1978: 143: freeze-up was about 20 September.
[371] Zuev 2009: 145.
[372] RGADA. F. 199. No. 528. Ch. 2. D. 3. L. 24 ob.; No. 539. Ch. 2. D. 6. L. 74, 78 ob.; KPTs: 180; Gurvich 1957: 246–252; Vdovin 1965: 76, 124; Zuev 1998: 40; 2009: 160. A. S. Zuev (2009: 160) suggests that on 22 March 1754 there were two attacks on the Russians, but Ya. M. Peresypkin tells, according to the documents, about one in which the Chukchi killed two Cossacks and one *raznochinets*; B. Kuznetskii, in

V. Shatilov described the general military situation on the Anadyr' in his report of 20 December 1751:

> . . . also the Chukchi arrive at the Anadyrsk stockade by water in baidary in summer time . . . and at this time if the Russian people show even a little negligence while fishing, the Chukchi kill them, taking captive wives and children, copper and iron kettles, axes, knives, and so on, since there is not thereof in their land. And in this regard, in the summer time at fishing, the Chukchi are extremely oppressive and dangerous.

In fact, from 1675 there was a prohibition to sell weapons to potential enemies of the empire that led to a deficit in iron tools and arms among the Chukchi. In his 1735 instructions, the head of Okhotsk, G. G. Skornyakov-Pisarev, specifically ordered the Anadyrsk commander not to sell to the Chukchi not just arms, but also other iron items.[373]

The Chukchi point of view on the situation on the Anadyr' is represented in a family tradition told to W. G. Bogoras by the Reindeer Chukchi Remkilen on the Omolon River in 1896:

> In the old days our people had no iron, and the Russians did not want to trade with them or sell them iron. And our people attacked the Russian people and Russian dwellings. Because of iron they killed people. And they took the kettles and knives. On the Anadyr' side there was the Chukchi Ken'yulin, a great warrior, a strong man. He lay by the road near the Russian village and kept watch on passersby. Whoever approached, Ken'yulin killed with an arrow from ambush, stripped the dead body, and placed it on the river ice; he left the clothing and sled by him, and he took with him the kettle and knife, the spear and axe. And Ken'yulin traded the iron among the Chukchi and became rich.[374]

Probably to this period of unrest, at the end of the 1740s–first half of the 1750s, also belong the reminiscences of the translator Cossack, Sotnik Andrei Kobelev (b. 1744), who told that he, as a boy, collected fallen Chukchi arrows during a Chukchi attack on Anadyrsk.[375]

---

giving his testimony, did not mention the deaths of two servicemen (RGADA. F. 199. No. 528. Ch. 2. D. 3. L. 24 ob.); he might not have known about it.

373 Quotation: Shashkov 1864: 67; 1735: Zuev 2009: 93.

374 Bogoras 1900: 392.

375 Argentov 1887a: 2.

3. Neither War nor Peace: Attempts to Establish Peaceful Relations between Russians and the Chukchi in the Mid-1750s to Mid-1770s

3.1. Activities of the Anadyrsk Commanders Ivan S. Shmalev and Semën Kekerov

On 7 December 1752 the Senate, based on a proposal by the Military College, approved a captain of the Shirvanskii (Shirvan's) Infantry Regiment, Ivan Stepanovich Shmalev, as commander of the Anadyrsk stockade with his promotion to the rank of second major. As early as 1749 I. S. Shmalev had presented to the Military College a project in which he presented four propositions on methods of bringing the Chukchi and Koryak into citizenship, in particular by assigning an experienced officer to the duty of the commander of the stockade, obviously hinting at his own appointment in this way. Action against the Chukchi was suggested in the document as of old: to bring them either in "tranquility" or in "complete ruin."[376]

For reinforcement of the Anadyrsk garrison, four officers and 257 soldiers of the Shirvanskii and Novouchrezhdennyi (Newly Established) Regiment were sent, together with Shmalev, from Tobol'sk, where the major set off from on 19 November 1753. He spent a year in Yakutsk and arrived at his place of assignment only on 1 February 1756. Together with the reinforcements that arrived, the garrison of the stockade now reached 700 men (in 1759)—the maximal size for its history.[377]

On 20 September 1755, a Chuvan woman and a Koryak woman, who had fled from Chukchi captivity, arrived at the Anadyrsk stockade and reported that "the Chukchi want to make peace with the Russian people and be under the High Autocratic Hand of Her Imperial Highness in citizenship and to pay tribute." They evidently transmitted the general

---

[376] Senate Archive. Vol. VIII. 1897: 741–744.

[377] AAN. F. 3. Op. 10. No. 137. L. 116; AII. F. 36. Op. 1. No. 643. L. 583; Alekseev 1958: 11–22; Zuev 2009: 157, 159, 428. Dates for the arrival of I. S. Shmalev in Anadyrsk differ in the documents: on 25 December 1755 (according to memories of G. Sheikin; according to T. I. Shmalev—simply in 1755): RGADA. F. 199. No. 539. Ch. 2. D. 6. L. 78 ob.), 1, 2, 6 February 1756. Such discrepancy is probably connected with the fact that the major sent reinforcements that were with him to Anadyrsk, in different groups.

mood of the Chukchi. Indeed, according to the observation of B. Kuznetskii, who lived among them at that time: "Those Chukchi, based on notes during his, Kuznetskii's, life there, are in fear of the Russian people, and talk more about peace than about war." In this regard, six Anadyrsk tributary Yukaghir, accompanied by a former captive Yukaghir woman, set off in the fall to the Chukchi to ransom wives and children taken captive. Upon arrival they stated directly to the Chukchi that "They were sent to them, the Chukchi, by the new commander, who had arrived in Anadyrsk, and he brought with him a decree in which it was ordered to reconcile with them."[378] However, the new commander I. S. Shmalev arrived in Anadyrsk only at the beginning of February 1756 and evidently was unable to give instructions directly to the local Yukaghir. The Yukaghir had probably heard something about some peace plans of the major, who was still on the way to Anadyrsk from Nizhnekolymsk where he had restored fortifications.

On 25 February 1756 in Anadyrsk, together with returning Yukaghir, a Chukchi embassy appeared that consisted of three people (the "best men" Khargigit, Amgovgov, and the wife of toyon Mego), to whom B. Kuznetskii was assigned as a translator. The negotiators had to find out in fact whether I. S. Shmalev intended "to reconcile" with the Chukchi and whether they would be able to pay as tribute one fox fur from each man without the taking of amanaty; they proposed to conduct the conclusion of peace and delivery of tribute later at the Chekaevo tract. At the beginning of March, after negotiations, two of the emissaries were sent back; Khargigit was left as a hostage. Sergeant Koshkarov, Corporal G. Tirskii, Cossack B. Kuznetskii, and translator-raznochinets (a meshchanin, or man from people of different ranks) N. Artem'ev were sent for negotiations with the Chukchi and reconnaissance. Beyond the Nerpich'ya (Kanchalan) River, Koshkarov met with the toyon Leucha of the Reindeer Chukchi, and then at the mouth of the Anadyr'—with Tevryu (Tatymkin), who confirmed their intention to pay tribute. On 21 April the servicemen who had been sent returned to Anadyrsk accompanied by five elders (led by Menigyt'ev) and two women. At a council of the officers with the emissaries, they agreed to meet in the Chekaevo tract on 20 June. Having

---

[378] Quotation: RGADA. F. 199. No. 539. Ch. 2. D. 6. L. 34; second quotation: Ibid. L. 34 ob.; third quotation: KPTs: 183.

taken Khargigit, Menigyt'ev and comrades left the stockade on 29 April.[379]

On 4 June 1756 I. S. Shmalev, with a detachment of 254 men in nine shitiki, went on the usual hunting expedition, during which he was also supposed to meet with the Chukchi. Before 1 July the detachment was occupied at the Bani tract by hunting reindeer, and then on two ships and one dinghy (a total of 70 men) arrived at the mouth of the Anadyr', where on 12 July there were two nomad camps of Reindeer Chukchi (of Tatymkin and Menigyt'ev), with whom negotiations were conducted. On 28 July Shmalev arrived at the tract on the Krasnaya River, 15 verst below which he set up a camp fenced with poles. Three-hundred Chukchi warriors arrived here in twenty-five baidary and stayed 350 m from the Russians. The Chukchi handed over one captive Russian and two Yukaghir women, requesting the return to them of their captives; they gave tribute from 107 men, obtained gifts, promised to pay again in the following year, but they categorically refused to give amanaty. On the stormy rainy night of 3/4 August the Chukchi unexpectedly left. On 4 August, receiving news of this flight, Shmalev assembled an officers' council, at which they decided not to initiate hostile activities and not to pursue the Chukchi, who had already given tribute. On 5 August one ship, with sotnik O. Nizhegorodov leading 25 servicemen and two translators (including B. Kuznetskii), was sent for clarification of the reason for the flight of the Chukchi. On 8 August, the sotnik returned, bringing Khargipin Kovrin, an elder of the Foot Chukchi, who explained the reason for the flight: on the stormy night it seemed to the Chukchi guard that the Russians wanted to attack them; the alarm signal was given, after which the Chukchi, breaking camp, fled in their baidary. On 9 August Shmalev set off for Anadyrsk, where he arrived on 17 August. After three days Lieutenant M. Bel'skii, who had been left by the major to hunt reindeer in Krasnyi Yar, also returned there.[380]

---

[379] RGADA. F. 199. No. 528. Ch. 2. D. 3. L. 26–26 ob.; No. 539. Ch. 2. D. 6. L. 34–35 ob., 78–82; KPTs 180–184; Vdovin 1965: 126–127; Zuev 2009: 160–165.

[380] RGADA. F. 199. No. 528. Ch. 2. D. 3. L. 27–28; No. 539. Ch. 2. D. 6. L. 35 ob.–37, 82–84 ob.; AII. F. 36. Op. 1. No. 643. L. 583–583 ob.; Vdovin 1965: 127–128; Zuev 2009: 165–168.

Thus, the Chukchi and Russians demonstrated their determination toward the establishment of relations without the aid of weapons. Incidentally, there was no written agreement, no pledges of allegiance were given, and no amanaty were produced. Mutual distrust was manifested, which was reflected not only in the stationing guards, but also in the unexpected panic and flight of the Chukchi under the imaginary threat of attack by the Russians—the Chukchi still remembered well the actions of Sotnik V. Shipitsyn in 1741 and feared repetition of the situation.

On 4 December 1756, 200 Chukchi carried out a raid on a Yukaghir camp in the Shchuch'ya River tract 25 verst from the stockade. Having learned of this, Shmalev sent Lieutenant Bel'skii with a detachment of 200 men on skis and dog sleds. Lieutenant Bel'skii pursued the Chukchi 15 verst, but was not able to catch them since they were traveling on reindeer.[381]

The commander of Anadyrsk had to respond to these hostile activities. During the summer of 1757 I. S. Shmalev carried out another retaliatory voyage-campaign against the Chukchi. Corporal G. Sheikin remembered the expedition in this way: ". . . Mr. major went with a detachment of two hundred on a campaign to the Chukchi in summer on shitiki ships on the Anadyr' River to Anadyr' Bay itself, as well as on the Belaya, Chernaya, and Anamena rivers; But since they always have dwellings on high cliffs and rocky hills, he returned, not inflicting on the Chukchi any vengeance."[382] Consequently, in July–August 1757 Shmalev, again with a detachment of approximately the same composition and probably in the same nine shitiki, sailed formally for action against the Chukchi, but in fact for the purpose of procuring food for the residents of the stockade. The servicemen went on the river to the Anadyr' Estuary, crossing to the Belaya, Tanyurer, and then Velikaya rivers. They usually hunted reindeer on these rivers, and the Chukchi might arrive here as well. The latter were not found, which the servicemen explained by the effectiveness of Chukchi surveillance. In general, this was the usual

---

[381] RGADA. F. 199. No. 539. Ch. 2. D. 6. L. 36–38 ob.; AII. F. 36. Op. 1. No. 643. L. 583 ob.

[382] AII. F. 36. Op. 1. No. 643. L. 583 ob.; Nefëdkin 2015: 186.

expedition from Anadyrsk down the river to hunt reindeer and search for the enemy.

At this same time there had been changes in the policy of Siberian and central authorities with regard to the Chukchi. The Irkutsk Vice Governor General Major I. Wulf, receiving I. S. Shmalev's reports about negotiations with the Chukchi, ordered to Shmalev by the decrees of 17 September 1756 and 14 May 1757 not to take amanaty from the Chukchi, to accept tribute willingly, to take an oath of allegiance from them as was their custom, to prohibit the Russians from going to the Chukchi in order to avoid unnecessary conflicts, and also to not interfere in Chukchi-Koryak relations. The Senate in a decree of 3 December 1758 generally supported the initiatives of the vice governor.[383] On the whole, it is possible to speak of a transition from a forceful system of relations with the Chukchi to the establishment with them of stable peaceful relations. The Russian authorities were even prepared in fact to leave their subjugated Koryak alone in their struggle with the Chukchi.

On 17 March 1758 in Anadyrsk Second Major I. S. Shmalev died after an illness. Lieutenant S. Kekerov, the senior officer in rank, again accepted command of the stockade before the assignment of a new commander. In 1758, severe flooding occurred on the Anadyr', as a consequence of which no fish were caught, and there was no reindeer hunt, and in 1759 famine broke out in the stockade, during which forty-five men died. Based on the archival information of A. S. Sgibnev, though not supported by other documents, a detachment of 200 Chukchi, returning at the end of spring 1759 from hunting in Kamchatka, having joined with other Chukchi, "laid siege" to Anadyrsk. However, it should be noted that the Chukchi did not lay siege to the stockade, which F. Ch. Plenisner pointed out several years later. If such "siege" did occur, it was the fact that the Chukchi set up camp at the stockade and threatened the residents who went beyond the walls. And this was possibly just another Chukchi threat to attack Anadyrsk, which in general was not realized; indeed, according to the documents it turned out that the next encounter, after 1756, of the Russians with the Chukchi was in 1760, whereas no Chukchi

[383] Vdovin 1965: 128–129; Zuev 2009: 170–171.

were encountered in the summer of 1759 when S. Kekerov sailed with a detachment to the Krasnaya River for reindeer hunting and fishing.[384]

Hunting expeditions to reindeer crossings on the lower Amur continued up to the liquidation of the stockade. In 1778, based on the preserved documents, sergeants at the Gizhiga fort compiled the following description of events:

> After the killing of some Yukaghir and after the death of Major Shmalev, the Chukchi met with the Anadyrsk Russian detachment that was there for the first time in the 760th year down along the Anadyr' for hunting swimming reindeer in the Chekaevo tract. At this meeting of the Foot Chukchi, the toyons Khargipin Kagrypov and Anyku Khanagarin and their relatives talked to the former commander of the Russian detachment Lieutenant Kekerov about the killing at the Anadyrsk fort of Yukaghir and their justification they stated that they, the Yukaghir, themselves deceiving, have killed their former best Chukchi men Khoyav'i and Levucha. And the Yukaghir, by the killing, did not permit the Chukchi time to meet with the Russian people and to pay tribute because of the danger.[385]

In general, it turns out that the Chukchi specially avoided an encounter with the Russians at reindeer hunts on the lower Anadyr', justifiably fearing punishment for raids on the Yukaghir, whom they accused of the death of two "best men," Levucha (possibly the already mentioned Leucha) and a Khoyav'i, that is, in fact the Chukchi pursued a blood feud. The Chukchi confirmed their desire to live with the Anadyr' people in peace and to pay tribute.

In the following years, the situation at the reindeer hunts on the lower reaches of the Anadyr' was peaceful:

> From this 760th, in the 761st and in the 62nd with commanders Lieutenant Kekerov, and in 763rd Mr. Colonel Plenisner, in 764th to 769th with Captain Peresypkin, every year down along the Anadyr' at the localities of Chekaevo and the mouth of the

---

[384] Famine: AAN. F. 3. Op. 10. No. 137. L. 132, 134–135; RGADA. F. 199. No. 528. Ch. 2. D. 3. L. 31; AII. F. 36. Op. 1. No. 643. L. 583 ob.–584; seige of the stockade: Sgibnev 2005: 81; Slovtsov 1886. Bk. 2: 79; Bogoraz-Tan 1934: 50; cf.: Zuev 2009: 174, Note 43; expedition of 1759: RGADA. F. 199. No. 539. Ch. 2. D. 6. L. 85 ob.; F. Kh. Plenisner: AAN. F. 3. Op. 10. No. 137. L. 8 ob., 82.

[385] Quotation: RGADA. F. 199. No. 539. Ch. 2. D. 6. L. 38 ob.; cf. S. Kekerov's report: Ibid. L. 39 ob.–40.

> Krasnaya River at the reindeer crossings [Chukchi.—A. N.] met the Russian detachment, and upon meeting before the removal of the Anadyrsk stockade they voluntarily paid to the treasury part of the tribute without amanaty. And they, the Chukchi, did not perpetrate murderous death upon this detachment or to other Russian people, either on the Anadyr' or at the Krasnaya and other rivers, and nowhere in travels by dry land, even up to 775th year.

These encounters on the lower reaches of the Anadyr' also had an important economic significance: after payment of tribute the Chukchi could officially trade with the Russians; however, part of the trade was illegal. It is known, for example, that in 1761 in the Chekaevo tract the Reindeer Chukchi elder Khargipin traded the Eskimo girl Itteni to the serviceman Aleksei Shipunov for a copper kettle.[386]

In 1781, Corporal G. Sheikin generally recalled that time, focusing, however, on the non-peaceful character of Russian-Chukchi relations in 1750–1763: "Captain Shatilov and Lieutenants Kekerov and Bel'skii went every summer in shitiki to the Chukchi at the reindeer crossing and farther, and killed some of the Chukchi, and at the end the Chukchi always fled from the Russian people; however, they were not at peace until the arrival of Mr. Colonel Plenisner himself." Having analyzed the documents and oral depositions of the servicemen, Lieutenant Colonel F. Ch. Plenisner thus characterized in his detailed report to the Siberian governor F. I. Soimonov of 31 October 1763 the campaigns of the Anadyrsk detachment down the river:

> And though the former commanders here, Lieutenant Kekerov, Captain Shatilov, and Major Shmalev, reported to the main command, that the indicated detachment here allegedly turned to campaigns every year . . . in fact, those campaigns, which were actually against the Chukchi and Koryak, turned out precisely to be: in the 731st, in the 744th, and in the 746th years on the Chaun River by Major Pavlutskii against non-peaceful Chukchi, in the 747th against Chukchi on the Orlova River, where Major Pavlutskii was killed, in the 749th and in the 750th by Lieutenant Kekerov against the Koryak traitors, in the 751st by Captain Shatilov at Taigunovskii Nos against the same traitorous Koryak, and of all the actual campaigns from the

[386] Quotation: RGADA. F. 199. No. 539. Ch. 2. D. 6. L. 85 ob.; about Itteni, see: KPTs: 186; cf.: Belov 1954: 183.

> beginning of Anadyrsk there turned out to be seven. And the others, though they were reported by these commanders, should not be considered genuine because these commanders wrote and considered this for a yearly campaign when the indicated detachment went on ships to the Chekaevo tract to hunt at reindeer crossings and fish and only protect themselves from the Chukchi at that time cannot be mistaken for a campaign or anything similar.[387]

Thus, the lieutenant colonel, sorting out the situation, directly stated that all the summer expeditions of the Anadyrsk detachment down the Anadyr' River cannot be considered genuine campaigns from the military point of view; indeed, the Anadyr' people sailed for procurement of provisions, and the servicemen just defended the hunters and fishermen from possible Chukchi attack.

3.2. Abolition of the Anadyrsk Stockade

After the death of Second Major I. S. Shmalev, the Siberian governor F. I. Soimonov began to look for a commander at the Anadyrsk stockade. In 1759, the prime major of the Yakutsk Regiment, Fedor (Friedrich) Khristianovich Plenisner (c. 1711–1778), passed through Tobol'sk, going to the Military College in St. Petersburg, bringing to the capital his request for resignation based on conditions of health. In Tobol'sk, the capital of Siberia, F. Kh. Plenisner met F. I. Soimonov, who persuaded the major to be the head of Anadyrsk. On 21 July 1760, the Senate confirmed the proposal of the governor and, assigning Plenisner as the Anadyrsk commander, gave him the rank of lieutenant colonel on 3 January 1761. On 7 November 1760 F. I. Soimonov sent to the Senate a report, in which he proposed orienting toward the peaceful convincing of the Chukchi toward Russian citizenship, and on 20 February 1761 he gave Plenisner 17-point instructions with the task of inspecting the condition of Anadyrsk, its detachment, profits, and expenses, as well as to find a way toward pacifying the Chukchi. The lieutenant colonel left the capital of Siberia on 23 February 1761 with a company of soldiers from Tobol'sk and Selenginsk. On 6 January 1763 Plenisner arrived in Anadyrsk

387 First quotation: AII. F. 36. Op. 1. No. 643. L. 584; Nefёdkin 2015: 186; cf.: RGADA. F. 199. No. 539. Ch. 2. D. 6. L. 38 ob.; F. Ch. Plenisner's quotation: AAN. F. 3. Op. 10. No. 137. L. 46 ob.–47, 148.

accompanied by 31 officers and soldiers; later 164 servicemen also arrived. The total in the stockade came to be 687 servicemen.[388]

Carrying out the instructions assigned to him, by 23 February 1763 Plenisner had sent the first report to Soimonov, in which he placed in doubt the appropriateness of the existence of the Anadyrsk party. To get the better view of the situation in the region, Plenisner decided to set off himself to the lower reaches of the Anadyr' for procurement of provisions and an encounter with the Chukchi. On 28 July 1763, with a detachment of 247 servicemen in 13 shitiki, among whom was also the baptized Chukchi N. I. Daurkin, sent from Tobol'sk to Anadyrsk as translator. On 5 August, at the mouth of the Krasnaya River, the lieutenant colonel pitched camp, surrounding it with a barricade. Forty verst (42 km) from the Russians the Chukchi set up their camp, deploying 60 baidary, which brought 600 to 1,200 people. To Plenisner's camp went 35 baidary, in which came more than 300 Chukchi men, women, and children, the presence of the latter clearly indicating the peaceful intentions of the Chukchi. During the stopover, the lieutenant colonel was occupied with collecting information. Thus, on 5 August he ordered the translator Sotnik N. Kurkin to ask the elder of the Foot Chukchi Khekhgitit about the residents of the islands in Bering Strait and Alaska. On 6 August, the Chukchi elders Petutin, Menyakhta, and Anyku with three relatives appeared before the lieutenant colonel and asked permission to hunt reindeer on the Krasnaya River. The Chukchi gave tribute and reaffirmed their desire to be Russian citizens without providing amanaty. During the stop, one of the Chukchi elders closely approached the barricade of the Russian camp, and, in order to keep him away, an officer struck him with a stick. Since for the Chukchi a strike is a terrible insult, they regarded this as a challenge to battle, putting on armor and taking weapons in hand. Only the intervention of Plenisner himself saved the situation, which did not turn into a clash. On 19 August, at the request of the Chukchi trading was initiated: the Chukchi gave furs, oil, thongs, and walrus tusks for tobacco, beads, needles, bells, and thimbles. As a precaution trade was carried out

---

[388] RGADA. F. 199. No. 528. Ch. 2. D. 3. L. 20 ob.; AAN. F. 3. Op. 10. No. 137. L. 118. (number of the detachment in 1763); Senate Archive. Vol. XI. 1904: 407; Sgibnev 2005: 81–82; Gol'denberg 1979: 171, 173; Zuev 2005a: 57–58; 2009: 176–182.

from behind the barricade—the servicemen did not leave the camp. On 20 August Plenisner sailed in three ships back to Anadyrsk, leaving a large part of the detachment (200 men) led by then still lieutenant Ya. M. Peresypkin in three places to hunt reindeer. On 1 September, the lieutenant colonel arrived in Anadyrsk. Without the chief commander, the servicemen had continued to conduct trade with the Chukchi. Corporal G. Sheikin recalled one incident:

> . . . in a dispute a Chukchi trading with one Russian man a mask for a small knife and, having quarreled between themselves, then the other Chukchi began to climb over the barricade to the Russians, but the Russians did not let them get there; then the Chukchi having dressed and quite ready for battle, cried to the Russians, "Tok-tok!" that is, "Let's play!";[389] however, lieutenant Peresypkin prohibited that anyone shoot, after which the Chukchi also walked away, not doing any harm.

On 2 September, the lieutenant terminated the hunting and on the 12th returned to Anadyrsk, bringing a thousand carcasses of procured reindeer. Before this, during the return, on 4 September, N. I. Daurkin fled to the Chukchi from the camp in the Chekaevo tract.[390]

Having become acquainted with the situation in Anadyrsk by means of personal observations, interviewing witnesses, and studying the archival documents, Plenisner compiled with German thoroughness an extensive detailed report (31 October 1763) for Soimonov, in which he gave answers to the question posed by him. Having considered the history of Anadyrsk and telling about his activities there, Plenisner formulated his main conclusion, namely, that it was necessary to abolish the stockade and disband the Anadyrsk party. The lieutenant colonel argued his position from the following circumstances: because of the remoteness the party was difficult to supply with provisions, which is why a large part of the campaigns of the servicemen (except seven—in 1731, 1744, 1746, 1747, 1750, and 1751 against the Chukchi and Koryak) were simply expeditions for procuring provisions for themselves; there was no one for the

---

389 Took (tog`ok or tok) is an imperative interjection corresponding to the Russian "Nu!" ["Well"], and meaning "Let's go!", "Let's start!" (Moll and Inenlikei 2005: 143).

390 RGADA. F. 199. No. 528. Ch. 2. D. 3. L. 28 ob.–29 ob.; No. 539. Ch. 2. D. 6. L. 85 ob.–87; Zuev 2009: 183–187, 364; depositions of Khekhgitita: KPTs: 184–185; quotation of G. Sheikin: AII. F. 36. Op. 1. No. 643. L. 584 ob.; Nefëdkin 2015: 186.

detachment to collect tribute from: the Yukaghir had been killed by the Chukchi, and the Koryak migrated to the south; the Koryak were able to defend themselves from the Chukchi by means of fortifying their camps, and a Russian detachment was not necessary for this; the Chukchi, because of the lack of governance among them and the remoteness of their places of habitation, did not need to be brought into citizenship by force; the Anadyrsk stockade itself did not bring in any income: based on Plenisner's calculations, every 1,000 rubles of income turned around to 47,372 rubles 50 kopecks expense (to the treasury and tribute population), and the total cost, based on documents for the known period (1710–1764), of maintaining the stockade and local population cost the treasury 1,381,007 rubles 49.5 kopecks. Plenisner proposed instead of the Anadyrsk stockade to use the Gizhiga fort, which was located in a more favorable environment and to which it was more convenient to bring provisions and goods by sea from Okhotsk.[391]

Receiving Plenisner's first, preliminary report of 23 February 1763, Soimonov, who was now in charge of Siberian affairs in the Moscow Department of the Senate, was inclined toward the idea of withdrawal of the bulk of the servicemen from Anadysk. The department sent Catherine II its report, based on which the Senate on 5 March and the Empress on 15 March 1764 ordered the Anadyrsk party generally abolished. This decree was sent to Plenisner on 4 May 1764. Plenisner's second detailed report, of 31 October 1763, together with several other Anadyrsk documents, also was sent to Soimonov. On 25 May 1764, the Moscow Department of the Senate again prepared the relevant report about abolition of Anadyrsk for the Empress, who approved it at a session of the Senate on 11 June of this same year. On 25 June 1764, the Moscow Department sent to Plenisner this decree, which granted him the rank of colonel and ordered him to withdraw the military from the stockade, however, leaving the civilians there. On 28 October 1766, in response to another of Plenisner's recommendations, supported by the new Siberian

---

[391] AAN. F. 3. Op. 10. No. 137. L. 1–71; cf.: Vdovin 1960: 46; Zuev 2009: 187–191. Regarding the conquest of the Chukchi, F. Ch. Plenisner wrote: ". . . for eradication of the Chukchi (based on the great difficulty through considerable distance to them through empty and treeless places during the former campaigns) and forcing them into citizenship does not provide any benefit" (AAN. F. 3. Op. 10. No. 137. L. 62–62 ob. = 169).

governor Major General D. I. Chicherin, a new imperial decree followed, which ordered the church to be dismantled and the stockade burned, that is, abolition of the settlement in general.

On 13 April 1764 Plenisner left Anadyrsk for Gizhiga accompanied by part of the detachment, leaving Lieutenant Yakov Merkur'evich Peresypkin as commander of Anadyrsk. Colonel Plenisner himself was head of the Kamchatka Region, to which both Anadyrsk and Gizhiga were subject, from 1765 to 1772. On 11 March 1765 Plenisner, being already in Okhotsk, received the first decree of the Senate (of March 1764), and on 11 July 1765, located in the same port—the second, then on 1 August 1767—the third. These decrees, along with Plenisner own orders, were sent to Lieutenant Ya. M. Peresypkin in Anadyrsk to be carried out.[392]

Though the stockade was in the phase of abolition, the Anadyr' people continued their annual summer hunting expeditions down the river. On 2 August 1764 Peresypkin set off to the Krasnyi Yar tract in 18 shitiki with 130 soldiers and Cossacks. They arrived at the place on 7 August. Then on 13 August N. I. Daurkin appeared there accompanied by the Chukchi elders Menyakhta, Nykangin, and Nivigin and their relatives. On 14 August, the lieutenant sailed in twelve ships to the mouth of the Krasnaya River to collect tribute, where, fearing attack, he made camp with a fortification of boards surrounded by a barricade. At the ensign's summons the Chukchi once again agreed to pay tribute, but not to give amanaty. Instead of giving them, the Chukchi proposed in subsequent years (1765–1767) that the Russians take their women as wives, which, incidentally, only one Cossack had done; also, the Chukchi refused to stop their raids on the Koryak. Peresypkin, accepting tribute from 77 people, arrived back at the Chekaevo tract on 16 August, where he stayed for ten days to hunt reindeer. On 27 August, having killed 650 reindeer, the detachment sailed back, arriving in Anadyrsk on 9 September 1764.[393]

Captain Ya. M. Peresypkin told about his summer expeditions in subsequent years in the following way:

---

[392] RGADA, F. 199. No. 528. Ch. 2. D. 3. L. 31 ob.–32; D. 9. L. 52; Vdovin 1960: 47; Zuev 2009: 192–198.

[393] RGADA. F. 199. No. 528. Ch. 2. D. 3. L. 29 ob.–30 ob.; Divin 1979: 323–324; Zuev 2009: 200–201.

In the 1765th year on the 2nd of August the indicated Captain Peresypkin set off from Anadyrsk by water in thirteen ships and one hundred fifty men, and on the 6th arrived in the Chekaevo tract, where the camp was located. And on the 8th the Chukchi elder Ivigin Inpinev arrived in two baidary with relatives, from whom tribute of eight red foxes was taken. He announced this: "The best elder Menyakhta, with relatives of many people, resides at the mouth of the Krasnaya River, and awaits the arrival of the Russian detachment for payment to the treasury of tribute." That is why on the 9th of August, he [Peresypkin] went with the whole detachment for a collection to the mouth of the designated Krasnaya River, where on the same day he arrived. There, from the willing Chukchi, on the 10th, 11th, 13th, 14th, and 15th days of the 765th year, tribute was collected for the treasury of red foxes from seventy-three one each, and from five two each (for a total of eighty-three), and six blue Arctic foxes from four men, and one white Arctic fox from one. And on the 16th the detachment set off to Chekaevo to hunt reindeer for food, where from the 17th to the 19th seven hundred fifty were procured. On the 21st he went to the fort and arrived safely on the 30th of August.

In the 1766th year on the 31st of July Captain Peresypkin set off from Anadyrsk by water in ships with a regular military and non-military detachment numbering a hundred forty men and met Chukchi elder Kherguv'yu with relatives in two baidary on the 2nd of August at the mouth of the Belaya River, who stated: He went to the fort for a visit; and the other Chukchi are at the mouth of the Krasnaya River in sixty baidary, who want to pay tribute. From where he [Peresypkin] went to the Chekaevo tract and arrived on the 5th, and on the 8th went with the whole detachment to the mouth of the Krasnaya River, and arrived on the same day. And being there, took from willingly arriving Chukchi on the 9th for the 766th year tribute of fifty-eight red foxes and eight blue and four white Arctic foxes. And on the 10th he went to the Chekaevo tract to hunt reindeer for food for the detachment, and procured four hundred. And on the 26th he set off to Anadyrsk and arrived safely on the 8th day of September.

In the 1767th year on the 3rd day of August the frequently mentioned Captain Peresypkin went from the former Anadyrsk by water in eleven ships with a hundred forty-nine men in the detachment and arrived on the 10th at the Chekaevo tract. And on the 11th, upon arriving at the mouth of the Krasnaya River, he collected from the Chukchi people for the

> treasury tribute of ninety red foxes, one marten, and fourteen blue Arctic foxes and two white ones. From where on the 18th he went to Chekaevo and arrived on the same day, and on the 25th he returned to Anadyrsk and arrived safely on the 1st day of September."[394]

In the summer of 1768, as a consequence of the gradual withdrawal of the detachment from Anadyrsk, no voyages for the purpose of hunting for the detachment were carried out. Only two ships of Anadyr' people sailed down the river and conducted trade with the Chukchi. Three guns were returned to the Russians. The guns had been seized by the Chukchi in the spring of the same year from the small group of sotnik Stepan Belyaev in the region of the Oklan River with a promise of return. In 1770, the Anadyr' people again sailed for new trade. The Chukchi, with the encounter, and based on current tradition, brought tribute, which incidentally the Anadyr' people did not accept. This was a voyage under the pretext of hunting reindeer, since the frequent trips to the Chukchi were prohibited, and they did not have permission from the commander of the stockade.[395] Moreover, Captain Peresypkin, leaving Ensign Mikhail Klyusov, as the penultimate commander of Anadyrsk, left to Gizhiga with 52 Cossacks on 4 November 1769 with a caravan carrying office and artillery supplies.[396]

The gradual withdrawal of the military and civilian personnel from Anadyrsk occurred from 20 November 1765 to 3 March 1771. At Plenisner's direction all amanaty were released from the stockade, 142 servicemen were forced to resign, and of them 90 went to Yakutsk; a large part of the detachment and residents—193 soldiers and Cossacks (including all the veterans of the Chukchi wars) and 378 inhabitants—went to Gizhiga, the smaller part—44 Cossacks and 184 residents—went to Nizhnekolymsk. On 6 October 1770, the Spasskaya church was dissembled and the logs shoved into the water, while the priest Prokopii Trifonov went to Nizhnekolymsk with the church paraphernalia. The heavy weapons (flintlocks, grenades, cannonballs), including five cannons, were buried at Soldatova Channel. On 3 March 1771, the

---

394 RGADA. F. 199. No. 528. Ch. 2. D. 3. L. 30; cf.: Zuev 2009: 203.

395 RGADA. F. 199. No. 528. Ch. 2. D. 9. L. 48, note 2. L. 51 ob.; No. 539. Ch. 2. D. 6. L. 88.

396 RGADA. F. 199. No. 528. Ch. 2. D. 3. L. 20 ob.–21.

fortification of the stockade, the public buildings, and the 75 homes of the local residents were set on fire by the last commander of Anadyrsk Lieutenant Pavel Mordovskii. This day was the last day of existence of the stockade. Only one blockhouse remained in the place of the stockade.[397]

3.3. Defeat of the Chukchi at Gizhiga (1775)

On 2 March 1775, the commandant of Gizhiga Captain Ya. M. Peresypkin sent Sergeant Anisim Maklakov at the head of 20 servicemen and three Koryak to the Kamenskii Outpost, where the settled Koryak lived, for collection of tribute. They were joined by N. I. Daurkin with the task of negotiating with the Chukchi, if the latter were encountered on the way. On the morning of the next day, on Osinovaya Creek (a channel of the Paren' River), a meeting with a group of 150 Chukchi occurred.

Here's how N. I. Daurkin himself described this meeting and the following events in his "Zhurnal" ["Journal"], which was subsequently sent to the Irkutsk provincial office:

> And while traveling on the way they stopped in one place because of a storm. As the storm passed they saw a multitudinous number of Chukchi no farther from them than 50 sazhen [106 m.—A. N.], and those Chukchi, getting off sleds, having dressed themselves in armor, and taking up spears and bows and arrows, came against the Russian detachment, not even permitting the detachment to make a fort from the sleds, or to remain with the sleds in order; seeing this, he, Daurkin, in accordance with Sergeant Maklakov, with the Cossack Purgin, who knows the Chukchi language, went rapidly from the detachment then to meet those Chukchi, in order to stop them and hold negotiations, and meanwhile to give the Russian detachment time from the attack to make a fort from the sleds. And then, having run up to them, he called out in Chukchi. And they, the Chukchi, inviting him, stopped and placed their weapons in the snow, then moved back. Upon arriving to them, he, Daurkin, talked for a quarter of an hour, and in that time the detachment made a fort, and then he, Daurkin, brought to the fort 7 leaders from the main Chukchi men, and behind them others also came and stood about the fort.
>
> Upon the arrival of the Chukchi he, Daurkin, had negotiations with them through an interpreter about them being

[397] RGADA. F. 199. No. 528. Ch. 2. D. 9. L. 52; Vdovin 1960: 46–47; 1965: 134; Zuev 2009: 198–199.

subjects of Her Imperial Majesty and paying tribute. To what, they, the Chukchi, also promised to pay tribute in Anadyrsk, as it was before, when of the Russian detachment people would sail up along the Anadyr' River in boats. With this condition, the Chukchi said that they would live peacefully with all Russians and foreigners, only with the Reindeer Koryak they would never to live peacefully, because they, the Chukchi, have few reindeer.

Among these negotiations and the Chukchi promises, he, Daurkin, gave both the leaders of the Chukchi, and also the other Chukchi, an official Chinese ball [tobacco.—A. N.] and his own spice cakes and Cherkessian tobacco. Then he, Daurkin, on a promise the Chukchi about being in citizenship, invited them to the Gizhiga Fort, where they, in agreement with the whole Chukchi army, went with him, Daurkin.

On 9 March, still not reaching the Gizhiga Fort, they stopped at a standing Koryak camp, from where the best elders, Menyakhta with his brother-in-law and nephew, arrived at the Gizhiga Fort; and those elders reporting to Captain Peresypkin declared their desire to be citizens. Captain Peresypkin asked the elders for amanaty, to which from them, Menyakhta, stated: "If his son was here, he would leave him as an amanat, but about the other Chukchi who arrived with him, he, Menyakhta, did not know whether the elders would give". However, for his part, this Menyakhta left his nephew in amant. With these negotiations Peresypkin gave those elders tobacco and regaled them with wine.

And this same day Peresypkin sent him, Daurkin, with the Chukchi elders to their camp for negotiations again about being in citizenship and about payment of tribute, to which they also had agreed; and at this, they requested to see their amanat left in the fort; and if he is alive, then they the Chukchi in truth will believe and can give another.

And meanwhile Captain Peresypkin arrived at the Koryak camp and brought the amanat that remained with him, and showing him to the Chukchi, asked for another—but they requested a little time and to go feed the reindeer—and for the elders to arrive in camp on foot for negotiations. But Peresypkin, without giving a truce, since he was drunk, began with anger and threat to announce to the Chukchi elders: If they do not give another amanat, then all will be killed. This is why the Chukchi began to leave, and Peresypkin sent a detachment and Koryak in pursuit, who began to shoot with guns at the

> Chukchi; but the Chukchi first bowed to appease, but then they also armed themselves and engaged in battle, and up to 50 of the Chukchi people were killed, and up to 40 women and children were taken captive.
>
> Moreover, as interpreter Cossack Pavlov translated for Captain Peresypkin the negotiation the Chukchi deliberately, that is, not what the Chukchi said; as the captain ordered to strike the Chukchi, he, Daurkin, much proposed so they won't be struck, since then they would retaliate."[398]

Such is N. I. Daurkin's version of events. However, clarifications of other members of the events, Captain Ya. M. Peresypkin and Lieutenant P. Mordovskii, permit making substantial additions to the above-mentioned statements: At the beginning of the winter of 1774 a group of Chukchi left the region of Kresta Bay. Along the way, it destroyed a Yukaghir camp, then moved toward Kamchatka, but because of deep snow it turned southwest toward the Koryak Kamennyi outpost, where two Chukchi were detained, who had gone to their relatives. The Chukchi began to search for the Reindeer Koryak, in order to take their reindeer for provisions, but encountered A. Maklakov's detachment. Through N. I. Daurkin's negotiations with his uncle Menyakhta and other elders the Chukchi and Maklakov's detachment peacefully parted and together with Daurkin went to Gizhiga. They set up camp 200 m from the Koryak camp, which in turn, was located 1.5 km from the Gizhiga Fort. In the Koryak camp was a detachment of a hundred servicemen led by Lieutenant Pavel Mordovskii. It had been sent there to defend the Koryak in the most dangerous time of winter. Seeing the Chukchi, the Russians and Koryak left the camp and began to prepare for battle, but Daurkin explained to Mordovskii that the Chukchi came in peace in order to accept Russian citizenship. The Chukchi, nevertheless, managed to drive off ten draft reindeer from the Koryak camp. During subsequent negotiations in the Gizhiga Fort with the commandant Captain Peresypkin, the Chukchi gave up one amanat, the nephew of Menyakhta, and they promised to give another later.

Arriving in the Chukchi camp, the captain, already being drunk, demanded giving a second hostage. The Chukchi requested time to think and, in addition, wanted to go to the pasture for reindeer a half kilometer

---

398 "Dve zapiski. . ." [Two notes. . .] 1873: 365–366.

from the Koryak camp. Peresypkin began to threaten the Chukchi, who started to retreat to the pasture. The captain ordered to begin battle, fearing an attack on the Koryak herds, which were at pasture on the same side. Mordovskii's detachment of 139 men with a cannon followed the departing Chukchi. The Chukchi began to prepare for battle. The lieutenant sent for Daurkin in order to continue negotiations. But at this time the Koryak began to shoot their guns without being given the order; the Chukchi shot from bows. After a volley from the guns the Chukchi threw themselves on the ground, in Daurkin's statement—in order to show their peaceful intentions, but based on indications of the interpreter S. Pavlov—in order to avoid losses from bullets. Having shot from the cannon, the detachment began to advance, the Chukchi jumped up, threw down their armor and bows and fled. The Koryak Prince Imleko with the servicemen pursued the Chukchi and gave battle, but was defeated by the Chukchi. Based on oral information collected by T. I. Shmalev, in this pursuit 20 men died (including the prince) and 18 suits of armor and several guns and spears were lost, and a few reindeer driven off. According to Peresypkin's report to the Kamchatka commander M. Bem, two Russians were killed in battle and twelve servicemen and Koryak wounded, whereas the Chukchi lost 54 people killed and 40 taken captive.[399]

Thus, the destruction of the Chukchi at Gizhiga on 9 March 1775 occurred in general as a result of a combination of circumstances. The Chukchi themselves arrived to adopt Russian citizenship, to agree to provide tribute, and even to give up amanaty, but due to the hasty actions of the drunken commander of Gizhiga, Ya. M. Peresypkin, were crushed. This event soon became known to the central authorities: the report of N. I. Daurkin of 29 September 1775 was sent to the capital to the highest name. On 26 January 1776, a decree from the Senate to the Irkutsk office followed with the requirement to examine the case.[400]

---

[399] About the battle at Gizhiga: RGADA. F. 199. No. 528. Ch. 2. D. 9. L. 4, 51 ob.; No. 539. Ch. 2. D. 6. L. 56–57 ob., 90–91 (losses in the retreat of the Koryak); Mamyshev 1809: 23–25; Shakhovskoi 1822: 305–306; Sgibnev 2008: 18; "Dve zapiski. . ." 1873: 360–370; Alekseev 1961: 61–62; Vdovin 1965: 97 (about the encounter with A. Maklakov); Zuev 2009: 217–230, 291.

[400] "Dve zapiski. . ." 1873: 361; Zuev 2009: 221–230.

4. Adoption of Russian Citizenship by the Chukchi

On 1 July 1776, being in service in Okhotsk, Captain T. I. Shmalev received an Imperial decree to investigate the Chukchi massacre of 1775 and to bring the Chukchi into Russian citizenship. On 7 August, the captain, together with N. I. Daurkin, arrived in Gizhiga. In 1778, having conducted the investigation, T. I. Shmalev found Ya. M. Peresypkin and the interpreters guilty of the massacre of the Chukchi. Upon his report of 13 July 1781 Catherine II ordered Peresypkin and P. Mordovskii demoted and dismissed from service, and the translators to exile for a year of penal servitude, which was carried out. This was the first order fulfilled. In parallel Shmalev also executed a second task.

Arriving in Gizhiga, Shmalev learned that the Chukchi woman Kytna, widow of the Koryak Akhiika Itchalina, mother of three grown sons, who lived among the Reindeer Koryak in Paren', wants to reconcile the Chukchi and Russians. Since Shmalev received through the Koryak a rumor that the Chukchi were preparing for a winter a raid on the Koryak and on the fort, so it had been decided to prevent the event in an assembled council of officers and Koryak elders on 28 November 1776 and to send Kytna to the Chukchi with her son and three more captured Chukchi women and Koryak for negotiations regarding accepting Russian citizenship. In answer to this embassy, in the fall of 1777 the head Chukchi toyon Amulyat Khergyntov gathered a council of Chukchi elders, in which it was decided to accept Russian citizenship, but the Chukchi wanted to turn over the tribute, as formerly, in Anadyr'. On 22 February 1778 Captain Peresypkin, who gave the order to attack the Chukchi in 1775, was removed from command of the Gizhiga Fort, and a triumvirate of three sergeants was installed in his place. On 25 February 1778, the caravan of the chief Chukchi toyon Amulyat stopped a verst from Gizhiga. Within the Chukchi delegation, which numbered 22 people, besides Amulyat, there were the toyon Aetkim from the Arctic Ocean and the toyon of the Foot Chukchi Cheikhulyan. On 28 February negotiations were conducted in the fort. On 3 March, in Gizhiga, the chief toyon solemnly accepted Russian citizenship with payment of tribute and giving of amanaty. The agreement was recorded in the office the next day; nine Chukchi brought tribute in the sum of 13 rubles 20 kopecks, and in response gifts costing 32 rubles 25.5 kopecks were handed out to them. In all, the Chukchi left six amanaty. The Russians returned 28 captives to the Chukchi, in exchange they were supposed to return Koryak prisoners. On 11 and 12 March the Chukchi left, accompanied by Kytna and her son Tekkev and the Koryak elder Kaikip Ivakin, who were assigned a reconnaissance mission.[401]

On 8 March 1779 Kaikip and Tekkev returned to the Gizhiga Fort and with them arrived two Chukchi toyons: the same Aetkim and Khegruv'ya sent from the Foot Chukchi by Amulyat, with ten accompanying men. The Chukchi stopped at the Koryak camp three verst from the fort. Those who came confirmed the need to comply with last year's agreements, and returned ten captured Koryak. In addition, nine Chukchi (including Amulyat, who sent a parka in absentia) gave tribute of fox furs in the sum of 11 rubles 60 kopecks and received gifts worth 1 ruble 75 kopecks. Having left amanaty, the Chukchi departed on 23 March. In 1781 Amulyat again arrived in Gizhiga, in 1783 Khamakhai came to a meeting with the Russians, and in 1784 Amulyat, Aetkim, and Khegruv'ya arrived. In March 1788 about 40 Chukchi, including fifteen or sixteen women (including captives) and as many children, went for a meeting with their relatives in Gizhiga, and on the way, they encountered the French envoy J.-B. de Lesseps (1766–1834). Then the Chukchi came to Gizhiga only in 1794 and in 1798.[402]

The Kamchatka commander M. Bem treated Shmalev's report about the induction of all the Chukchi into citizenship in 1778 skeptically, though the central government supported Shmalev's actions. According to an imperial decree of 11 November 1799, the Chukchi were released from tribute for 10 years because of an epidemic and a reindeer die off; they were permitted to conduct unhindered trade with the Russians; Chukchi elders were given plaques with the Russian coat of arms, which indicated that this territory belonged to Russia, to set up near certain places on the coast.[403] The coats of arms were handed out in 1781 to Amulyat. Properly speaking, the annexation of Chukotka to Russia can be dated to the time

---

401 RGADA. F. 199. No. 539. Ch. 2. D. 6. L. 1–17 ob.; Alekseev 1958: 55; Vdovin 1965: 137; Zuev 2006a: 26–37; 2009: 226–244.

402 RGADA. F. 199. No. 539. Ch. 2. D. 6. L. 41–53 ob.; Lesseps 1801: 101–111; Zuev 2009: 244–251.

403 Arkhiv Gosudarstvennogo soveta [Archive of the State Council]. 1869. Vol. 1. Pt. 2: 259–260; Alekseev 1961: 63; Zuev 2009: 242–243, 248.

of publication of this decree. Although the toyons, who agreed to Russian citizenship, were able to answer only for their clans, the other Chukchi under their influence nevertheless also gradually began to adhere to these agreements. The definite haste on the Russian side in adopting the resolutions was clearly dictated by the fact that European penetration into the region was begun and territories could go to other European powers.

5. Russian-Chukchi Relations at the End of the 18th Century

After formal accession of the region to Russia a secret geographic and astronomical expedition (1785–1794) was organized and led by a shipmate of J. Cook, Captain of 1st Rank J. Billings (ca. 1758–1806). The task of the expedition was the investigation of the newly adjoined land. Therefore, it was necessary to notify the Chukchi in advance about the peaceful character of the expedition. In 1786, Captain Shmalev met with the Chukchi at the Kamennyi outpost (fort) of the Koryak and reported to them about the arrival of an "embassy" from the Empress. In 1788, the Yukaghir Prince Sozykin, of the Khodynsk and Chuvan clans, and the Gizhiga Koryak Petushkov were sent on behalf of the Zashiversk police chief I. Banner to the Chukchi, including to convince them of the peaceful character of the expedition. T. I. Shmalev himself (1787), as well as the translators N. I. Daurkin and I. Kovelev, participated in the maritime part of the expedition, which contributed to the success of the undertaking. At the end of the expedition Billings and a few companions, accompanied by Chukchi, crossed the Chukchi Peninsula on land (in winter 1791/92).[404]

After official permission to trade with the Chukchi, active development of commercial connections ensued, that is, the Chukchi gradually became economically linked to Russia. After cessation of raids on the Koryak, the Chukchi were in extreme need of tobacco and metal items, and for their part, they were able to offer furs, walrus tusks, and baleen. By 1781 the so-called Chukotsk fair began to operate. It took place in February in a field 100 to 250 verst northeast of the Kamennyi stockade,

[404] Sauer 1802: 69; Slovtsov 1856: 19; "Dve zapiski. . ." 1873: 371; Grenader 1975: 94–95; Divin 1979: 398–399.

at the mouth of the Penzhina; the turnover of the fair reached 25,000 rubles.[405]

Another place of annual meeting of the Russians and Chukchi became the Angarka River—a tributary of the Bol'shoi Anyui. In 1788, the Zashiversk local government police chief Ivan Banner, to whom the Kolyma Oblast' [district] was then subject, sent the Yukaghir Prince Sozykin and the Koryak Petushkov to the Chukchi, in order to invite the Chukchi to a meeting on the Angarka River in 1789 or 1791. From 1789 to 1794 the fair with the Chukchi took place in the small Angarsk Fortress, and then moved to an island in the Bol'shoi Anyui and was named Ostrovnaya [Island] fair; there it continued until 1848. The stockade at Ostrovnoe was occupied only during the time of the fair at the beginning of the year. The trade itself took place after the Chukchi gave tribute (at one fox pelt per man) and received Russian gifts. To this trading came the northern Chaun Chukchi who, in particular, lived closer. By 1788 Banner introduced the tradition of granting Chukchi toyons scarlet tunics and daggers with the imperial monogram on the belt, and in 1791 they began to give neckbands with silver and copper medals to toyons, that is, the elders now stood out by their status insignia of distinction. In this same year, at the decision of the Irkutsk office, Khamakhai became the chief toyon, receiving a special document confirming his status. In February 1791, two Nose toyons and twenty prominent Chukchi arrived at Ostrovnoe, and a total, as Banner counted, of 581 people were brought into citizenship. In 1794, 307 Chukchi arrived at the fair, including four toyons and two American Eskimos; in the following year—six toyons and 62 more people; in 1796—the same number of toyons and 120 Chukchi; a year later—70 people; and in 1799—five toyons and 277 men. After this, reports of the Irkutsk office about the fair are no longer received in the Senate; relations with the Chukchi occurred entirely within the competence of the local authorities.[406] From this time the fair at Ostrovnoe was the primary location of Russian and Chukchi meetings.

---

[405] Shakhovskoi 1823: 38–39.

[406] KPTs: 188–189; Arkhiv Gosudarstvennogo soveta. Vol. 1. Ch. 2. 1869: 260; Slovtsov 1856: 19–25; Sgibnev 2008: 25; Argentov 1886: 12; Bogoras 1902: 154; Bogoraz-Tan 1934: 53–55; Vdovin 1965: 138–141; Divin 1979: 399; Zuev 2009: 250–253; 2009a: 86–87.

Relations between the Russians and Chukchi developed basically on a peaceful course, though excesses nevertheless happened. In 1785, the Kamchatka commander F. von Reineke sent Cossacks to the Penzhina River in search of a favorable place for construction of the city of Aklansk. However, the prospect of construction of a Russian fort in their territory did not please the Koryak, who persuaded the Chukchi, led by toyon Khegruv'e, who was visiting the Koryak, to attack the Russian caravan, which consisted of nine people, and seize the booty of tobacco and iron implements. As a result, five Cossacks were killed and three women and a juvenile were taken captive, which brought condemnation among the northern Chukchi. The generally complex situation in the region at the end of the 18th century was characterized by the exiled Polish brigadier Iosif Kopets (1758–1827). Traveling from Kamchatka to Gizhiga at the end of 1796, he noted that he "had to go near the land of the Chukchi, who constantly attack and rob Muscovites and Siberians."[407]

---

[407] 1785: Lesseps 1801: 95, 101–103; Sgibnev 2008: 24–25; cf.: Chernenko 1957: 136; EM: 164 (in 1784 the Anadyrsk Chukchi killed the "retired Cossack Dolgopolov and comrades" on the Penzhina River); quotation: Kopets 1896: 1004.

CHAPTER 5

INTERTRIBAL RELATIONS IN THE 18TH INTO THE FIRST HALF OF THE 19TH CENTURY

1. Chukchi-Koryak Wars

1.1. The First War

In the 18th century, along with the regular Russian opponents of the Chukchi, were the Koryak, with whom over the course of almost the whole century there was war for possession of herds of reindeer and women. The Chukchi-Koryak hostility has roots in the pre-Russian past, possibly in the time of separation of the two peoples. I. S. Vdovin suggests that the Chukchi and Koryak were separated in the 14th–15th centuries by a Yukaghir wedge along the Anadyr', after which two separate peoples were formed. However, it is quite clear that the thin Yukaghir population could not break the connection of the Chukchi with the Koryak. In the mid-17th century B. O. Dolgikh counted fewer than 5,000 Yukaghir in the territory from the Lena to the mouth of the Anadyr'.[408]

According to legends of the North Kamchatka Koryak, there were two wars with the Chukchi. N. N. Beretti provides these stories: "The legends, told to me by old natives, say that the earliest inhabitants of the region located north and northeast of Gizhiga were the Koryak. The Chukchi arrived here much later, evidently from the north. At the time of their arrival the Koryak aleady possessed large herds of reindeer, which the Chukchi did not have.

[408] Vdovin 1972: 109; 1987: 45; 2010: 513–514; pro: Leont'ev 1976: 177–178; Lebedev and Simchenko 1983: 97; contra: Vol'fson 1991: 35–37; Nefëdkin 2003: 131; about the number of Yukaghir: Dolgikh 1960: 554.

The Chukchi attacked the Koryak for the purpose of taking their herds and prevailed over them as the consequence of the unpreparedness of the Koryak for military activities." This war is attested by the name of the tributary of the Talovka River—Imlana ("Broken"), where the arrows and spears of the enemies were broken. Having lost the war, the Koryak began to train their young, and in particular to teach the young people to artfully dodge arrows. "The second Chukchi-Koryak war arose again at the initiative of the Chukchi, who evidently were not satisfied with the herds they took in the first war. This war took place during the time of the first appearance of Russians and foreigners in Kamchatka, since some Koryak at that time already had firearms; the Chukchi did not have them and had never even seen them. The second war ended quickly, the Chukchi were defeated. The main reason for their defeat was panic caused by the firearms. The old natives say that the Chukchi believed the shots from the guns were thunder in the heavens and the wounds from the bullets injuries from lightening."[409]

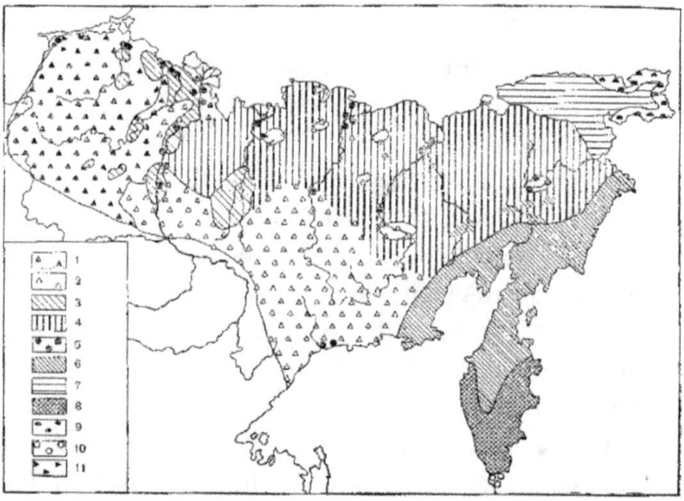

Figure 30. Settlement of the peoples in northeastern Siberia at the end of the 17th century. 1—Tungus (Evenk); 2—Lamut (Even); 3—Yakut; 4—Yukaghir; 5—Russians; 6—Koryak; 7—Chukchi; 8—Kamchadal (Itel'men); 9—Eskimos; 10—Kuril'tsy (Ainu); 11—Tavgiitsy. (Reproduced after Gurvich 1966: 267).

Beretti translates information of a Koryak historical legend applied to local conditions, and along the way inserting his remarks in it

[409] Beretti 1929: 5–7. About the first Chukchi-Koryak war: Neiman 1872. No. 1: 34.

with a significant degree of confidence, which, of course, should be distinguished from the general flow of information. It turns out that the Koryak were the autochthonous population in the Gizhiga District and in northern Kamchatka, where the Chukchi arrived much later. During the Koryak-Chukchi wars, in the 1750s–1770s, only short-term raids were carried out in these territories, and the Chukchi began to resettle to the south across the Anadyr' only after the abolition of Anadyrsk in 1771. Consequently, it can be supposed that this is about the resettlement of the Chukchi to the region. In the legend, it is noted that the Koryak already possessed large herds, which the Chukchi did not have and which served as the reason for war. Based on G. F. Müller's report, at the end of the 1730s the largest herds of reindeer in Siberia were possessed by the Nenets-Yurak and the Chukchi, who had herds numbering 200 to 300 head. It seems that the story talks about an earlier period of history, as Beretti himself thinks, judging by a further narration. J. Lindenau also notes that before the arrival of the Russians "the Koryak were occupied in large part by robbery and waging war on the reindeer Tungus, Lamut, and Chukchi."[410]

The second war played out through one or several generations, when the Koryak could teach their youth the military arts. The use of firearms serves as a rather clear chronological indicator. The large number of firearms that appeared among the Koryak could scarcely be viewed as trophies taken from the Russians in the course of uprisings from 1745 to 1757. They most probably came to the Koryak legally. Indeed, in June 1763 the ban on the sale of firearms to tributary peoples was officially removed, even though later powder and lead were sold under the guarantee of the chiefs.[411] Among the Chukchi themselves guns began to spread only from the middle of the 19th century. Although the Koryak attributed their victory over the Chuchi to firearms, the end of the war with the Chukchi (at the end of the 1770s) was not due to the effective use of them by the Koryak.

---

[410] About the southern, Vaezhskie Chukchi: Vdovin 1962: 154; Nuvano 2008: 81; about the herds: Müller 2009: 224; quotation: Lindenau 1983: 105; cf.: Neiman 1872: 34. In the first quarter of the 19th century among the poor Chukchi there were up to 100 reindeer, among the rich—to 1,000 (Kiber 1827: 197; Vdovin 1987: 50).

[411] Alekseev 1996: 62.

Firearms produced a large psychological impression on the Chukchi. However, it was evidently not the Koryak who used them on the field of battle, but the Russians, as was done, for example, at Gizhiga in 1775. At the end of the 18th century, according to C. Merck, the Koryak used guns only for defense, for shooting from the yaranga.[412]

The Ural mining engineer and writer N. R. Mamyshev (1777–1840) also gives an artistic retelling of the legend. He received his information from the son of the commandant of Gizhiga (possibly the son of Captain Yakov Merkur'evich Peresypkin) and described the wars of the Koryak with the Chukchi in this way:

> It is not known precisely why and when hostility began among these two peoples; it might be that both leading a nomadic life, there was a dispute over land; reliable only is that their hatred was irreconcilable, so to speak, with the blood being transmitted from father to son, from generation to generation, from clan to clan. Neither the Chukchi nor the Koryak presently know its reason, but are bitter enemies of each other.
>
> . . . Being everywhere pursued by their hostile enemies, the Koryak decided to give a last desperate battle, which they still tell about with horror even today: but their furious bravery was not able to overcome the multitude of enemies and served only that they all fell . . . in the place of battle; scarcely a few dozen families were left of this once numerous people. . ..
>
> These few remaining departed into the impenetrable dense forest, yielding their sweet homeland to their bloodthirsty pursuers. Much time passed; finally, they increased and began to be here and there with their herds. But the Chukchi did not forget their enmity, and everywhere afflicted them and everywhere pursued them as a true prey. At this time, the Russians appeared near their possessions. The Koryak were joyous for defense by the winners of Siberia. The Chukchi were humble before their powerful patrons. But hatred and vengeance seethed in the hearts of both. The Chukchi, relying on their strength, the Koryak, on the strength of their defenders, often, where possible, attacked each other, chased off the herds of reindeer, and took captive the wives and children of the enemies.[413]

---

[412] Merck 1978: 120.
[413] Mamyshev 1809: 21–22.

According to this already elaborated variant of the legend, the events of the first war appear this way: the Chukchi attacked the Koryak, forcing them to accept general battle. The battle, because of the numerical superiority of the enemy, was lost by the Koryak, in spite of their bravery in battle. The remaining Koryak retreated into the forests, that is, to the south of the Anadyr', into the forest zone. But after the lapse of a long time the second war occurred, when the Koryak had regained their numbers. Again, clashes began to emerge between the nomads, and the Chukchi were again the attacking side. At this time, the Russians appeared in the region and took the Koryak under their protection, with which the Chukchi reconciled themselves.

Again, the time of the first war in the narrative of Mamyshev is vague. Indeed, by the end of the 18th century even the Koryak and the Chukchi themselves did not remember its beginning. The second war was conducted by the Koryak with the aid of the Russians, who actually possessed firearms, which clearly points to events of the 18th century. This version of the legend looks more plausible than the legend recounted by Beretti.

The Koryak story "The Settled People and the Reindeer Herders" also probably belongs to this first war, at the beginning of which it says that all the Chavchuven were annihilated as a result of the wars, with the exception of seven nomad camps. This generally correlates with Mamyshev's information about the remaining few dozen families.[414]

In Chukchi folklore, as far as I know, information about this first war with the Koryak is not clearly presented. However, in the legends told by Rol'tynto from Meinypil'gino, descendant of Evyskyv (1980), a cousin of Kunlelyu, it says: "After the first war, when Kyyty lived, everyone was at peace. There was plenty of room. They lived well." If this fight is understood literally as the first war of the Chukchi with the Koryak, then it is possible to calculate the approximate chronology of events. One camp of Makchymkat, father of Vyttegin (Vytegyn, later nicknamed Kyyty), moved over the ice to summer pasture from the north to the south shore of the Anadyr' Estuary. Vyttegin predicted that it was necessary to turn back; part of the camp returned, while those remaining on the right bank of the

[414] Menovshchikov 1974. No. 149: 471–472; Lebedev and Simchenko 1983: 127–128; Nefëdkin 2003: 288–289.

Anadyr' were destroyed by the Tanity. Rol'tynto himself represented the tenth generation of Kyyty's descendants. Calculating, on average, one generation as 25 to 30 years, we determine that the event can be dated 250 to 300 years ago (1680–1730). The narrator believed that Kunlelyu was the grandson of Kyyty. And Kunlelyu's father was married to a Koryak woman, to whom he proposed rather then captured. Since Kunlelyu actively fought in the second half of the 1750s, the time of Kyyty's crossing, taking in consideration his youthful age, can be determined as about 1690. However, according to the careful genealogical calculation of Ivan Korav'e, the descendant of Kunlelyu's cousin Evyskyv (Eviiskiv), the latter was the grandson of Magchymkat and son of Kymtek.[415] In this case, the events of the story about Vyttegin are dated only to thirty years earlier than the activities of Kunlelyu, that is, to 1720, which seems less probable from the historical standpoint. It can be supposed that Makchymkat's crossing to the north shore of the estuary occurred at the height of the war. Therefore, the threat by the Koryak was entirely real. The events themselves can be dated to about 1700.

Researchers usually date the first Koryak war with the Chukchi to deep antiquity, before the appearance of the Russians in the region.[416] According to Koryak legends, the first Koryak war with the Chukchi was before the arrival of the Russians in the region, which can be supposed a time even earlier than when the "Yukaghir wedge" divided the two peoples along the Anadyr', that is, according to I. S. Vdovin, after the 13th–14th centuries, when the Yakut began to exert pressure on the Tungus, and the Tungus on the Yukaghir, forcing them to move to the east. But based on Koryak legends themselves, the time of the first war is not clear. If we turn to Chukchi folklore, the first clashes at the mouth of the Anadyr' can be dated to the boundary of the 17th and 18th centuries, when the Chukchi began to go to the south shore, crossing to summer pasture on the last ice. Indeed, it was only in 1695 that rather regular tribute collection from the

---

415 Lebedev and Simchenko 1983: 95–100 (Rol'tynto quotation—p. 98); cf.: Menovshchikov 1974: 298–299 (courtship of Kunlelyu's father); Leont'ev 1976: 152–154 (genealogy of I. Korav'e—p. 154); 1983: 129; Kytgaut 2003: 34; for a similar story about fording the Amguema in an attempt to escape the Russians, see: Bogoras 1900: 335–336.

416 Jochelson 1997: 223; Tan-Bogoraz 1930: 63; contra: Vdovin 1944: 259–260.

Koryak began, and they in this way came under the protection of Russian authority.[417] Thus, the third force appeared in Chukchi-Koryak relations.

It can be supposed that contact was not just in the area at the mouth of the Anadyr'. Even if the disputed identification of the Koryak discovered by M. V. Stadukhin during the voyage of 1649 is not considered, it is possible to think of several other possibilities for such Chukchi-Koryak contacts. In the formal reply of the Kolyma steward I. Dorofeev of 20 June 1686 to the Yakutsk governor it is mentioned that the tributary Yukaghir of the Nizhnekolymsk winter quarters requested servicemen sent against the Chukchi and Koryak oppressing them. However, the steward indicated in the document that there was no one to go in campaigns against the Koryak, Chukchi, and Khodyn people because of the lack of servicemen on the Kolyma. In 1689, the Cossack M. Zinov'ev, sent to three Kolyma winter quarters for tribute collection, explained to the Yakutsk governor that the shortage of tribute for the current year was because the tributary Yukaghir "were killed by the non-tributary Chukchi and Koryak." Thus, the Reindeer Koryak had appeared in the territory of the Kolyma winter quarters in the second half of the 1680s, probably when the number of the Yukaghir population had been reduced due to epidemics and military clashes. First, the Koryak threatened the Yukaghir of the Verkhnekolymsk, but, judging by the first formal reply, they reached much farther north, to the territory controlled by the Lower winter quarters. According to a petition to the Yakutsk governor from the Kolyma tribute collector, boyar's son Andrei Bulygin, in 1662 it became known to the latter that "on the Kovyma [Kolyma] River they live beyond the Kovyma at Kamen', migrate to the tributary Yukaghir, Anauli, and Koryak, and do not pay tribute, and with them come forty and fifty men in a clan."[418] Consequently, from beyond the mountains whole clans of Koryak migrated to the territory of the Kolyma Yukaghir, probably to winter camps. On 11 September 1739, the detachment of Lieutenant D. Ya. Laptev met at the mouth of the Indigirka a newly baptized Koryak, who was hunting there. If compared with later

---

[417] Dolgikh 1960: 425, 555.

[418] Document of 1686: AAN. F. 21. Op. 4. No. 32. L. 130–131; document of 1689: MIYa. Ch. 3: 994; petition of 1662: DAI. 1851. Vol. IV: 381; cf.: Ogorodnikov 1922: 268; Dolgikh 1960: 554, 562.

times, in the Chukchi territories the Koryak went along the rivers in the Chaun region, as they did in their campaigns in 1768 and 1769. In 1765, the Chukchi believed that the Koryak poisoned members of the Russian expedition with the juice of buttercup at the Chaun River. At the beginning of the 1790s, the boundary between the Chukchi and Koryak went along the Rauchuanskii and Ilirneiskii ranges, and the Reindeer Koryak themselves clashed with the Yukaghir arriving in summer on the Omolon to fish. Even later, in 1852, in A. I Argentov's Chaun parish 94 Chukchi families, Chuvan people, and "a small number of Koryak" (35 people) were counted.[419] Thus, it seems that precisely in this western contact zone there could have been clashes of the Koryak with the Chukchi.

1.2. Chukchi-Koryak Conflict in the 18th Century

Now we turn to historical events recorded in the Russian written sources. In 1699, the Koryak began to be listed as tributaries of the Anadyrsk department; they took active part in the campaigns of the servicemen against the Chukchi in 1702, 1730, 1731, 1744, 1746, 1747, and 1775. In this same 1699, the Koryak began to complain of Chukchi attacks. The Moscow historian G. A. Leont'eva, studying the servicemen of Eastern Siberia, unfortunately without indicating the original sources, writes that in the fall of 1699, at the request of the tributary Koryak from Anadyrsk an expedition was formed against the Chukchi who had arrived at the Anadyrsk stockade and wanted to drive off the Koryak's reindeer. The campaign was successful, and the Chukchi gave tribute and even provided amanaty.[420] As can be understood from the paraphrasing of the message, there was only the threat of driving off the reindeer from the Koryak, which, in fact, did not happen. The Anadyrsk servicemen acted sufficiently promptly in this case by sending out a detachment that brought the Chukchi in the Anadyr' area into formal citizenship by means of tribute acquisition and the taking of hostages, which at other times the Chukchi categorically refused to do. If this account is true, it is the first known

---

[419] Argentov 1857: 89; 1857b: 19; cf.: Sementovskii 1852: 143 (according to A. I. Argentov, in the Chaun parish were 101 "yurts"—730 people); 1739: Okhotina-Lind and Møller 2015: 362; 1765: Makarova 1968: 63–64, note 35; about the boundaries of the 1790s: Sarychev 1952: 259; fishing: EM: 27.

[420] Koryak of 1699: Dolgikh 1960: 425; Leont'eva 1997: 109; cf.: Zuev 2002: 67, 234; 2009: 43; Nefёdkin 2003: 24, note 2.

campaign in defense of the tributary Koryak, from whom the Chukchi could drive off the reindeer herds.

In the summer of 1716, according to depositions given in Anadyrsk by the Reindeer Koryak Beloglaz, the Chukchi arrived "with battle" onto the Pakhachinsk Koryak and slaughtered their reindeer,[421] but did not drive any off—in general, uncharacteristic behavior during a raid, which could be explained by some exceptional circumstances, in particular striving to deprive the enemy of an economic basis for existence.

Based on information from the newly baptized Yukaghir Stepan, at the first snow (September–October) of 1716 the Yukaghir of the Lower Khodynsk clan and the Chukchi went on a raid for Koryak reindeer. Since the Chukchi captured the edible fish supplies of the Yukaghir (that had been taken from the Aklansk Koryak), the Yukaghir abandoned their allies. On the Paren' River the Chukchi massacred the Koryak and stole the herds.[422] It is worth noting that in this campaign against the Koryak, the Khodyn Yukaghir of the Anadyr' came out as allies of the Chukchi, and probably showed the Chukchi the way to the Koryak.

In 1720, the Koryak again turned to Anadyrsk with a complaint about the Chukchi, who stole their reindeer. I. S. Vdovin is even of the opinion that it was from this raid that Chukchi raids on the Reindeer Koryak began,[423] though, as we see, the raids began earlier.

In 1725 the Reindeer Koryak, whose nomad camps were located near Anadyrsk, notified the Russians that many of them were killed by the Chukchi, and asked that 50 servicemen be sent to them for protection of their herds. And even some tributary Koryak seceded from the Russians, seeing the small number of servicemen and impunity of the Chukchi raids.[424]

In 1728, Koryak came running to Anadysrk and reported an attack by the Chukchi, who had killed part of the Koryak, destroyed eleven yarangas, drove off eleven reindeer herds, and took sleds and property.

---

[421] RGADA. F. 199. No. 539. Ch. 2. D. 6. L. 26; Zuev 2002: 248 (with incorrect dating of the event).
[422] RGADA. F. 199. No. 539. Ch. 2. D. 6. L. 26; Zuev 2002: 265.
[423] Nul. . . 1866. No. 17: 4; Vdovin 1950: 84; 2010: 513; pro: Belikov 1960: 274; Gurvich 1966: 104; contra: Gurvich 1982: 202.
[424] Sgibnev 1869: 27; 2005: 49; Gurvich 1966: 104.

The Koryak men, killing their families, saved themselves by skiing to the stockade.[425]

In 1730, a new invasion of the Chukchi across the Anadyr', deep into Koryak territory, followed. The Cossak leader A. F. Shestakov, moving from Gizhiga in conquest of the Penzhensk and Alyutorsk Koryak, on the Tylka (Tylkhaya) River received from reconnaissance information that the Chukchi had destroyed four Koryak "yurts." Shestakov moved in pursuit of the departing Chukchi with a detachment of 150 soldiers. On 14 March at the Egacha River (later, the Shestakovka) the Chukchi defeated this detachment, with the Cossack leader himself being killed. The Chukchi went to their camp, taking trophies and reindeer. According to the report from Anadyrsk to Yakutsk by the serviceman Ivan Batankuev (1 April 1730), between the Paren' and Oklan Rivers "a considerable number of Reindeer Koryak were killed, and 8 herds of field and draft reindeer were taken, and the aforementioned Chukchi took a considerable number of wives and children of the Reindeer Koryak captive." The Anadyrsk commissar Petr Kolesov reported the following about this March raid: ". . . the non-peaceful Chukchi arrived in a multitude in a campaign with military weapons in hand, killed the Reindeer Koryak Yallakh and relatives, and took the reindeer herds and yurts and all their household belongings, wives and children captive," killing 40 people. On 30 November 1730, the tributary Koryak who lived at Olyutorsk appeared in Anadyrsk with a complaint to D. I. Pavlutskii that the Chukchi had been attacking their relatives between the Paren' and Penzhina Rivers, killed 35 people (including Settled Koryak), took wives and children captive, and stole the reindeer herds; they asked for protection by the Russians and offered to assist. A. S. Sgibnev through his archival data counted up to a hundred Reindeer Koryak in the spring of 1730.[426]

---

425 Nul. . . 1866. No. 17: 4; cf.: Vdovin 1965: 65; Zuev 2009: 73.

426 Quotations: AAN. F. 21. Op. 4. No. 33. L. 254 (depositions of the Tungus in Okhotsk, 22 May 1730), 263 (report of I. Batankuev, 1 April 1730), 264 (report of I. Ostaf'ev: the Chukchi "killed many Reindeer Koryak, as well as a large number of Settled); Zuev 2009: 67 (communication of Kolesov); also see: AAN. F. 21. Op. 4. No. 33. L. 241 ob.; RGADA. F. 199. No. 539. Ch. 2. D. 6. L. 26; Müller 1996: 64; Sgibnev 2005: 49; Vdovin 1965: 65; Gurvich 1966: 104, 113; Zuev 2003: 134 (request of the Koryak to D. I. Pavlutskii about help); 2009: 67, 73.

In general, this is the picture we get. In March 1730, the Reindeer Chukchi carried out another raid on the Koryak for booty. The precise number of attackers was not known. The attack was conducted on the Chavchuven, who migrated between the Paren' and Penzhina and even farther south, and on the Settled Koryak, probably the Kamen' people, where five people perished. Possibly on the return trip Yallakh's camp, which consisted of four yarangas, suffered destruction, where 35 people were killed—probably all or almost all the male population of the site who engaged in battle. It must be kept in mind that in Koryak yarangas there were a larger number of *pologi* [sleeping compartments, tents] than in those of the Chukchi: K. Kekketyn mentions a yurt with 14 pologi in a nomad camp that consisted of only three yurts. Yallakh himself was possibly not killed, since in the depositions about the death of D. I. Pavlutskii, among the other Reindeer Koryak princes tributary to the Anadyrsk stockade there is a Yallakh Tinivgin.[427] The Chukchi drove off eight herds—evidently each Koryak yaranga on average pastured its reindeer in two herds; if it is assumed there was one herd, this will mean that not just Yallakh's camp suffered in the raid. Since Shestakov's detachment was nearby, he took the fleeing Koryak under his protection and began to pursue the Chukchi, who were substantially superior to him in number. The Chukchi apparently retreated northeast because of the threat of an encounter with the Russians. The Cossack leader, probably not having precise information on the number of the enemy, but hoping for the effectiveness of firearms, engaged in battle. The Chukchi attacked and won the battle, using their numerical superiority and quick attack with transition to hand-to-hand fighting. Encumbered with booty, they did not pursue the enemy but rather continued on home.

In the spring of 1731, in a retaliatory campaign of Captain D. I. Pavlutskii, more than a hundred and fifty Koryak took part. During the campaign 42 captive Koryak were freed. However, after Pavlutskii's departure to Yakutsk in August 1732, the Chukchi in this and the following year continued hostile activities against the Russians, involving also the allies of the latter—the Koryak.[428]

---

[427] Kekketyn 2010: 68; AAN. F. 21. Op. 4. No. 33. L. 254; KPTs: 169.

[428] 1731: Nefëdkin 2003: 25; March 1733: RGADA. F. 199. No. 539. Ch. 2. D. 6. L. 26.

Then, after some pause (which perhaps results from our lack of information), the raids on the Koryak continued at the end of the 1730s. At the very end of March 1738, the Chukchi killed five tributary Koryak. On 21 February 1739, at the Anadyrsk tribute house, the tributary Reindeer Koryak Yaglakh Tyganov and his comrades gave a detailed description of a raid in December 1738:

> The past year 738, in mid December, the perpetual villains of Koryak, the Chukchi, in a multitude on a campaign attacked them suddenly at night time, on nine yurts and destroyed their yurts, and the tribute prepared for payment to the treasury of Her Imperial Highness in year 738 and their other clothing and possessions they robbed, and both wives and children and relatives of the male and female gender sixty-six people they took captive as well as twenty-one thousand two-hundred draft and non-draft karginy. And after this, on the next day assembling, they, the Koryak, came to rescue Prince Tamali and comrades, since he was by them, the Chukchi, under siege; and with this they, the Chukchi, killed some of them, the Koryak, and precisely: the best Prince Etel' Lekhtalov with comrades, fifteen people, and three non-tributary people. And in year 738, in the last days of March, they, the Chukchi, killed of the tributary Koryak five people.

Slightly earlier, in January 1739, the Koryak told of a raid to the Anadyr' Reindeer Koryak Aiga Plaksin in slightly less detail:

> In the past 1738th year, during the Filipov fast, non-peaceful Chukchi suddenly arrived and ran off ten herds of our reindeer, and killed the Reindeer Koryak toyon Etel', and with him killed thirteen people of the tributary Koryak, and the tribute coffers from the tributary Koryak that toyon Etel' himself for the year 739 gathered and was taking with him to the Anadyrsk stockade, and these coffers the aforementioned non-peaceful Chukchi captured leaving nothing and took with them.

The Irkutsk office reported to the Siberian command in mid December 1738 about the campaign

> in Koryak land on the Olyutorka River of the non-peaceful Nose and River Chukchi in multitude with a militant hand, and about the killing of 25 Koryak people and about no small number of their wives and children taken captive with household belongings, and about driving off eleven herds of their reindeer without remnant to their Chukchi land, and that they boasted of being a second time in spring of the year 739 on

> the last route in multitude and by a militant hand laying ruin and devastation to the Koryak land and the Anadyrsk stockade.

S. P. Krasheninnikov also described in a letter to I. G. Gmelin and G. F. Müller of 5 August 1739 the raid of the Chukchi in winter of 1738–1739: "... the present winter the Chukchi came to the Koryak, and killed a small number of them, the Koryak, among whom the best Koryak man Etel' was killed, and they drove away and stabbed to death a herd of many reindeer."[429]

Comparing the evidence, we get the following picture: in mid December 1738 on the Okyutorka River the Chukchi attacked at night a camp of nine yarangas of the "best prince" Etel' Lekhtelev, whom they killed along with fifteen Koryak, took 66 women and children captive, and drove off eleven herds for a total number of up to 21,000 reindeer (that is, an average of 1,900 head per herd). The Chukchi also took the tribute acquired by the prince, the yarangas, sleds, and other belongings. Further, the Chukchi went to the camp of another prince, Tamali, who defended himself in his dwelling, and only the help of the assembled Koryak saved the situation on the next day. Probably a total of 25 Koryak were killed in the raid. This raid can generally be viewed as a kind of winter migration of the Chukchi to southern districts north of Kamchatka, into the land of the Olyutorsk Koryak, carryied out by both the Chukchi of the Chukchi Peninsula and by the Anadyr' River Chukchi. Since the tribute collected for the Russians was stolen, this touched the immediate interest of the Russian authorities in the region, and the situation was examined in the Senate and had more repercussion than other raids.

In the following year, 1739, the Chukchi began to bring their threat to the Koryak to life—making another raid on the Apuka and Khatyrka Koryak, leading again to the death of the men, taking away the women, and driving off the reindeer herds.[430] Krasheninnikov noted concerning the consequences of the recent Chukchi raids:

---

[429] AAN. F. 3. Op. 10. No. 137. L. 16 ob.–17 = 95–96 (first quotation); Pokrovskii 1941: 230 (second quotation); Senatskii arkhiv [Senate Archive]. Vol. II. 1889: 35–36, 535; KPTs: 160 (third quotation); Krasheninnikov 1949: 597 (fourth quotation); also see: Sgibnev 2005: 63; Gurvich 1966: 105, 113; Zuev 2009: 101; Okhotina-Lind and Møller 2015: 365, 370 (deposition of the Ustyug merchant A. Verkhotin: 19 people killed, 50 taken captive).

[430] Gurvich 1966: 105; Zuev 2009: 101.

> . . . and if it was not for protection of them from Anadyrsk, the Chukchi would destroy them to the ground, and would make serfs of masters deprived of their herds, forcing them to live in earthen yurts and to nourish themselves on roots and fish like the Sitting [Settled] Koryak, as they perpetrated on the Katyrka and Apuka Koryak in 1738 and 1739.[431]

Consequently, after the raids of 1738–1739, the Apuka and Khatyrka Reindeer Koryak, having lost their livestock, were forced to settle down and, so to say, "downgrade" their social status, turning from nomadic into settled.

On 24 October 1741, the Anadyrsk commander sotnik Vasilii Shipitsyn informed translator Yakob Lindenau that on 23 July of that year the Reindeer Koryak of the Anadyrsk Department, Nanylan Tygakov and Uchivanin Pavokov, handed him a petition, in which was reported:

> In the past month of March the non-peaceful Chukchi arrived from Anadyrsk Nos and from the mouth of the Anadyr' River into Koryak land on the Enigu River, called Talovka by the Russians, and suddenly attacked their best tributary Koryak prince Tamole Kuyukin and killed him, and with him they killed twelve tribute-paying Koryak people; they also killed a large number of their wives and children, and took others captive, and five of the field and draft reindeer herds they drove to their Chukchi land.

On November 5th, 1741, Captain M. Lebedev sent to the Irkutsk office a report in which there is reference to a message of the Reindeer Koryak:

> In the first days of the past m[onth] of March, of this year, the non-peaceful Chukchi arrived on the upper reaches of the Talovka River to them, the Koryak, in a multitude; they suddenly attacked with weapons and killed their best Koryak Tamolyu and with him eight people of the tributary Koryak, stabbed to death their Koryak wives and children, and took others captive with them, as well as all their household belongings from the yurts, and they took seven herds of reindeer, and they killed all the remaining reindeer.[432]

---

[431] Krasheninnikov 1949: 451–452; cf.: Ditmar 1855: 56, 59; Bogoras 1921: 47; Stebnitskii 1938: 133–139; Antropova 1957: 176; Vdovin 1973: 99.

[432] Report of V. Shipitsyn: AAN. F. 21. Op. 5. No. 142. L. 31; report of M. Lebedev of 5 November 1741 and of 22 March 1742: RGADA. F. 199. No. 528. Ch. 2. D. 3. L. 13 ob.; cf.: RGADA. F. 199. No. 539. Ch. 2. D. 6. L. 26; Vdovin 1965: 120; Gurvich 1966: 105; Zuev 2009: 101.

Thus, at the beginning of March (Old Style) 1741 the nomadic Chukchi from the Chukchi Peninsula and the mouth of the Anadyr' carried out a raid on sleds. They moved to the Talovka in the north of Kamchatka, where they attacked a camp of the Chavchuven and killed the prince and eleven men. The Koryak, despite the suddenness of the attack, probably tried to resist. The women and children, according to the document, were in part killed and in part taken captive. However, it is possible that the Koryak attributed the murder of non-combatants to the Chukchi—the Koryak women might have killed their children and themselves. The five (or seven) stolen herds could have belonged to this one camp, the other herds evidently were not found by the Chukchi. It is noteworthy that, as in 1716, they killed the reindeer they were not able to drive off, in order to inflict material damage to the enemy and undermine its economic base.

Figure 31. The Koryak in ancient armament pose for ethnographer W. Jochelson in 1900–1901. (Reproduced after Jochelson 1997: 101, Fig. 53).

In response to this Chukchi raid the Koryak prepared a response campaign in July to the mouth of the Anadyr' with a detachment of 60 men, in order to return their loved ones and get back the herds. Four princes of the Koryak in their petition urged the Russians from the Anadyr' to help them in the campaign. In answer to this request the commander of the stockade V. Shipitsyn sailed on 6 August in ten shitiki (with 70 servicemen) down the Anadyr' to hunt for reindeer and simultaneously to search for the enemy Chukchi. During the sotnik's negotiations with twelve toyons, the latter were slaughtered and the remaining Chukchi fled.[433]

At the beginning of 1742 about 400 Chukchi approached Anadyrsk, and near the Baranov Kamen' tract they killed tributary Koryak (among whom were eight chiefs) and drove off the reindeer. On 28 February 1743, the Chukchi carried conducted a raid on the Koryak on the Orlovka River (a tributary of the Main) and stole the herd. On 15 March a detachment of 40 Russian and Koryak led by Ivan Modin was sent from Anadyrsk to pursue the enemy. In the battle that ensued on this same river the superiority at first was with the Russians, who killed three of the enemy, but reinforcements that arrived to the Chukchi forced Modin's people to make a "saneburg" [sled fort] fortification and only then to retreat. The campaigns of Pavlutskii against the Chukchi that followed in 1744 and in 1745 stopped the active attack of the latter, and probably even disrupted their intended river campaign in baidary against the Anadyr' in the spring of 1745.[434] At this same time, because of the Chukchi threat, the Reindeer Koryak of the Anadyrsk Department began to resettle to the south on the Akalan, Penzhina, Gizhiga, and Pakhacha rivers, where in 1745 they became the initiators of the big Koryak uprising of 1745–1757.[435]

On 12 March 1747 about 500 (more precisely, between 400 and 600) Chukchi attacked the Koryak at the mouth of the Orlova River, not far from Anadyrsk, took seven women and one juvenile captive, and drove away seven herds (up to 20,000 head) that belonged to five princes, including two herds of Ivaka and two of Eigel' Lekhtelev, as well as draft reindeer of the Anadyrsk party. On this same day Major Pavlutskii, with a

---

[433] AAN. F. 21. Op. 5. No. 142. L. 31–31 ob.; Gurvich 1966: 105; Zuev 2009: 102–103.

[434] AAN. F. 3. Op. 10. No. 137. L. 103; RGADA. F. 199. No. 539. Ch. 2. D. 6. L. 26 ob.; KPTs: 166; Zuev 2009: 119, 124.

[435] Zuev 2004: 68–71.

detachment of Russians and Koryak mounted on reindeer, went in pursuit of the enemy and on 14 March died in a battle with the Chukchi on the Orlova River. After the death of the major, Chukchi raids on the Koryak continued. In 1747 "the Chukchi arrived to the Kamchadal at a Kamchatka settlement and to the Koryak living on the Ukinskoe coast and carried out deadly murder and pillage." That is, during the winter the Chukchi reached the eastern coast of Kamchatka, to Nizhnekamchatsk, where they plundered property of the Itel'men and settled Karagin Koryak people. On 15 April 1748, 200 Chukchi suddenly attacked the Chavchuven beyond the Penzhina River and killed thirteen herders and drove off seven herds of reindeer. In the spring of 1749, according to depositions of the Koryak prince Avui Amilin, a large group of Chukchi arrived on the Itkana River, where they "killed the Koryak chief Molokh and relatives in two yurts, and they chopped off the head of this Molokh and took away eight herds of reindeer." Thus, the Koryak Itkan people on the east of the Taigonos Peninsula were subjected to a raid. According to depositions of the Koryak woman Telna who fled from Chukchi captivity: "In that 1749th year, in summer, the enemy Chukchi went in baidary on the Khatyrka River . . . against the Settled Koryak in a campaign." I. S. Vdovin suggests that this was a trading expedition, since other documents do not confirm such battle action. However, the absence of other evidence is not proof, especially since Kerek folklore confirms these raids as Chukchi raids in 1738–1739. In March 1750, the Chukchi conducted the next raid on the Koryak on the Talovka; Lieutenant S. Kekerov rushed to the aid of the latter but did not find the Chukchi—they had managed to escape.[436]

After the death of D. I. Pavlutskii the Anadyr' people continued to carry out regular annual maritime expeditions to the mouth of the Anadyr' for the purpose of hunting reindeer to supply the garrison of the stockade with food and to resist the Chukchi. Such river expeditions followed in 1749–1754. This contributed to some reduction in the military activity of the Chukchi on the Anadyr'. However, in 1755 the Chukchi had

---

[436] Campaign of 1747: KPTs: 169–170; Vdovin 1965: 63 (quotation); 92; Zuev 2009: 139–140; Nefëdkin 2003:245; 1748: RGADA. F. 199. No. 539. Ch. 2. D. 6. L. 26 ob.; Sgibnev 2005: 65; Zuev 2009: 140; 1749: Zuev 2009: 140; 1750: RGADA. F. 199. No. 528. Ch. 2. D. 3. L. 21 ob.; on the Kerek in 1749: Vdovin 1965: 63–64; Zuev 2009: 141; Kerek folklore: Leont'ev 1976a: 236; 1983: 18, 96; Tyneru 2005b: 88.

again attacked the Koryak. In depositions of the Cossack B. Kuznetskii who returned from Chukchi captivity we read: "In the past year of 755, in the fall time, they, the Chyukchi, told me that a multitude of them gathered and went on a campaign to the Koryak, and then the master of him, Kuznetskii, stated to him that he and his relatives do not go to the Koryak, but he goes to ransom his son at the Anadyrsk fort and takes him with him, and wanted to redeem him." The total gathered for the campaign was a huge number of Chukchi warriors—1,500–2,000 men.[437] The subsequent action of this sizeable mass of troops is unclear.

On the background of Russian-Chukchi and Chukchi-Koryak confrontation a Koryak-Russian conflict also emerges—the powerful uprising of the Koryak in 1745–1757, in which, along with its primary members—various groups of Settled Koryak—nomads also took part. Those who rose up tried to enlist the support of the Chukchi, which was known to the Russian authorities in 1749; indeed, in 1746, with preparation for the uprising, the rumor spread among the Kamchadal that the Chukchi, having reconciled with the Koryak, had joined with them against the Russians and gathered to come to their aid; this was reported in an interrogation. According to Kuznetskii's report, in 1755 to the Chukchi

> arrived one man from the Koryak, and he doesn't know what he was called, and brought two wives with him: one a captive Chukchi and the other Koryak, with this declaration that the Koryak want with them, the Chukchi, to live together among them and to go in common against the Russians in a campaign; to which those Chukchi announced that they did not wish to reconcile with the Koryak, but rather they want to make peace with the Russian people and to pay tribute, but without amanaty being taken from them, the Chukchi.

This proposal was brought to the Chukchi by the Koryak Reprigin. Reprigin, who had lived as a captive from childhood among the Chukchi, brought the proposal along with two women. It is possible that the Chukchi sent the Koryak who had grown up among them to ransom captives. The Koryak brought a negotiator accompanied by a Koryak woman who was supposed to seal a political alliance with marriage. However, the Chukchi

---

437 Campaign of 1755: Vdovin 1965: 124; Zuev 2009: 169; depositions of B. Kuznetskii: KPTs: 183 (quotation); Zuev 2009: 160.

did not accept the peaceful proposition of the Koryak—their blood enemies. And in 1768 the Chukchi, during an encounter with the Russians on the Anadyr', accused the Koryak of instigating an attack on the Russians.[438]

In 1756–1758 there were peaceful negotiations between the new Anadyrsk commander Second Major I. S. Shmalev and the Chukchi elders. The Chukchi were permitted not to provide amanaty and to bring tribute themselves, and it was recommended to the Anadyrsk commander not to intervene in relations of the Koryak and Chukchi, limiting himself only to passive defense.[439]

Later Captain Shmalev wrote that in 1756 the Chukchi closest to the Anadyr' asked permission of him to settle on the Anadyr' in winter time, since they needed firewood, which it was possible to get on the river. The Chukchi were permitted to cross the river. Besides, at the end of the 19th century the Reindeer Chukchi usually migrated to the Anadyr', where they spent the winter.[440] Indeed, somewhat earlier, in 1750, the reasons for the hostility of the Chukchi and Koryak were indicated in a memorandum to General Major Ch. Th. von Kindermann, who was in command of the border line:

> And they, the Chukchi, are always looking to eradicate all the Koryak; and from the beginning of their lives always plaguing them with quarrels because there is scant reindeer food in all their Chukchi land and there is not any kind of forest, and in this they are in extreme need. . . . And among the tributary Koryak the land is free to live in and for contentment of the reindeer, both forests and food; and for this, those Chukchi look for their, the Koryak's, places to seize.[441]

In general, the document, based on local observations of the Anadyr' people, transmitted the Chukchi point of view on the reason for the

---

438 1746: KPTs: 97 (but cf.: KPTs: 101, 107–108); Okun' 1935: 53; 1749: Senate Archive: Vol. VIII. 1897: 741–742; B. Kuznetskii: KPTs: 183; cf.: RGADA. F. 199. No. 528. Ch. 2. D. 3. L. 25 ob.; No. 539. Ch. 2. D. 6. L. 79 ob.; Zuev 2009: 161–162; 1768: RGADA. F. 199. No. 539. Ch. 2. D. 6. L. 89 ob.

439 Zuev 2009: 160–173; cf.: Sgibnev 2005: 81.

440 Vdovin 1962: 154; 1965: 129; about the Chukchi of the end of the 19th century: Gondatti 1897a: 85, 95; 1897b: 169.

441 Divin 1979: 58; cf.: RGADA. F. 199. No. 539. Ch. 2. D. 6. L. 89 (speech of the Chukchi in 1768 on the Anadyr').

conflict, which consisted of territorial claims of the Chukchi on territory of the Koryak rich in moss and woods.

In the second half of the 1750s, on the background of the Russian-Chukchi peace negotiations, there was another escalation of Chukchi-Koryak conflict. The events that occurred interested local historian T. I. Shmalev, and in 1778 he described them in detail, as much as he could learn, in an essay "A Note Made by Captain Timofei Shmalev Due to the Ancient Rancor Taking Place among the Koryak with the Chukchi and the Campaigns from Both Sides, Deadly Murders, Plundering, and Especially from the Incident in the 757th Year from the Koryak to the Chukchi Deception and Murder of Their Best People Tatymkin and Arelpuga and Their Relatives." This composition was then sent by the captain to G. F. Müller and is now preserved in the so-called "Portfeli Millera" [Portfolios of Müller] (fond 199) in the RGADA in Moscow.[442]

The events, according to Shmalev, developed in the following way. In 1756 the elders of the Reindeer Koryak from the Velikaya and Khatyrka Rivers agreed with the Chukchi from Kanchalan to mutual marriages:

> . . . in the fall Ivaka Lekhtelev with his wife, as she was of the Chukchi race, went to Chukchi land and concluded peace with the Chukchi, and for steadfastness of the peace, that the Chukchi married their children to Koryak, and the condition that the Koryak take Chukchi in marriage. . . . Ivaka persuaded the best people of the Chukchi, Tatymkyn and Arelpuga with their clans with wives and children and reindeer herds, to go with him to Koryak land.

In March 1757, it became known to the Russians in Anadyrsk that "the Chukchi in forty-three yurts migrated beyond the Anadyr' and to the Khatyrka River with the Koryak to reconcile and intended to summer over there." However, then Ivaka stated that the Chukchi "came only to spy on their land" and were going to attack the Koryak. They decided to kill the Chukchi. The Chukchi and Koryak herds were specially mixed. During the separation of the mixed herds "they killed [the Chukchi], and then plundered the yurts, and split up the reindeer herds." There were up to 5,000 reindeer. Having learned about this, the Chukchi "decided at a

---

[442] RGADA. F. 199. No. 528. Ch. 2. D. 9. L. 45–56; Nefëdkin 2016a: 152-183.

council that for such deception, whatsoever it will take, to eradicate the Koryak so that not even a memory of them is left."[443]

In general, it is clear that the increase in the number of reindeer among the Chukchi, which happened as a a result of stealing herds from the Koryak and Yukaghir, and because the natural reproduction of the animals required new pastures. Therefore, the southern Chukchi from Kanchalan agreed to peace with the nomadic Koryak, which, based on tradition, was supposed to be reinforced by mutual marriages. Evidently, an agreement was concluded between the old enemies of the adjoining Chukchi-Koryak territories, who knew each other. In the winter of 1756/57 two Chukchi clans crossed the Anadyr' on the ice and went south to Khatyrka. Forty-three yarangas carried out the migration, that is, at a minimum of two hundred Chukchi. Those who arrived planned not only to winter over, but also not to return to their pastures in summer. Then the Koryak, not without basis, decided that they want to take their pastures, as a consequence of which the Chukchi were killed, the yarangas plundered, and the herds divided. Shmalev dates this event to 1757, whereas some modern historians (I. S. Vdovin, A. S. Zuev)—to the following year. According to the legends, the killing occurred in winter (1757/58). Learning about the incident, representatives of the southern Chukchi assembled a council, at which they decided to conduct total war against the Koryak. Even the Chaun Chukchi, in the words of the Nose Chukchi in the summer of 1760, were going to take revenge and to go to Anadyrsk in order to kill the Russians and Koryak there, especially Ivaka Lekhtelev.[444]

T. I. Shmalev did not know all the details of the origin and course of the conflict, which became clearly reflected in the cycle of stories (in essence, the heroic epos) about Kunlelyu, the hero of the Tel'kep Chukchi.[445] If one tries to summarize the various versions of the story

[443] RGADA. F. 199. No. 528. Ch. 2. D. 9. L. 46–46 ob.; Vdovin 1962: 154; 1965: 69, 129; 1973: 258; Zuev 2009: 169, 173, 175; cf.: Nefëdkin 2003: 28; 2016a: 156-157.
[444] RGADA. F. 199. No. 539. Ch. 2. D. 6. L. 39 ob.–40.
[445] Texts about Kunlelyu, see: Baboshina 1958: 238–240, 243–245, 247–251; Kostenetskaya 1973: 172–185; Menovshchikov 1974: 298–306; Leont'ev 1976: 152–153; 1983: 129; Belikov 1982: 183–197; Lebedev and Simchenko 1983: 98–100; Kymytval' 1987: 87–94; Nutekeu 2005: 12–14; interview of A. P. Al'ya in August 2013 in the village of Khatyrka, see: Nefëdkin 2016: 144–146.

about this hero, the following picture is obtained. After a long war the elders of the Koryak and Chukchi decided to reconcile. In order to reinforce peace, it was decided to intermarry. Two girls were sent from the Chukchi side, the father of one of them, by the name Arel'pyno (Arepu), went to the Koryak with his daughter named Kytgy (Cheivyne). The former enemies began to live in the tundra in one winter camp by Lake Ëmragytgyn (80 to 90 km from modern Khatyrka, on the right bank of the river) or, according to another story, to the north, on the upper reaches of the Vaamochka River; in summer, the camps migrated to the sea. Someone by the name Motlynto (Mollinto)—a Chukchi (sometimes referred to as a Koryak), who lived then among the Koryak, then among the Chukchi, for selfish reasons decided to embroil the two peoples. He incited the Koryak to attack the Chukchi, saying that he had heard about preparations for an attack by the latter. According to another variant, a dispute occurred because of the division of a herd, which was also rigged by Motlynto. Indeed, in dividing up the mixed herd it was possible to obtain a few of the others' reindeer. In the ensuing quarrel the Koryak, superior in number, led by Olyalek killed the Chukchi. Arel'pyno managed to flee at night with his daughter, but he died en route, while his daughter was able to escape.

Arel'pyno had a younger cousin named Kunlelyu, whose camp had migrated to the lower reaches of the Velikaya River on the south shore of the Anadyr'. Kunlelyu found Kytgy among the Kerek, who told him about the incident. The Koryak sent a detachment to the Kerek on reindeer sleds, which at night was destroyed by Kunlelyu's warriors. Fearing retaliation by the relatives of the slain Chukchi, Motlynto and the Reindeer Koryak migrated farther south; according to one of the versions, even to Kamchatka. Kunlelyu set off with a small group for revenge. After various adventures the people of Kunlelyu found the dwelling of Motlynto on a steep hill (Atagrymvyke, or Achepotskaya Hill), which had only one road up, bathed in water in winter. At night, with the help of riding reindeer sent up the hill, to which thongs were attached, the warriors one after another climbed this hill. Finding Motlynto's dwelling, the Chukchi challenged him to battle. After killing his family, he went out, but lost the battle and was captured. Later, the Koryak from the camp of the elder Olyalek joined the battle. In the battle, the Koryak spent all their arrows and perished together with their families, including Olyalek's family. The

Chukchi drove off the Koryak herds, and during the trip Motlynto was with them as a shepherd. They killed him just when the Chukchi returned home. Then the daughter of Arel'pyno, Kytgy, began to develop relations between the warring sides. The information from the stories about the cessation of warfare after the murder of Olyalek can be also taken as the standard end of a legend, as well as a real historical fact about subsequent termination of the war between the Koryak and the Chukchi.

As we see, the folklore data overall not only does not contradict the description of T. I. Shmalev, but rather substantially add to it. The initial reconciliation of the Koryak and Chukchi occurred because of significant human losses. The decision about peace was undertaken by the elders of both peoples. They gave their daughters as wives to their former enemies, in particular Arel'pyno did this and probably Tatymkyn, not mentioned in the story. If, in the captain's description, the instigator of the subsequent dispute appears the former peacekeeper himself, Ivaka Lekhtelev, brother of the "main Koryak prince Eigel'," then in the folklore it is a Chukchi reindeer herder Motlynto (literally, "Bleeding"). This seems to be a contradiction, but generally is not. Motlynto acted on behalf of Ivaka Lekhtelev (who in the folklore could have been transformed into the Koryak leader Olyalek).[446] The dispute between the Koryak and Chukchi occurred because of the mixing of the reindeer herds and conflict with subsequent division of them. The small-numbered Chukchi were killed. The relatives of the latter began to retaliate; in the folklore it is done by the small, eight people, group of Kunlelyu, consisting predominantly of relatives.

If the comparison of Ivaka with the hero of the epos is continued, it is possible to establish that he passed from life between 1759 and February 1768, that is, at the time when he gave an amanat in Anadyrsk. However, in this regard, among the rich herd-owners on the Nakhayan River, falling into Gizhiga Bay, only his sons Kaipik and Evlokh are mentioned.[447] In case of such comparison, the campaign of Kunlelyu

[446] The name Alalyk (Olalyk) is also encountered in Russian sources. Thus, on 25 March 1777 the Chukchi toyon Amulyat attacked the Reindeer Koryak Olalyk on the Pakhacha River, killing eight people, taking four women captive, and driving off the herd (RGADA. F. 199. No. 528. Ch. 3. D. 9. L. 44 ob.; Shakhovskoi 1822: 306–307; Zuev 2006a: 30; 2009: 234–235).

[447] RGADA. F. 199. No. 528. Ch. 2. D. 3. L. 42; D. 9. L. 48 ob.

against Motlynto can be dated to the 1760s. According to Russian documents, the Chukchi woman Kytna, widow of the Koryak Yakhyik Itgalin, offered herself to establish relations between the Russians and the Chukchi. She, together with her young, but already adult, son Igilev went to the Chukchi in 1776–1778 and conducted negotiations.[448] This is evidently Kytgy of Chukchi folklore.

Thus, after the murder of Arel'pyno and Tatymkin, warfare between the Chukchi and the Koryak was resumed with renewed force: practically annual Chukchi raids on the Reindeer Koryak were begun. In April 1759, a Chukchi group numbering up to 200 people arrived at the Uka shore of Kamchatka, along the way robbing and killing the Cossack Chudinov, who was going from Nizhnekamchatsk to collect tribute from the Alyutor people. The attackers robbed seven camps of Reindeer Koryak, killed 30 men and 20 women, and took nine women and four men captive, including children, as well as drove off, according to Koryak claims, 50,000 reindeer. And, as T. I. Shmalev correctly noted about the women's deaths, the Koryak women committed suicide upon threat of falling captive, that is why losses among the women were only a third fewer than the men.[449]

In 1760 the Chukchi destroyed seven camps of the Koryak who lived on the Krestova River, a tributary of the Omolon River; 25 men were killed, 20 women died, three women and seven men were taken captive, and up to 1,900 head of reindeer driven away. The Chukchi, moving south to the Gizhiga River, clashed with the Koryak fortified in a "posada" 50 verst from the Gizhiga fort, but were forced to leave without success and booty. This was probably done by the Chaun Chukchi, who, according to the words of Chukchi toyons, gathered in a campaign on the Anadyr' in the summer of 1760. In 1761 the Reindeer Koryak on the Oloi River suffered an attack by the (Chaun?) Chukchi: their yarangas were plundered, the men killed, and the women, juveniles, and children taken captive. In the following year the Nose Chukchi set off on a campaign to the upper reaches of the Opuka River; they plundered two camps of

---

[448] RGADA. F. 199. No. 528. Ch. 2. D. 9. L. 22 ob., 24, 28 ob., 42, 43 ob.; No. 539. Ch. 2. D. 6. L. 1 ob.–15, 45, 46, 48–48 ob., 51 ob.; Zuev 1997: 77; 2009: 232–237.

[449] RGADA. F. 199. No. 528. Ch. 2. D. 9. L. 30 ob., 47–47 ob. (the opinion of T. I. Shmalev); Zuev 2009: 173–174; cf.: Sgibnev 2005: 81; Gurvich 1966: 106.

Reindeer Koryak, where 16 men and 17 women were killed, 13 men and two women taken away captive, and 5,700 reindeer driven off. In the winter of 1763 the Chukchi toyon Amulyat went to Koryak lands, where he traded with the Settled Koryak and robbed the Reindeer Koryak. In 1765 another Chukchi raid followed to northern Kamchatka, during which seven households were destroyed, 40 men and 24 women died, eight women and two children were taken prisoner, and 33,700 reindeer were run off.[450]

By the end of the 1750s the Anadyr' Koryak had migrated south to the region between the Gizhiga and Olyutorka Rivers—on F. Plenisner's map (1764) the northern edge of settlement of the Koryak ran south of the Khatyrka. However, while the Anadyrsk stockade existed, part of the Reindeer Koryak remained under its protection. The Koryak supplied the garrison with their reindeer for campaigns and supplied auxiliary contingents. After abolition of the stockade the Koryak migrated to the protection of the Gizhiga fort, which had an established garrison of 200 men. At the fort in 1767 there were sixteen clans of reindeer people, with 319 male tribute payers. In the most dangerous time—fall and winter—when there was snow cover and the Chukchi could move quickly on sleds, the Koryak went to be near the Gizhiga fort and kept only the draft reindeer there, whereas their herds were pastured to the south, in the river valleys "on the Okhotsk side." In summer the Koryak, not fearing attacks from the now immobile Chukchi, migrated far from Gizhiga.[451]

In 1763, still being at Anadyrsk, Lieutenant Colonel F. Ch. Plenisner conducted negotiations with the Koryak elders and recommended that in winter they create large camps for safety and defend themselves against the Chukchi with firearms; to ease management of the Koryak, Eigel' Lekhtelev, after providing amanaty, was assigned as the main prince, who then on behalf of the Empress received a sword as a sign of his rank. In a report to F. I. Soimonov of 31 October 1763, Plenisner

---

[450] Campaigns of 1760–1765: RGADA. F. 199. No. 528. Ch. 2. D. 9. L. 30 ob.–31, 47 ob.–48; cf.: RGADA. F. 199. No. 539. Ch. 2. D. 6. L. 40 (notification of the Chaun raid in 1760); campaigns of 1760–1763: Zuev 2009: 174, 208.

[451] About the Koryak clans: Dolgikh 1960: 557; migration to the south: Zuev 2009: 176, 195; method of migration at the Gizhiga: Kosven 1962: 282, 287; map of F. Plenisner: Orlova 1951: Appendix; Efimov 1964: No. 126.

thus described the position of the Reindeer Koryak and the reasons for their defeats:

> . . . concerning the previous attacks, formerly from the Chukchi against the loyal Koryak, they followed from nothing else but disagreements with their Koryak relatives, who do not always listen to their chiefs and elders, and thus they put up their yurts and live in one or two, and as many as three together. And if every chief and elder would live with his relatives and each would not be fewer than twenty and thirty yurts together, then in such considerable number of yurts the Chukchi will never dare to make raids. For the Koryak always in battle, as before, and now, predominate in opposition to the Chukchi, because they have sufficient firearms and armor, of which the Chukchi do not have any. And the Chukchi make raids from extreme poverty, so that they obtain from the Koryak some booty, and especially iron items, that is, axes, flint, and kettles, which the Koryak receive quite sufficiently from the Russian traders.

Later, in 1777, Captain T. I. Shmalev in a similar way reported:

> When the Anadyrsk residents were taken to the Nizhnekovymskoe [Nizhnekolymsk] and Izhiginskoe [Izhiga] forts, the Yukaghir were almost all eradicated as a result of war with the Koryak, so for the Chukchi there was nowhere left for them to obtain needed . . . things. As I heard from captive women, they [the Chukchi] regret very much the removal of the Anadyrsk stockade, and therefore, they will be forced to go on more campaigns against the Koryak for robbery.[452]

After abolition of the Anadyrsk stockade no one hindered the Chukchi from crossing to the south shore of the Anadyr', and now the Chukchi no longer feared attacks by the servicemen. What is more, earlier, in the 1750–1760s, the Russian authorities did not require the Chukchi to cease their raids, believing that the Koryak, congregating in larger camps, could protect themselves. However, the number of the Koryak themselves, as researchers propose, decreased by half in the first half of the 18th century because of wars with the Russians and Chukchi.[453]

---

[452] Treaty with the Koryak: Zuev 2009: 191–192; AAN. F. 3. Op. 10. No. 137. L. 48 ob.–49 ob., 151–152; quotation of F. Kh. Plenisner: AAN. F. 3. Op. 10. No. 137. L. 48ob.–49 = 151–152; quotation of T. I. Shmalev: RGADA. F. 199. No. 528. Ch. 2. D. 9. L. 52; sword to the Koryak princes: AAN. F. 3. Op. 10. No. 137. L. 155.

[453] Vdovin 1962: 154; about the reduction in number of Koryak: Dolgikh 1960: 560; Gurvich 1966: 107; Zuev 2004: 72; 2009: 174.

On 3–4 February 1768 a group of Chukchi numbering up to 300 attacked the camp of the Koryak Tykka located 5 verst from the Gizhiga fort, where three men were killed, four women taken captive, and up to 400 reindeer stolen. A total of 13 households suffered in the raid on the Gizhiga Reindeer Koryak, five men were killed, ten women and two boys taken captive, and 46,000 reindeer driven off, based on information obtained from the Koryak, probably exaggerated. At the same time, nine Chukchi died in clashes with the Koryak, evidently primarily in the battle at Tykka's camp. Judging by the losses, the Chukchi did not particularly engage in the fighting, preferring to steal the reindeer from the pastures. Upon their return, on 18 February, the Chukchi again attacked the reindeer camp at Gizhiga, where a skirmish occurred. To help the Koryak, the detachment of Ensign Pavel Mordovskii with a hundred soldiers and Cossacks came from the fort. After the Russians' first volley (shot), the Chukchi threw themselves into the attack, incurring with this losses from the dense fire. Ten of the "main and best people" were lost and a mass of other Chukchi—a total of 34 people. The Chukchi retired. The servicemen did not pursue the enemy, but managed to recapture half the stolen reindeer.[454]

Evidently, the encounter on the Oklan River of the Chukchi with three servicemen led by the sotnik Stepan Belyaev, who were carrying a case with documents from Gizhiga, belongs to the beginning of this campaign of the Chukchi against the Koryak. The Chukchi took the servicemen's guns, arguing that they needed the weapons for an upcoming campaign against the Koryak, and promised to return the weapons in the summer at the meeting when hunting on the Anadyr', which was done.[455]

The Koryak, inspired by the defeat of the Chukchi at Gizhiga, decided to carry out a raid on their enemy. Captain Shmalev, in the words of the main Koryak elder (in 1777), thus described this raid:

> In this same 768th year in the month of March, after the Chukchi were there, the Reindeer Koryak Tykka Engylekhtenov went with his relatives, a total of 15 people, on

[454] RGADA. F. 199. No. 528. Ch. 2. D. 9. L. 31, 48–48 ob.; Sgibnev 2005: 85; Vdovin 1965: 135; Zuev 2007a: 292; 2008: 40; 2009: 206–207 (according to A. S. Zuev, there were two clashes of the Gizhiga detachment with the Chukchi); Nefëdkin 2016a: 160.
[455] RGADA. F. 199. No. 528. Ch. 2. D. 9. L. 48, Note 2. L. 51 ob.; No. 539. Ch. 2. D. 6. L. 88; Vdovin 1965: 68; Zuev 2009: 204; Nefëdkin 2016a: 176, n. XI.

> a campaign to the Chaun; and, spending the summer and fall there searching for the Chukchi, arrived at the first snow back at the Gizhiga fort. And there in the month of September of this summer they villainously took three Chukchi women who were going in the tundra to dig roots, and received as booty up to 400 reindeer, since there was no one of the reindeer people as shepherds for the reindeer. And since they were in small number, after the capture they hurriedly returned, and arrived safely.[456]

The Koryak campaign was apparently organized in March, quickly, and immediately after the defeat of the Chukchi in order to take advantage of the confusion of the latter. One of the elders of the Gizhiga Reindeer Koryak, Tykka Engylekhtenov, did not assemble a lot of warriors, but rather took only 15, that is, a third of the men of his clan (46 men in his clan paid tribute in 1777)—probably those who were migrating nearby. They were evidently in a hurry to leave mounted on reindeer over the snow cover still being on the ground. This was not a true military campaign, but, judging by the number of participants, a simple raid designed as a surprise attack. The direction of the expedition toward Chaun can clearly be explained by the fact that the Chukchi, who carried out a raid also in 1760, went there. Tykka obviously wanted vengeance for the raid of 1760, when several people of his group were killed by the Chukchi and the herd driven away, as well as for the recent February raid, in which three men of his camp were killed and four women (including his daughter) taken captive. He probably wanted to free his relatives; indeed, later, in the fall of 1774, Tykka went with a group of five people again to Chaun in order to exchange his daughter for a captive Chukchi woman. Though the campaign of 1768 lasted half a year, from March to September, its success was modest; indeed, Tykka's small group did not engage in open battle but only seized three Chukchi women, attacking them suddenly; and in addition, the Chaun region was not thickly populated.

According to A. S. Sgibnev's data, in 1769 the Chukchi again drove away Koryak reindeer, killing several herders. In March 1769 four princes of the Reindeer Koryak, without the knowledge of the Gizhiga

---

[456] RGADA. F. 199. No. 528. Ch. 2. D. 9. L. 48 ob.; Nefëdkin 2016a: 160; cf.: Ibid. L. 47 ob.–50; Zuev 2009: 207–208.

administration, again set off on a campaign, which T. I. Shmalev briefly tells about in the following way:

> In the 769th, in the month of March, to this same Chaun went an expedition of the Koryak of Lelu Medvedev, Umevi Yallakhov, Poitelo Umevin, and Umya Aukevgon with their houses and with their relatives, up to 80 people, including those raised by them, by the Koryak, six people of the Reindeer Kamen', the Paren', and the Itkan' Settled Koryak. And in the month of May of that year to the Chaun island they made a crossing on reindeer over the ice, and by spying on those Chukchi in the night destroyed one yurt, and killed 9 Chukchi men, took 8 women and one youth captive, took two Chukchi reindeer herds, in which there were up to 1,000 reindeer, based on division by those Koryak.[457]

This campaign is also probably in response to the last year's raid of the Chukchi on the Gizhiga nomadic Koryak; indeed, the expedition of the Koryak in 1768 did not bring substantial success. Now four clans were united, mobilizing a large part of their men, to which were joined six volunteers from the settled Koryak, in general composing a sufficient force by local standards—80 warriors. The Reindeer caravan moved in March to the Chaun, where it arrived in May. Going over to Aion Island, the Koryak attacked at night and destroyed the yaranga of the nomadic Chukchi, killing the men, taking the women captive, and driving away the herd of reindeer—this was the expedition's only success.

Starting from 1769, during the period from 1 February to 1 May—in the most dangerous time due to raids—on F. Ch. Plenisner's order, the Gizhiga commander was to send a detachment for protection of the Koryak camps; however, because of the slowness of the arrival of dog sleds, bringing three riders each and provisions, aid from the servicemen did not arrive in time in the most distant sites.[458] This is attested by Chukchi raids in the following year.

W. G. Bogoras thought that the southern Tuman' Chukchi were the "wildest and most indomitable, and in former times they were the

---

[457] Sgibnev 2008: 16; quotation: RGADA. F. 199. No. 528. Ch. 2. D. 9. L. 49–49 ob. ; Nefёdkin 2016a: 161; cf.: Sgibnev 2005: 85; Vdovin 1950: 88; Zuev 2009: 208.

[458] RGADA. F. 199. No. 528. Ch. 2. D. 3. L. 37 ob.; D. 9. L. 52–53 ob.; No. 539. Ch. 2. D. 6. L. 90; Gurvich 1966: 110; Zuev 2009: 207.

initiators in wars against the Koryak."[459] It is possible to speak about this situation only for the second half of the 18th century, when the Chukchi crossed the Anadyr' and began to settle among the Koryak. As we saw, the Chukchi carried out raids essentially from two sides: the Chaun Chukchi conducted them toward the Okhotsk coast, whereas the Nose—more often in the Kamchatka direction.

In April 1771, the Chukchi (evidently the "Nose") carried out a raid on the Reindeer Koryak on the Opuka River, where they ravaged seven camps of two owners, killing four shepherds; eleven women were killed and 4,000 reindeer stolen. On the way back, the Chukchi were attacked by the Koryak at the Parapol'skii Dol. The Koryak killed some of those who had attacked, captured women and juveniles, and recovered two herds (probably 2,000 head). In the following year there was a larger Chukchi raid on the Koryak who lived on the Opuka and Lesnaya Rivers. Five households were ravaged, fourteen shepherds and nine women died, eight men and four women were seized, and 45,000 reindeer stolen. Eigel' Lekhtelev, assigned by the Russians as the main prince, threw himself with his elders and warriors in pursuit of the retreating Chukchi for the purpose of returning the herds. After killing four Chukchi, the Koryak were able to return only a tenth part of the reindeer. In 1773 the Chukchi, in their turn, attacked the camp of Eigel' himself, which was migrating on the Talovka. Since among the latter at this time were sixteen Settle Koryak with guns, they were able to repel the attack, killing one Chukchi. However, four herds of the prince, with 10,000 head in each, were stolen, three shepherds who were in the pasture were killed, and nine women and a young son of Eigel' were taken captive. Also, 3,000 reindeer were chased away from the camp of the Koryak Notaime.[460]

At the end of 1774 eight Chukchi toyons leading 150 warriors went from Onemenskaya Bay on the right side of the Anadyr' for the next raid on the Reindeer Koryak. Along the way the Chukchi destroyed a camp of the tributary Yukaghir. First, the group set off toward Kamchatka, but because of deep snow the reindeer were unable to go, and the Chukchi turned to the west to the Kamen' stockade of the Koryak, in which two

---

[459] Bogoraz-Tan 1937: 13; cf.: Antropova 1957: 176.

[460] RGADA. F. 199. No. 528. Ch. 2. D. 9. L. 31, 45; Ch. 2. D. 3. L. 37 ob.; No. 539. Ch. 2. D. 6. L. 90; about the raid of 1771: Gurvich 1966: 110.

Chukchi were held. On 2 March 1775, on the Osinova River, a Chukchi group encountered the detachment of Sergeant A. Maklakov, before whom the Chukchi peacefully dispersed. But on 9 March, approaching Gizhiga, the Chukchi were defeated by a detachment of a hundred men led by Ensign P. Mordovskii and the allied Koryak joined to him, who had a camp one and a half verst from the fort for protection against a probable Chukchi raid. According to Ya. M. Peresypkin's report to the Kamchatka commander M. Bem, 54 Chukchi were killed, 40 women and children taken captive, and a herd of reindeer taken; on the Russian side, two Russians were killed and twelve servicemen and Koryak were wounded.[461]

After the defeat at Gizhiga, Menyakhta threatened to attack the Koryak in the winter of 1775/76 and in general destroy the Gizhiga fort. However, the Chukchi came for revenge only in March 1777, led by toyon Amulyat. T. I. Shmalev briefly described the route of this campaign:

> . . . the main Chukchi toyon Amulyat Khergyntov and his brother Petutim Khergyntov with 180 people came to the Apukha and Pakhacha Rivers in the month of March in search of the Koryak. The Koryak Nutelly had 28,000 reindeer driven off; from the Ellansk outpost on the Khatyka, they took one Koryak juvenile and gave him for ransom. On the 19th of March, they arrived at Aputsk outpost of the foot [unmounted] Koryak and, treating the Aputsk [Apuka] Koryak in a friendly way, in addition conducting trade, finally the Aputsk from the fort killed with a gun a Chukchi by the name Iatyrgin. Then they arrived at Pakhacha outpost, but there, other than discussion, nothing occurred. And going up the Pakhacha River, they found a yurt of the foot Koryak, which the group plundered and took two girls captive. On 25 March, close to the Pakhacha, one yurt of the migrating Reindeer Koryak Alalyk Akhyitkalov, a total of eight people killed and 4 women taken captive; also, reindeer, from both Alalyk and another Koryak Tynanpii, a total of two herds driven away, obtained as booty, and returned to their land.[462]

---

[461] Shakhovskoi 1822: 305–306; "Dve zapiski. . .," 1873: 360–371; Zuev 2009: 217–221.

[462] Quotations: RGADA. F. 199. No. 539. Ch. 2. D. 6. L. 5 ob.; cf.: Ibid. L. 16 (dating), 44 ob.; Shakhovskoi 1822: 306–307; Gurvich 1966: 115; Zuev 2009: 232–234; threat of Chukchi in the winter of 1775/76: "Dve zapiski. . .," 1873: 370; cf.: RGADA. F. 199. No. 539. Ch. 2. D. 6. L. 2 ob.

Before us is a brief report on the typical raid at the end of the winter season lasting less than a month. The campaign was purely for plunder, and directed at those areas that were not protected by the Gizhiga fort. The efforts of the attackers (180 people) were not especially great, though they were led by "the chief Chukchi toyon" Amulyat Khergyntov and his brother Petutin. Evidently, they did not count on a direct clash with the Russians. They drove off herds from the Reindeer Koryak, killed men, and took women captive; they did not storm the outposts of the Settled Koryak, which were defended with firearms, preferring to trade, though disputes emerged here as well, which led to the death of one Chukchi; though finding a single dwelling of the Settled Koryak, the Chukchi destroyed it and took the women captive. Taking two large herds, the Chukchi returned home. This raid was unexpected for the Koryak; it occurred without large battles and serious losses, but brought booty—this is a typical example of a successfully conducted raid.

Chukchi relations with the Kerek, the approach to whom from land was covered by mountains, were not developed in such a confrontational manner. In 1769 the Khatyrka people (Kerek) told the Russian tribute collectors that they "have never had enemies, and though they hear about the non-peaceful Chukchi and Koryak, they have never seen them themselves," therefore, they did not fortify their settlements. In 1777 T. I. Shmalev writes directly about the friendly relations of the Khatyrka Koryak (Kerek) with the Chukchi; in 1775 the Kerek arrived as guests to the Chukchi on the right bank of the Anadyr'; in the following year Purekhom Menyakhtin with his clan visited the Kerek. After 30 years the Chukchi raid of 1739 was probably forgotten and the residents, remaining without reindeer, began to live settled. They did not possess reindeer, for which the Chukchi conducted their raids, but rather the Luoravetlany preferred to trade with the Settled for mutual benefit. However, conflicts happened, as they did in 1749. Based on documents, it is also known that at the beginning of 1782 eight Chukchi killed ten Khatyrka Kerek. Preserved in the epos of Kunlelyu is the legend that Lame Kerek, who was a reindeer herder but now living in a pit house, fought with Kunlelyu, but lost the fight. The Russian document and Chukchi folklore possibly tell of the same event. Kerek legends colorfully depict conflicts with the Chukchi and the Koryak. During the winter raids, small groups of attackers approached the Kerek pit houses at night and threw in through the smoke hole tow soaked in oil, forcing the residents either to burn up or run out. If the Kerek learned about a raid beforehand, they hid in caves that were in the high shore of Pekul'neiskoe Lake and at Cape Navarin. Such attacks on the Meinypil'gyn and Khatyrka Kerek can be dated to the 1770s.[463]

In general, for the period of 1730–1770s more than 50 Chukchi attacks on the Reindeer Koryak were recorded by Russian authorities; in two last decades, they occurred almost annually. During the period from 1759 to 1774, based on documents known to T. I. Shmalev, the Chukchi killed 138 Koryak (also 103 women died), 34 men and 52 women were taken captive, and 239,300 reindeer were stolen (though the captain himself supposed that in actuality only a fifth part of that number were stolen—48,000); in the campaigns, 63 Chukchi died (besides the loss of the clans of Tatymkin and Arelpuga—probably unknown to Shmalev) and 12 were taken captive.[464]

After negotiations in Gizhiga in March 1778 the main Chukchi toyon Amulyat and toyon of the Arctic Ocean Aetkim began to strictly maintain peaceful relations with the Russians and with the Koryak. In 1781, when once again the main Chukchi toyon Amulyat Khergyntov came to Gizhiga, he agreed to stop raids on the Koryak. In January 1784 Amulyat even discouraged the elder Kamukha of the Arctic Ocean, who had already set out on the expedition, from a further campaign against the Koryak.[465] After this, Koryak-Chukchi relations developed primarily along a peaceful course. In 1785 the Chukchi, with the support of the Koryak, killed on the Penzhina five Cossacks, who had arrived for the purpose of surveying the place for the construction of the town of Aklansk.

---

463 Quotation of the deposition of the Khatyrka people: Zuev 2009: 208; 1775: "Dve zapiski. . .," 1873: 370; T. I. Shmalev: RGADA. F. 199. No. 528. Ch. 2. D. 9. L. 51, 54, Note 2; Zuev 2009: 232; conflict of 1782: Zuev 2009: 257; about Kerek (the individual): Baboshina 1958: 242–243; Kerek folklore: Leont'ev 1983: 96; Tyneru 2005b: 88; evacuation to caves: Leont'ev 1976: 148; 1976a: 236.

464 RGADA. F. 199. No. 528. Ch. 2. D. 9. L. 31, 49–49 ob.; Vdovin 1950: 95; 1965: 18, 65; Antropova 1957: 175; Nefëdkin 2003: 25; Zuev 2009: 368.

465 Amulyat and Aetkim: RGADA. F. 199. No. 539. Ch. 2. D. 6. L. 49–49 ob.; treaty of 1781: Shakhovskoi 1822: 289; Vdovin 1965: 69–70; Zuev 2009; 257; 1784: Zuev 2009: 258.

Later, 500 Chukchi arrived at Olyutor for trade with the Koryak, which was carried out without any conflict. To notify the Chukchi about the future expedition of J. Billings from Gizhiga, sotnik I. Kobelev was sent to negotiate with the Chukchi. He arrived on 24 December 1789 to the Koryak camp of elder Ammat Yakhvikov, where on 8 March of the following year Reindeer Chukchi peacefully came, led by Imlerat; the latter together with I. Kobelev and N. Daurkin went to Kolyuchin Bay. Somewhat later, in 1791, C. Merck thus described relations of the former enemies: "... the Chukchi who go to Gizhiga have their friends among the Koryak, to whom, upon arriving, they give gifts in the form of parkas of martens and foxes and others, for which the Koryak exchange with them some of the tired reindeer for the subsequent trip home, as well as give the Chukchi meat for the road, killing reindeer especially for this."[466] In general, from the end of the 18th century relations between the southern Chukchi and the Koryak bore predominantly a peaceful character—mutual interest in economic relations began to prevail.

2. Chukchi War with the Anadyr' Yukaghir

At the boundary of the 17th and 18th centuries, when the Chukchi began to appear in the Anadyr' basin, the ethnographic situation there was changed: The Settled Anauly Yukaghir on the middle course of the river had practically died out due to Cossack pogroms and epidemics, and only the Reindeer Yukaghir hunters were left—the Chuvan people and Khodyn people; the first lived on the upper reaches of the Anadyr', the second on the middle course of the river. The herds of reindeer among these groups of Yukaghir were small and the animals were used primarily as a means of transportation, initially as draught animals. G. F. Müller in particular noted that among the Anadyr' and Kolyma Yukaghir there were usually no more than 20 head of reindeer (1730s).[467]

In 1680 the Yukaghir of the Anadyr' area became tributary, that is, they adopted Russian citizenship.[468] Besides giving tribute, they were supposed to be guides and translators during expeditions, to supply reindeer for transportation, and be auxiliary contingents during campaigns

[466] 1785: Sgibnev 2008: 24–25; 1789: EM: 161; quotation: Merck 1978: 145.
[467] Ethnography: Dolgikh 1960: 418–442; opinion of G. Müller: Elert 1999: 122.
[468] Ogorodnikov 1922: 331.

against the Chukchi, Koryak, and Itel'men. In particular, the Yukaghir were mobilized in campaigns of the servicemen against the Koryak (in 1691, 1708–1709, 1713–1714, 1719–1720, 1732, and 1751), to Kamchatka against the Koryak and Itel'men (in 1696 and 1697–1699), and against the Chukchi (in 1702, 1731, 1744, and 1746). On the other hand, the Russian authorities were supposed to protect their subject Yukaghir, on whom attacks by the Chukchi from the north and Koryak from the south became stronger from the 1680s, which led to military conflicts between the Russians and these ethnic groups.

The steward of the Anadyrsk stockade, desyatnik Vasilii Tarasov, thus described to the Yakutsk governor the complaints of the tributary Yukaghir with their payment of tribute (1685):

> And the foreigners that pay the great monarch tribute, and those paying tribute say: you ask tribute from us, but you do not defend us from non-tributary Koryak and Chukchi; those Koryak and Chukchi kill many of our relatives on the sable hunts, and wives and children, and take our relatives captive, and steal the reindeer; they knock down our dwellings; and they do not let us hunt sables for the great monarch in tribute; and there is no one to defend against those non-tributary Koryak and Chukchi; the service people are few, and no guns, powder, and lead.

In general, it turns out that because of too few Cossacks and supplies at Anadyrsk the servicemen were not able to protect the tributary Yukaghir (in 1685 there were 117 men) from the Koryak and Chukchi; primarily the Koryak, who attacked the Yukaghir hunters on hunts, in the camps, and individual outposts, displacing them from their traditional hunting places. From the Yukaghir they took women and children captive and drove off reindeer, leaving them without family or transportation.[469]

For several years the situation on the river remained the same. The next steward of Anadyrsk, Andrei Tsypandin, in 1688 explained to the Yakutsk governor the shortage of tribute in the following way: ". . . in past years many tributary foreigners died by the will of God, and others by non-peaceful non-tributary foreigners, the Chukchi and the Koryak, were killed on sable hunts and in their dwellings, and others . . . went to distant

[469] Quotation: DAI. 1867. Vol. X: 351; cf.: Zuev 2002: 66; tributary: Dolgikh 1960: 424.

unknown places. . . ." As we see, the steward explains the reason for the reduction in tribute collection by the fact that some Yukaghir died from epidemics, others simply migrated away from the winter quarters, escaping the illness, and some were killed while hunting and in their camps. This was again done by the Koryak and Chukchi. Judging by the tribute books of the Siberian Office, in 1688 154 men (that is, Yukaghir) numbered among the tribute payers, whereas 94 old and 8 new payers delivered tribute, 35 "left" and 14 died.[470]

During the uprisings against the Russians the same Yukaghir tried to enlist support from the enemies of the Russians and their subject Chukchi. In 1696, 60 servicemen and hunting people, accompanied by 60 Khodyn people and Chuvan people, set off to Kamchatka, led by V. V. Atlasov. In 1697, during the campaign, the Yukaghir led by the Chuvan Oma rebelled and engaged in battle with the Cossacks on the Palana River. They coordinated their actions with preparation for an uprising by the tributary Yukagir Pochin, who intended to unexpectedly attack Anadyrsk. The Yukaghir themselves "called upon the Chukchi to take the Anadyrsk stockade and kill the servicepeople, and retrieve the amanaty. And if they are not able to take the stockade, then they will burn it with fire." However, the Chukchi did not come to help and the surprise attack on the stockade was unsuccessful, and Oma fled.[471] In general, the attempt at coordination of actions between the Yukaghir and the Chukchi against the Russians was not successful.

In 1701 the Khodyn Yukaghir Nekras with relatives delivered to the Anadyrsk steward S. Chernyshevskii a petition with a complaint about the Chukchi and a request that "to send servicepeople with them from Anandyrsk against the non-peaceful Chukchi to the Anandyr' Nos, whose mortal kills and robbery interfere with them, the Yukaghir, in the hunt for reindeer." In response to this petition, in 1702 a large expedition was organized against the Chukchi; Aleksei Chudinov was sent with a detachment consisting of 24 servicemen and townsmen and 110 tributary

---

[470] Quotation: MIYa. Pt. 3: 978; about the tributary: Dolgikh 1960: 424.

[471] Quotation: Belov 1957: 102–103; events, see: KPTs: 26; Gurvich 1966: 22; Zuev 2002: 232.

Yukaghir and Koryak. This detachment destroyed a Chukchi camp of thirteen yarangas at the mouth of the Anadyr'.[472]

We note that now the Yukaghir complained not about hindrance by the Chukchi in the sable hunt, but about conflicts in the reindeer hunts on the Anadyr'. Apparently at this time, during the summer, the coastal Settled Chukchi and Reindeer Chukchi sailed in baidary of the Settled into the lower Anadyr', where they themselves hunted reindeer. During the hunt competition arose, with the more militant and probably more numerous Chukchi killing and pillaging the Yukaghir, taking captive the women and children.

In 1710 the boyar's son Vasilii Ignat'ev and his colleagues in Yakutsk described for Governor D. A. Traurnikht the situation on the Anadyr' in this way: ". . . and from the Anandyrsk stockade the servicepeople and tributary Yukaghir sail along the Anadyr' down for reindeer hunting; and the non-peaceful Chukchi kill these servicepeople and tributary Yukaghir and take the wives and children captive." Since provisions had not yet been delivered to Anadyrsk from Yakutsk, the servicemen used the Yukaghir methods of preparing provisions for themselves: hunt the reindeer swimming across the river going to summer pasture, and then to winter pasture. In the list of deceased and killed Anadyr' Yukaghir of 1704,for the few preceding years were indicated: thirteen men killed by the Chukchi, six by Koryak, three by Lamut, and three in Kamchatka.[473] In general, it turns out that the Chukchi had become the most dangerous enemies of the Anadyr' Yukaghir.

The Yukaghir, who knew the Chukchi language, especially the more reliable baptized ones, were actively involved by the Anadyrsk administration as envoys to the Chukchi. Thus, in 1709 the steward of the stockade pyatidesyatnik Efim Petrov sent the newly baptized Ivan Tereshkin with the Khodyn Pelech and the serviceman Ivan Lokosov to the Chukchi for the next call to them to pay tribute. Returning, Tereshkin reported: ". . . having gathered they, the Chukchi, in a multitude, were

---

[472] PSI. Bk. 2: 524–525; KPTs: 155–156 (quotation); Müller 1996: 38; Bogoraz-Tan 1934: 44–45; also see: Dolgikh 1960: 433; Gurvich 1966: 112; 1985: 54; Zuev 2002: 235–236; Nefёdkin 2003: 24.

[473] Quotation 1710: KPTs: 155; cf.: Gurvich 1966: 74; list of 1704: Dolgikh 1960: 425, 428; Gurvich 1966: 74.

going to fight at the Anandyrsk stockade; and he, Ivan, meeting on the Anadyr' above the Chukochya River close to the Yukaghir reindeer hunts and after talking to them the Chukchi turned back; and for faith they, the Chukchi, sent from themselves to Anandyrsk one person"; Tereshkin agreed to meet with the Chukchi in the coming year at the same place to call for tribute payment.[474] Consequently, in the words of the Yukaghir, he arrived to the Chukochya River (Kanchalan), where the Chukchi probably were hunting reindeer at Yukaghir hunting places, and not only brought in tribute payment one Chukchi, but also dissuaded the Chukchi from a campaign against the stockade. Incidentally, it is impossible to say exactly, whether the Chukchi were gathered to sail to the stockade or only threatened to do this, as was often the case.

At the end of 1713 the Anadyrsk steward Afanasii Petrov, with a detachment of servicemen and Yukaghir (Khodyn and Chuvan), set off on a campaign against the well-fortified Koryak Olyutorsk stockade. During the siege Petrov took bribes, took the procured game, and finally, requisitioned reindeer from the Yukaghir. On 2 December 1714, during the return of Petrov's detachment, the Khodyn people led by Oma, Pochina, and others attacked the servicemen in a blizzard on the upper reaches of the Talovka River, killed Petrov and 29 servicemen, and took the Kamchatka tribute collection. The surviving Kamchatka stewards, V. Kolesov and I. Eniseiskii, fled into the Aklansk stockade to the Koryak, whom the Yukaghir besieged. The latter, having taken the draught reindeer collected by the Russians from the Koryak, boasted of going to take the Anadyrsk stockade and steal the reindeer of the servicemen. In fact, on the night of 16/17 December the Yukaghir stole the Anadyrsk herd, which the Koryak were grazing. In February 1715, the Aklan Koryak, under pressure from the Khodyn people, killed ten Russians invited to a meal, and both Kamchatka stewards who were with them were wounded. They burned them in the house. During the course of these events there turned out to be former enemies in the union: The Reindeer Koryak and Settled Aklan Koryak, on the one hand, and the Khodyn people and Chuvan people on the other. When the allies dispersed, the Khodyn people thought of "going to the non-peaceful Chukchi people and calling them to themselves in evil

---

[474] PSI. Bk. 1: 405–406; cf.: PSI. Bk. 2: 530; Andrievich 1889: 122; Bogoraz-Tan 1934: 45; Gurvich 1966: 112; 1985: 54; Zuev 2002: 68, 245.

intent in a campaign on the Russian people at the Anandyrsk stockade, sometime in the summer, during the hunt for food." However, after returning to their district the Khodyn people did not continue hostile actions, they released prisoners and sent tribute to Anadyrsk, and in July 1715 the Chuvan people also relented, giving up the Kamchatka treasury collection and delivering tribute.[475]

According to the Yukaghir, the Chukchi, who would arrive in summer at the reindeer hunt on the lower Anadyr' were supposed to sail with them to Anadyrsk. This, incidentally, did not happen. Nevertheless, a union with the Chukchi was created. According to the newly baptized Yukaghir Stepan Ach'yakh, during the first snow (September-October) of 1716 the Yukaghir of the lower Khodyn clan led by Onesko and a multitude of Chukchi made a raid on the Reindeer Koryak. But along the way, the Chukchi stole some of the Yukaghir reindeer, and then stole the supply of Koryak fish stores. Because of this the Yukaghir left the Chukchi, who later set off to the Paren' River to oppose the same Koryak.[476] In general, this temporary union was clearly not durable and the allies—formerly enemies—did not especially trust each other.

In the end of 1718, at the call of the Khodyn Sopina Yukol'nikov who remained loyal to the Russians, five "Nose" Chukchi led by the prince Idoga voluntarily came to the Anadyrsk stockade to deliver tribute, in order to be subjects of Russia,[477] that is, the Yukaghir continued to be intermediaries in Russian-Chukchi relations.

Attesting to rather durable Yukaghir-Chukchi trade connections in this period was the presence of Yukaghir trading sites (to the south of the Etel'kuyum River, which empties into Kresta Bay), along which the army of D. I. Pavlutskii passed in August 1731. In their 1763 reminiscences about this campaign, servicemen refer to the market place as the "former," that is, no longer existing in 1763, but according to the description of J. Lindenau, the fair was still held in the fall in the beginning of the 1740s. The Yukaghir were intermediaries in the trade between the

---

[475] PSI. Vol. 2: 53–57, 73–75, 88–95 (quotation—p. 88); RGADA. F. 199. No. 539. Ch. 2. D. 6. L. 27; Maidel' 1894: 538; Elert 1996: 257–259; Zuev 2002: 257–261; 2004: 62.

[476] RGADA. F. 199. No. 539. Ch. 2. D. 6. L. 26; Zuev 2002: 265.

[477] AAN. F. 21. Op. 4. No. 33. L. 134–135; cf.: Müller 1996: 40–41; Efimov 1948: 224; Vdovin 1965: 12; Zuev 2002: 268.

Russians and the Chukchi; they sold Russian and Yakut goods (Chinese knives, spear points, axes, pal'ma (cutlass on shaft), kettles, flints) to the Chukchi, which the Russians themselves often could not sell to the Chukchi because of government prohibition.[478]

In 1753 a captive Koryak indicated that he had heard what the leaders of the rebellious Koryak discussed about the fact that the "Yukaghir betrayed the Anadyrsk party and communicated with the Chukchi, and now many among them, joined in conformity, and were going to murder Russian people, and are unchangeable in their malicious intention."[479] However, whether such scenario was realized is unclear, but probably not. It seems that these were only rumors about the intentions.

Meanwhile, Chukchi raids continued. In 1742 Lindenau, describing the geography of the Anadyr', noted that on the right side of the middle course of the Anadyr' there was a Yukaghir channel, "formerly Yukaghir lived here, who were killed by the Chukchi." However, the details of this warfare were not preserved in Russian sources. Such strategy was in fact probably characteristic for Chukchi raids of the mid-18th century. Here is how the Chuvan tradition describes the war with the Chukchi in the presentation of A. E. D'yachkov: "The Chukchi, knowing the recklessness of the Chuvan people, were planning to kill them by cunning, attacking by surprise or at night, or when they were noticed in small parties. . . ." Attacking and massacring all, the Chukchi hid before the Russians arrived to help.[480]

It is known that in 1753, near the Chekaevo tract, the Chukchi attacked a camp of Ivan Efremov, a tributary Yukaghir of the Khodyn clan, killing four men, taking captive seven men and seventeen women, and stealing reindeer and property.[481] This probably occurred during a reindeer hunt.

Winter attacks also occurred—standard for Chukchi predatory raids. According to depositions of the Cossack B. Kuznetskii, in March 1754 "on the upper reaches of Nalucha Creek (which is no greater distance

[478] Campaign of 1731: KPTs: 159; cf.: Vdovin 1965: 74–75, 77; Gurvich 1966: 15; Yakutsk goods: Elert 1999: 129; description: Lindenau 1983: 163; cf.: Dolgikh 1960: 436.
[479] KPTs: 127.
[480] Lindenaau 1983: 161; D'yachkov 1992: 186.
[481] Vdovin 1965: 76, 124; Zuev 2009: 160.

than 35 verst from the Anadyrsk fort), eight yurts of the tributary Yukaghir stood on a rock . . . in which there were up to 50 Yukaghir of the male sex, and the total, including the female sex and the young, he doesn't remember to indicate." A detachment of half a thousand Reindeer Chukchi fell on these Yukaghir, killing 17 men and three women, capturing 25 men and 56 women, and stealing 912 reindeer. As the Anadyrsk commander F. Plenisner noted in his report, only seven men escaped by fleeing. Preserved in the Yakutsk archive is "Reestr o yukagirskom pograblennom i vzyatom v plen nemirnom chukotskim narodom v 1754 godu pri urochishche rechke Naluche" [The Registry of the Yukaghir Pillaging and Capturing by Non-Peaceful Chukchi Peoples in 1754 in the Nalucha River Tract].[482] According to this registry, 25 households were destroyed—19 Chuvan and 6 Khodyn. The Chukchi took all valuable property: weapons, tools, kettles, tobacco, clothing, head gear, boots, tent coverings, and skis. Seized were 275 rideable buck reindeer, 120 draft females, 180 females, 386 calves, and 176 open and 212 covered sleds. The total value stolen was estimated in the inventory as the very large sum of 8,113 rubles 77 kopecks. Among those of the Chuvan people robbed were two princes (Yadacha and Nayama Sobol'kov) and one "best Yukaghir" (Ononda Chernousov), which might point, for example, to the robbery of two clans. A camp on the mountain, where B. Kuznetskii spent the night, consisted of eight *urasy* [a summer dwelling], in which 50 adult Yukaghir and youths lived, that is, an average of eight people in one "yurt." As follows, 25 pillaged households belonged to one camp, and in each urasa was on average three households.[483] In the registry only 16 different surnames are mentioned, the others being repeated: Sobol'kov, Alinov, Eremkin—in three households each, and Chernousov, Zykov, and Perebyakin—two each. The Yukaghir lived in large family groups, consisting of separate families, in one urasa; that is, one camp was pillaged.

What happened to the heads of the households is not indicated in the document; it is said only that the Khodyn Tynachin Perebyakin was

---

482 Deposition of B. Kuznetskii: KPTs: 180; losses of the Yukaghir: Vdovin 1965: 76; Zuev 2009: 160; F. Plenisner: AAN. F. 3. Op. 10. No. 137. L. 47 ob.–48 = 150; registry: Gurvich 1957: 246–262; cf.: Dolgikh 1960: 431; Gurvich 1966: 75–76.

483 G. F. Müller (2009: 242) noted that on the Anadyr', Alazeya, and Kolyma five to six families of Yukaghir lived in one "yurt."

killed by the Chukchi. The registry itself was compiled based on depositions of three Yukaghir princes (F. I. and S. I. Shmalavyi and Z. V. Krasovskii), and there is nothing at all about their households in the document. Evidently, those who compiled the registry knew the Yukaghir households of the camp well but did not know the fate of the heads of the households. The deposition of B. Kuznetskii informs us about this: in the fall of 1755 to the Chukchi

> arrived six Anadyr' tributary Yukaghir people . . . and they, the Yukaghir, arrived to ransom their captive wives and children taken from them, Yukaghir that those Chukchi accepted honestly and did not offend, and those Yukaghir ransomed their wives and children from these Chukchi, giving tobacco, kettles, axes, knives, spears, bows, arrows, and corals, and ransomed of the male and female sex in all nine people. The Chukchi gave as gifts to these Yukaghir reindeer parkas, white fawn, white footwear, and parkas of musteline and red fox.[484]

In captivity among the Chukchi were the wives and children of the Yukaghir, part of whom they managed to ransom, though one woman ran off by herself to Anadyrsk on 20 September 1755. Probably during the course of a year and a half the Yukaghir had collected ransom. As ransom they brought primarily Russian wares, receiving furs as reciprocal gifts for these. In total, six Yukaghir whose wives and children were being held captive came to the Chukchi. These six (evidently including the prince Yudacha Sobol'kov as well) were probably the surviving heads of the twenty-five households that were pillaged. The rest perished.

I. S. Gurvich suggests that, together with the Chukchi, the raid on the Yukaghir was conducted by the Koryak; indeed, in this registry it is indicated that the Yukaghir Patog Chichakhov "was robbed by the Koryak." However, it is possible to view this in another way: the reference to the Koryak in the list is present because just this household was robbed by Koryak taking advantage of the moment. In addition, it is known that in this same year, 1754, the Reindeer Koryak killed two Yukaghir princes who lived among them—the Chuvan elder Oblika and the Khodyn Nogoda.[485]

---

[484] KPTs: 183.

[485] Registry: Gurvich 1957: 247; events of 1754: RGADA. F. 199. No. 528. Ch. 2. D. 9. L. 45 ob.–46; Nefëdkin 2016a: 155-156.

In 1755, because of Chukchi raids, 65 tributary Anadyr' Yukaghir became impoverished and were left without reindeer or even without urasy. This also explains the desire of the Yukaghir to ransom not only their captured relatives but also the reindeer, necessary for migration. In the summer of 1756, together with the Anadyrsk detachment of Second Major I. S. Shmalev, who set off to the lower Anadyr' to hunt reindeer, sailed four Yukaghir men to ransom their relatives from the Chukchi. Two of the Yukaghir returned to Anadyrsk with the major, and two (the newly baptized Nikolai Dmitriev and the baptized Chukchi woman Katerina Ivanova with child) remained to visit the Chukchi elder Khargipin, who was supposed to return them to Anadyrsk on the first snow. In the fall of that year the Yukaghir prince Ivan Novikov, hoping for peace with the Chukchi, sent eight people to the latter for purchase of reindeer. However, after the Chukchi raid on the Yukaghir in December 1756, N. Dmitriev and the eight Yukaghir sent were killed.[486]

Captain Shmalev, who obviously relied primarily on oral evidence, presented 20 years later in detail in his work about Chukchi-Koryak relations, the background of the Chukchi raid of 1756. In the fall of 1751 two Yukaghir clans (the Chuvan led by Oblik and the Khodyn led by Nogoda) "betrayed" and went from the Anadyrsk stockade to the Reindeer Koryak, who at this time also rebelled against Russian authority.[487] In January 1753 the Yukaghir (25 people), together with Koryak (275 people), laid siege to the recently founded Gizhiga fort. In spring of the following year the Yukaghir princes decided to repent, to go to the Anadyrsk fort and again accept Russian citizenship. The Koryak, fearing that the Yukaghir might then lead the Russians to their land and give the latter reindeer for this campaign, staged a dispute, during which they killed all the resisting Yukaghir and divided their wives, children, and reindeer among themselves. Some of them, approximately 70 men, saved themselves because the son of Nogoda Eegoi and his relatives gave a deposition in Anadyrsk on 1 May 1754. In the summer the Yukaghir

---

486 1755: Vdovin 1960: 43, 76; 1756: RGADA. F. 199. No. 539. Ch. 2. D. 6. L. 37–38 ob.
487 According to the report of Captain V. Shatilov, 49 Yukaghir led by Andrei Alin separated (Gurvich 1966: 75), that is Andrei Oblik (RGADA. F. 199. No. 539. Ch. 2. D. 6. L. 29).

captives, who lived in the yaranga of the sister of the main Koryak prince Eigel' Lekhtelev and his brother Ivaka, waited until the men left on a hunt, killed the sister of the prince and several women, and fled to Anadyrsk. The brothers Lekhtelev, having discovered what had happened, pursued the fugitives and, overtaking them on the Main River, killed them all.

Not satisfied with this revenge, in the fall of 1756 Ivaka concluded peace with the Chukchi and provoked them into killing six Yukaghir who had gone to the Chukchi to ransom captives, even though in 1755 the Yukaghir had peacefully arrived to the Chukchi and ransomed those captured. Ivaka pointed out that it was ultimately necessary for the Chukchi to kill the Yukaghir so that the Chukchi and Koryak could live safely. Based on A. S. Zuev's assumption, the Koryak wanted to provoke worsening Russian-Chukchi relations; indeed, they knew about the negotiations of the Russians and Chukchi on the Anadyr' and possibly establishment of peaceful relations between them. However, as we see, at this time the Yukaghir themselves were formally not Russian citizens, and were mutinous. In addition, the hostility of the Koryak toward the Yukaghir could have emerged as a consequence of Chuvan aid to V. Shatilov's expedition against the Taigonos Koryak in 1751; as compensation for this aid the Chuvan people received about 120 Koryak reindeer.[488]

The Anadyrsk commander Second Major I. S. Shmalev described in detail in his report of 23 October 1757 to the Okhotsk commander the Chukchi raid that followed in December 1756: As early as August 1756 Shmalev, fearing an attack by the Chukchi, warned the Yukaghir princes Ivan Novikov and Semen Ivanov, who were on the upper reaches of the Chernaya River, not to send anyone to the Chukchi and at the first snow come to Anadyrsk. On 9 November the princes set up a camp of five urasy 25 verst from the stockade on the Shchuch'ya (Pickerel's) River tract. The major insisted that the camp be well guarded and observers were sent to watch the enemy. On 1 December the observers reported that down the Anadyr' "vapor" rising from people was noted. Learning this, Shmalev proposed to the princes to enclose the camp with a barricade and to accept

---

[488] Essay of T. I. Shmalev: RGADA. F. 199. No. 528. Ch. 2. D. 9. L. 45–46 ob.; Nefёdkin 2016a: 154-157; opinion: Zuev 2009: 168–170; campaign of V. Shatilov: KPTs: 115–120; Gurvich 1966: 75.

a Russian detachment of 50 servicemen, which the elders refused. As it turned out, the Yukaghir simply wanted to trade with the Chukchi and therefore decided not to accept the Russians. Such trust of the Yukaghir can be explained by the fact that the Chukchi had formally gone to Anadyrsk to deliver tribute (as F. Plenisner and G. Sheikin tell about this), to which, strictly speaking, they had agreed with Shmalev the previous summer. In addition, in April 1756, returning from negotiations in Anadyrsk, the Chukchi elder Minigit'ev gave the Yukaghir 100 reindeer as a sign of the beginning of friendship. The Yukaghir did not carry out Shmalev's order to come to the fort because of a freeze that was happening. Then on 4 December an old man from the Yukaghir camp arrived in the stockade on captured Chukchi reindeer to report the attack. Early on the morning of 4 December about 200 Chukchi had unexpectedly attacked the camp. In the course of this attack the prince I. Novikov and 17 other Yukaghir were killed and 14 were wounded. Twenty-three men, 58 women, and 39 children survived, as well as two people who were away. The Chukchi, judging by Shmalev's report, only took property and stole reindeer. According to I. S. Vdovin, 120 surviving Yukaghir were taken captive. Shmalev sent from Anadyrsk Lieutenant M. Bel'skii with a detachment of 200 soldiers and Cossacks, moving on skis and dogsleds, in pursuit of the Chukchi. They followed the enemy for 15 verst but were not able to catch the Chukchi, who were on reindeer teams. The reason for this, it seems, unprovoked Chukchi attack on the Yukaghir is unclear. A Yukaghir woman, freed in this same year from Chukchi captivity, informed the Russians that the Chukchi wanted to annihilate the tributary Yukaghir so that the latter did not lead servicemen to them. Ivaka also pointed this out to the Chukchi. However, this could simply have been a threat. Later, in 1760, this same toyon Khargipin explained to Lieutenant S. Kekerov, that "they, the Yukaghir, being deceiving, them, the Chukchi, previously killed Khoyav'ya and Levucha, the best of their Chukchi men," that is, it was in fact a matter of blood revenge for the Chukchi.[489]

[489] RGADA. F. 199. No. 539. Ch. 2. D. 6. L. 36–38 ob.; also see: Ibid. L. 84 ob.–85; L. 38 ob., 39 ob. (report of S. Kekerov); RGADA. F. 199. No. 528. Ch. 2. D. 9. L. 46–46 ob.; AAN. F. 3. Op. 10. No. 137. L. 48 = 150 (report of F. Plenisner); AII. F. 36. Op. 1. No. 643. L. 583 ob.; Nefëdkin 2015: 185 (deposition of G. Sheikin); RGADA. F. 199. No. 539. Ch. 2. D. 6. L. 29 (deposition of Eegoi). Opinion: Vdovin 1965: 76, 124; 2010: 515; about the numerical forces at Gizhiga: KPTs: 122; gift of Minigit'ev:

In the opinion of Corporal G. Sheikin, being in service at Anadyrsk in the 1750s, in this raid "a whole clan of Yukaghir was eliminated," and "a small number, about ten souls, was left of Yukaghir women and all those they took to Anadyrsk. And as a result, a whole clan of Yukaghir did not now dare to come anymore to the Anadyrsk stockade." However, in 1758 the newly baptized Chuvan prince Semen Ivanov with a group of Yukaghir, who survived the Chukchi rout in 1756, was at Anadyrsk.[490]

In 1763, in his report F. Plenisner, explaining the incident with the Yukaghir, noted that before 1751 there were 65 Khodyn and 87 Chuvan men in the Anadyrsk district; these Yukaghir lived between the mountains in the territory from the Yablonnaya River (a right tributary of the upper Anadyr') to Anadyrsk and were exterminated by the Chukchi in the time of Captain V. Shatilov and Major I. S. Shmalev (1750–1758): "of these (except the prince Pattoga Semeikin of the Chuvan clan with relatives of seven people presently being in residence up the river in Nizhnekovymsk), all died in past years, and others were killed and ravaged in the time of the Chukchi campaigns against them, and now at present there is no one, and those dwellings, where they were, all stand empty;" "one prince, the above-mentioned with his relatives, was left (a total of seven people), who, from hunger, migrated a few years ago back to the vicinity of the Nizhne stockade, and presently lives near the mouth of the Anyui River and pays tribute at the Nizhnekovymsk fort, and not at Anadyrsk."[491] In general, it turns out that Plenisner attributed the final disappearance of the upper Anadyr' Yukaghir to two Chukchi raids (1754 and 1756) and to epidemics. Only the Chuvan chief Pattoga with relatives (a total of seven tribute payers, that is, about 30 people) moved to the Anyui after these raids and began to pay tribute at Nizhnekolymsk.

In 1777 Shmalev noted about the Yukaghir: "Those who presently remain are quite few in number, so that of both the Chuvan and the Khodyn clans there are 4 men each, but they have become settled and they have

---

Vdovin 1965: 78, 127–128; Zuev 2009: 165; deposition of the captive Yukaghir woman: Vdovin 1965: 72.

[490] Quotations: AAN. F. 36. Op. 1. No. 643. L. 583–583 ob.; Nefёdkin 2015: 185; 1758: Gurvich 1966: 76.

[491] AAN. F. 3. Op. 10. No. 137. L. 119 (quotation), 149–150 (second quotation), 173.

residence with the Omatsk clan on the Anyui Rivers, which fall into the Kolyma River right at the Nizhnekolymsk fort." Folkloric data also confirm this: according to the stories of the Anadyr' Lamut told to G. Maydell in 1870, a Chuvan woman with a child fled from a Chukchi pogrom to Angarka; in spite of the Chukchi pursuit, she managed to escape.[492]

Travelling through the Anyui stockade in February 1792, C. Merck noted: "Around the small stockade on the Angarka . . . lives a clan of Yukaghir, the number approximately twenty people, who call themselves Chuvan and are the remains of those who lived earlier on the Anadyr'. They speak their own special language, though now, when they are baptized, also Russian; they have only a few reindeer; they eat from hunting; therefore, they constantly change place of residence."[493] The doctor evidently had in mind that in all (and not just the men) 20 people migrated around Anyuisk, who were Chuvan who had fled there from the Anadyr'.

However, some groups of Yukaghir still remained in eastern Chukotka. In the winter of 1774–1775, before their appearance at Gizhiga in March 1775, a Chukchi group of 150 warriors, who came from Kresta Bay, attacked a camp of tributary Yukaghir, killed the men, took the women and children captive, and stole the reindeer; and from those taken captive, after the battle remained a Yukaghir girl. Even much later the Yukaghir remained among the Chukchi; in fact C. Merck noted: "Also encountered among the Chukchi are Koryak and individual Yukaghir as laborers," probably having in mind those same captives of 1754.[494]

Later, G. Maydell provided a legend of the Anadyr' people about the disappearance of the Yukaghir in the following way: "The Chuvan people themselves attribute their annihilation exclusively to the Chukchi and especially point to the fact that they had to endure all the force of revenge of the Chukchi after the death of Major Pavlutskii, of whom, they,

---

[492] Quotation of T. I. Shmalev: RGADA. F. 199. No. 528. Ch. 2. D. 9. L. 45 ob., note; Lamut folklore: Schiefner 1874: 358; Burykin 2007: 383.
[493] Merck 1978: 151; cf.: Kiber 1823: 128; Ogorodnikov 1922: 260.
[494] 1775: Gurvich 1966: 76; Zuev 2009: 218; quotation: Merck 1978: 121.

the Chuvan, were the main allies."[495] In the beginning of the 1820s F. P. Wrangell also heard that "a large part of the Yukaghir and almost all the Chuvan fell in bloody battles," and the remaining Chuvan arrived on the lower reaches of the Kolyma, "driven by the Chukchi from the banks of the Anadyr'."[496] Consequently, the Chuvan on the Anyui, even after seven decades, preserved in historical memory a clear idea of their origin in the region of the Kolyma.

In general, clashes of the Chukchi with the tributary Anadyr' Yukaghir in the first half of the 18th century occurred because the latter supplied guides-interpreters and auxiliary contingents for Russian expeditions (end of the 17th–middle of the 18th century), though earlier peaceful relations had prevailed between the Yukaghir and Chukchi. The Yukaghir, who knew the Chukchi language and localities well, were actively used by the Russian administration as negotiators with the Chukchi up to the establishment of stable peaceful relations in the region and the opening of the fair on the Angarka in 1788. The Chukchi attacked the Yukaghir for the purpose of robbery, took captive the women and children, and drove off the reindeer. Both the tributary and non-tributary Yukaghir were exposed to rout, though regarding the latter, the events simply did not fall into the area of attention of the Russian administration and evidence about them is absent. The other side of the relationship was trade—the Yukaghir were intermediaries who sold Russian and Yakut wares to the Chukchi, part of which the Russians themselves did not generally sell to the Chukchi because of government prohibition against the trade of iron. From this arose some ambiguity and practically simultaneous hostility and trade relations between the ethnic groups. After the abolishment of Anadyrsk the remaining Chuvan people went to Angarka and changed to the district of the Nizhnekolymsk stockade. At the same time, the remains of the Omok Yukaghir still lived on the Anyui Rivers, who in large part died from smallpox and migrated at the end of the 17th century beyond the sea. The upper-Kolyma Yukaghir, Lamut, and even the Chuvan people moved to the lower course of the Kolyma.

---

[495] Maidel' 1894: 64; cf.: Neiman 1872: 35; Maidel' 1925: 24; Antropova 1957: 180. As C. Merck noted (1978: 120), the Chukchi "in earlier times also came out against the Yukaghir, who with their help were almost annihilated."

[496] Wrangell 1948: 218, 395, note 3.

Another part of the Chuvan people who survived went to the Reindeer Koryak, where they were assimilated by the Koryak.[497] It can be supposed that war with the Yukaghir was not a significant event for the Chukchi, such as the struggle with the Koryak or Russians—it is not reflected very clearly in the folklore. At the end of the 19th–20th century the old men of the southern Chukchi thought of the Yukaghir as brothers, and in their turn, the Kolyma Yukaghir viewed the Chukchi as relatives and allies.[498]

3. Warfare in Bering Strait

3.1. Hostilities between the Chukotkans and the Alaskan Eskimos

In the written sources, relations between the Chukchi and Alaskans are recorded from the beginning of the 18th century—these are both military and trade relations, dominated by the former. In the conflict between the residents of Siberia and Alaska the attacking side, as a rule, were the former. The American anthropologist W. Sheppard, who specially examined warfare on the basis of Alaskan folklore, found only one mention of a raid by Alaskans to the west.[499] Even W. G. Bogoras correctly compared raids of the Siberians with pirate raids.[500] Large combat operations were conducted in the summer on the water, though raids could also have occurred in winter when the strait could be crossed on the ice on sleds, as was done by the hero Elendi of Chukchi legend.[501]

Of course, members of conflicts primarily were those in the territories most closely located to each other: Cape Dezhneva in Siberia and Cape Prince of Wales in Alaska, situated at a straight distance of about 100 km. The folklore of the latter tells precisely who the enemy of the Americans was. In Alaskan legends from Wales, the land of the attackers is called Nyvkvak (Newkwak), and its residents—Nyvukagmit (Nevuqaġmit), the Nyvuqantsy being residents of the village of Nyvukan

---

497 Wrangell 1948: 218, 229–230; departure of the Chuvan people to the Koryak: Maidel' 1925: 24.

498 Chukchi ideas: Lebedev and Simchenko 1983: 132; cf.: Bogoraz-Tan 1934: 175; Yukaghir: Jochelson 1898: 260; 1900: 210–211; 1900a: 186; Bogoras-Tan 1928: 241; Gurvich 1966: 53; 1982: 197.

499 Sheppard 2009: 48, 57; cf.: Shvaittser and Golovko 2001: 30.

500 Bogoraz-Tan 1934: 79; cf.: Sheppard 2009: 52, 57.

501 Bogoras 1899a: 356–358; cf.: Gondatti 1898: XI.

(in Chukchi), that is, Naukan. In the legends of Kawerak (in the interior of the Seward Peninsula) opponents are also called Nuuvaġmiut. This evidence is also supported by the folklore of the Asiatic Eskimos: in the Naukan legend about Ekëtamyn, it is the settled residents of Naukan and Uelen—the easternmost villages of the Chukchi—who set off to make peace with the American Eskimos.[502] And that the Reindeer Chukchi participated in the maritime campaigns of the coastal residents also speaks a document of the 18th century, which depicts in detail the collaboration of the migratory and settled residents of Chukotka:

> The Reindeer Chukchi arrive to the Sitting Chukchi on reindeer and, incite the Sitting on their reindeer against the Koryak in winter campaigns, and, for their part, the Sitting Chukchi in summer take the Nose Reindeer [Chukchi], and those living in places in the tundra farther from the sea, in their baidary over the sea and along the rivers and give them their baidary in mutual friendship, and from them, the Reindeer, they take in place of their baidary different kinds of reindeer hides to dress themselves.[503]

The pyatidesyatnik V. V. Atlasov in his "story," given in 1701 in Moscow, revealed: "And the Chukchi foreigners there, who live near this nose [cape] and at the mouth of the Anadyr' River, told that opposite this indispensable nose there is an island; and from this island in winter, as the sea freezes, arrive foreigners; they speak their own language and bring scrawny sable, like the ferret animal."[504] Since the pyatidesyatnik had never been to Chukotka, he retold information provided by the coastal residents of Chukotka, who by the end of the 17th century had trade relations with the Eskimos, who spoke a different language. It could have been a matter both about American Eskimos (indeed, at that time Alaska was considered by Asians as an island) who brought the skins of some "scrawny sable" to trade, and also about the residents of Ratmanova [Big Diomede] Island, with whom peaceful relations had already been established at this time. We note, based on the description, that the

---

[502] Garber 1940: 55–59, 169–172; Burch 1998: 118; Sheppard 2009: 48; about Ekëtamyn see: Menovshchikov 1988: 214–215; Nefëdkin 2003: 312–314. About the names of Naukan: Leont'ev and Novikova 1989: 269.
[503] Vdovin 1965: 41; Nefëdkin 2003: 197; cf.: Burch 1998: 115.
[504] Berg 1946: 49; Polevoi 1967: 108.

Eskimos arrived in winter over the ice, and not as usual, in summer in baidary, which more probably indicates the residents of the Diomede Islands.

In 1711 the coastal Chukchi Makachkin told the serviceman P. I. Popov about Alaska, where he repeatedly went on campaigns, as a result of which about ten captive Alaskans were in his camp. The serviceman thus briefly laid out his story about the mutual relations of the Siberians and Alaskans: ". . . and for a long time and still now they, the Nose Chukchi, do not have peace between those island people and themselves; they go to battle with each other." In the words of Makachkin, it turns out that both Siberians and Americans went on offensive campaigns. Makachkin was not able to tell the time when conflict with the Alaskans began, but it lasted "for a long time" (that is, at least from the 17th century) and continued to 1711. However, Makachkin himself was repeatedly in Alaska: ". . . and he, Makachkin, was on this island in many years in campaigns, and they, the Chukchi, call this island Bol'shaya zemlya [the great land]."[505] In general, based on this description, the picture of mutual relations in Beringia turned out to be anything but peaceful, as in the almost contemporary account of V. V. Atlasov.

In 1718 the tributary Settled "Nose Chukchi" stated in Anadyrsk that

> opposite this their Chukotskii Nose there is an island small in form and treeless; and on this island live people of a different clan, not Chukchi, and they have their own language; and the people of this island are at peace with them, the Chukchi, and those island people are few in number. . . . And from this same island there is a large land beyond the sea . . . and they, the Chukchi, paddle in baidary in calm weather from this small island to that land in one day; and on this island live people similar to them, the Chukchi, but the language of those people is their own . . . and battle often exists between them, the Chukchi, and those people, and they battle with bows, but fire[arms] battle they do not know.[506]

In general, we see that by 1718 peace had already been established between the coastal residents of Chukotka and the Eskimos of Ratmanova

505 KPTs: 157.

506 AAN. F. 21. Op. 4. No. 33. L. 134–135; Nefёdkin 2016a: 9-10; cf.: Efimov 1948: 224.

Island, whereas with the residents of Alaska (the Great Land) they conducted frequent wars—there is no talk of trade. About the number of residents of the Gvozdev [Diomede] Islands it is possible to cite the later information of I. Kobelev: in 1779, in the two villages on Ratmanova Island lived 203 men and 195 women, and in 1791, 300 men and up to 150 women, whereas on Kruzenshterna [Little Diomede] Island—85 men and 79 women, and in 1791—45 and 55, respectively.[507]

Later, during the first campaign of D. I. Pavlutskii in 1731, Eskimo warriors from the islands in the Bering Strait helped the Chukchi, which was reported to M. S. Gvozdev by the islanders themselves in the following year. In addition, the servicemen found among the slain enemy after the third battle with the Chukchi one or two Eskimos, who were recognized by the labrets in their lips. Based on the assumption of the Russian authorities themselves, not only did the Eskimos from the islands help the Chukchi against the captain, but also some from Alaska.[508] Of course, primarily the islanders were allies of the Chukchi, connected with them through trade relations, but it is impossible to exclude the fact that even Alaskans sent aid to their opponents-partners against the new threatening enemy.

As early as 1741 the "island Chukchi" were allies of those on the mainland and gathered with them in the campaign against the Russians.[509] Participant of the Second Kamchatka Expedition, G. W. Steller, noted in the same year of 1741: "I know from quite reliable sources that the Chukchi trade with America from the second Chukchi island. Though right now because of misunderstanding they have refrained from this trade for several years, those who live on the islands continue it."[510] Consequently, the coastal residents of Chukotka traded with the Alaskans on Kruzenshterna Island. But in 1741 some conflict occurred because of

[507] 1779: Efimov 1948: 229–230; 1791: EM: 162–163.

[508] Answers of the Eskimos to M. S. Gvozdev: Polonskii 1850: 399, 400; Sokolov 1851: 94, 96; Efimov 1948: 240–241; Gol'denberg 1984: 129; Krasheninnikov 1949: 178; Eskimos among the bodies: Efimov 1948: 225; Zuev 2001: 28; assumption of Russian authorities: Belov 1956: 324, 330. About the arrival of "Toothed" in Chukotka: KPTs: 158; Okhotina-Lind and Møller 2015: 369.

[509] RGADA. F. 199. No. 528. Ch. 2. D. 3. L. 18 ob.; No. 539. Ch. 2. D. 6. L. 68 ob.

[510] Steller 1995: 74.

which the "Chukchi" stopped going for trade to the Gvozdev Islands. Probably, conflict arose between the coastal residents and the islanders.

The Cossack B. Kuznetskii, who was a captive of the "best Reindeer man" Mego from March 1754 to February 1756, thus described relations in Bering Strait:

> And when he, Kuznetskii, recognized conversations and somewhat learned to speak, and then in those their conversations he heard repeatedly that they fight not only against the Russians and Koryak, but there is an island directly in the sea from Chukotskii Nose, on which live Chukchi, being called toothed, with whom they had war before this, and now they live in friendship, and from this island there is another island, which they call the Great Land, and they reckon so many people as there is sand, which people they call Kykhmyshchi. . . . To this Great Land they go for war through the bays in leather baidary, not seizing that island where the toothed live, but in dry places on foot and they carry the baidary in their hands, while in winter they do not go.[511]

Consequently, in the mid-1750s the residents of Chukotka were at peace with the Eskimos of Ratmanova Island, which was established at the beginning of the 18th century; however, in 1741 this peace was interrupted by some incident, which is clear from the indications of G. W. Steller. Attesting to this conflict were two captives from Ratmanova Island who arrived in 1761 in the Krasnaya River tract together with the Chukchi elder Khargipin.[512] In the middle of the 18th century the war with the Eskimos of Alaska was in full swing—the serviceman does not even mention trade. The Cossack notes that they fight only in summer, sailing to America in baidary, that is, rather large campaigns took place. In general, there is no mention in this description of trade in Beringia—only confrontation.

In 1756 the toyon of the Reindeer Chukchi Tevryui told the Russians:

> There is a Great Land beyond the sea . . . and they, the foot [Settled] Chukchi, have war with these people every summer; and they, the people of this land, fight naked with the foot Chukchi. . . . And now with these people the river foot Chukchi are at peace and friendship, trading, barter for marten parkas.

---

[511] KPTs: 182.

[512] Belov 1954: 183.

> Those people reassure the Chukchi: "When the Russian people will look for them, then they should let them know quickly, because they will be in every readiness to aid them."[513]

In spite of the similarity of this description with Kuznetskii's story, there is an important difference: before the mid-1750s the settled residents of Chukotka carried out raids each summer in Alaska, but now peace had been established, and even military alliance. On the one hand, it is possible to think that in the translation the Great Land (Alaska) was simply confused with Ratmanova Island, but, on the other hand, the presence of marten parkas and the method of fighting nude clearly attest to Alaskan Eskimos. Incidentally, perhaps Kuznetskii and Tevryui tell about different periods in the relationship. Then at some point peaceful relations were established between certain communities in Chukotka and Alaska.

In the summer of 1755, having migrated to the coast, the Chukchi together with Mego arrived to the relatives of the latter—the Settled Chukchi. Kuznetskii met two women and one man there who four years before had been taken from the Great Land and who, in fact, told the Cossack about Alaska.[514] Subsequently, one of the raids on Alaska occurred in the summer of 1751.

In depositions, given immediately upon return, the Chukchi Cossack N. I. Daurkin related, among other things (1764):

> And there is trade between the Chukchi and the people living on those islands of Imyaglin and Imyalin for marten, fox, wolverine, lynx, wolf, and otter, because those two islands are not a great distance from the Chukchi cape. . . . And the third island, Akibyan, those living on this one are in agreement with the Great Land. And when those Chukchi go in baidary to Okibyan' and to the Kynmyt people on a campaign they take from them wives and children captive, also musteline parkas and other things.[515]

Consequently, in this period the coastal residents of Chukotka traded on both Gvozdev Islands and were in a peaceful state. Military activities were conducted against the residents of Alaska and their allies from King Island.

---

[513] RGADA. F. 199. No. 539. Ch. 2. D. 6. L. 81 ob.–82; cf.: RGADA. F. 199. No. 528. Ch. 2. D. 3. L. 26 ob.

[514] KPTs: 182–183.

[515] RGADA. F. 199. No. 528. Ch. 2. D. 3. L. 34 ob.; cf.: Gil'meren 1876: 462.

Figure 32. Alaskan Malamute Eskimo with labrets (1865). (Reproduced based on Whymper 1869: 159).

N. I. Daurkin also briefly wrote in his report (1765) that the Chukchi, with the residents of the Great Land (Alaska), "live unpeacefully from a long time ago, since they alternately go to battle each other. They fight with spears and bows, and on their arrows, as also on spears, the points are of crystal, which is very hard and which they poison with the toxin of the plant (buttercup) so that, if a wound made by them is not immediately sucked, the wounded would die in 24 hours."[516] In general, Daurkin retold information that was given to him by the Chukchi. It turns out that the hostility between the Asians and Americans was long and they conducted war on each other from both sides, that is, it was not just the residents of Chukotka who were the aggressors. However, the information about mutual attacks could simply be a generalization of a single raid.

In the "explanation" to his 1765 map this same N. I. Daurkin signed near the mouth of the Kheuveren River: ". . . the fort was built up, and the structure of wood; and among them there is an elder called by them by the name Inakhlun, who is short and fat, but strong; who arrived not long ago in years from a distant land, and precisely, at my request, they tell about him that in1761 he arrived and built the fort."[517]

Formerly researchers compared the Kheuveren River with the Yukon; however, more convincing seems the opinion that it is said about the locality of Port Clarence, about Toksook Channel between Grantly

---

[516] Pallas 1781: 248; see: Alekseev 1961: 18–29.
[517] Alekseev 1961: 33–34; Fedorova 1964: 98; Ray 1964: 5.

Harbor and Imuruk Lake, perhaps about the village of Kawerak located farther toward the center of the Seward Peninsula. Based on the story of the Chukchi to Daurkin, a new elder, who acquired authority as a consequence of his strength, arrived from afar in this Eskimo settlement. It is he, according to Daurkin, who in 1761 initiated the construction of a wooden fort, which clearly was not the usual affair among the local residents; this is how the information appeared. In an illustration on a map a fortification is represented, defended on one side by a river and on the other by a palisade, and having one highly raised observation platform, on which are warriors.[518] The structure itself, clearly a fort, attests to active warfare in the region. And since the fortification, judging by the map, was erected not far from the coast, the primary enemy against which it was erected was maritime, primarily the residents of Chukotka. We note that in the 19th century Kawerak was a large village, in which were seven large houses where three hundred residents lived.[519]

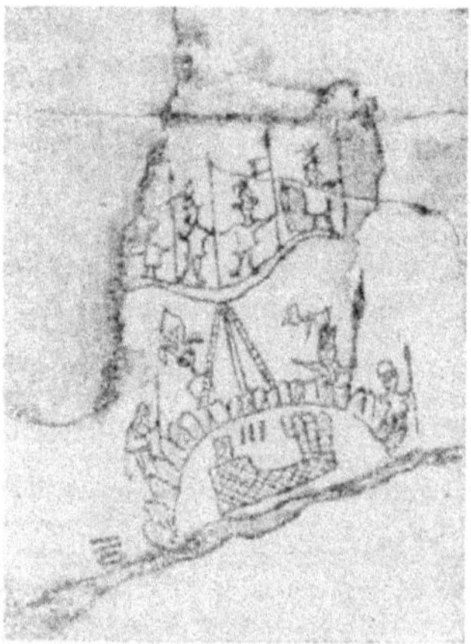

Figure 33. Illustration of a wooden fort at the mouth of the Kheuveren River on the map of N. I. Daurkin (1765). The palisade of the fort abuts the river; four warriors with headgear decorated with feathers stand on an observation platform holding spears in their hands. (Reproduced based on Alekseev 1961:39).

---

[518] Kheuveren—Yukon River: Efimov 1948: 148; Markov 1948: 33, 43; Kuzitrin River: Fedorova 1964: 104, 110; 1971: 163; Port Clarence: Alekseev 1961: 74; Kawerak: Ray 1964: 5–10; Koyuk River: Sverdlov 1992: 67; mouth of the Anadyr' River (Kagyrgyn): Burykin 2013: 384–389. Illustration of the fort, see: Alekseev 1961: 39; Fedorova 1964: 98, Fig. a; about this fortification, also see: Efimov 1948: 138; Medushevskaya 1954: 118; Fedorova 1960: 99 (construction of the fort for the purpose of defense against the Chukchi).

[519] Ray 1964: 10–11.

On the other hand, in Daurkin's time the residents of Chukotka had stable peaceful relations with the Eskimo of Ratmanova Island. Daurkin himself arrived in October 1763 on the island on reindeer sleds over the ice with only one companion, a Chukchi. The Eskimos, according to his report, accepted him hospitably, gave him a parka of fur of sable and marten in exchange for a few leaves of tobacco. Also, it was on this island, according to Daurkin's map, that Eskimos of Alaska ("Kmiug`mylsy") arrived for trading furs.[520] Obviously, the selection of the place for trade was favorable to both sides, Asian and American, since the island had a convenient location in the middle of the strait, and, evidently, the islanders had more or less stable relations with the two adversaries. In general, the information of Daurkin, who personally visited the island is one of the earliest pieces of evidence for trade between the residents of Alaska and Chukotka, though obviously the dominant type of relations were hostile. Probably toward the middle of the 18th century the Eskimos of Ratmanova Island were at peace with the Alaskans.

On 5 August 1763 the Settled Chukchi "best man" Khekhgitit arrived at the mouth of the Krasnaya River for a reindeer hunt; he told this about the wars with the Alaskans:

> Those same people of the Great Land being mentioned have little accord with the Chukchi who live on Chukotskii Nose and with toothed people on the above-mentioned Large island, because those Chukchi and the toothed people occasionally go to them on campaigns and conduct battle among them from both sides and shoot from bows with arrows and fight with spears; they, the Great Land people, have bows of spruce backing, and some, of the reed tree, and when they engage in battle with the Chukchi or with the toothed people, then they undress for ease of quick avoidance of the arrows in their nakedness. . . .[521]

Consequently, it is clear that at the beginning of the 1760s the residents of the Chukchi Peninsula and Ratmanova Island fought with the Alaskans. It was the Asians who went on campaigns to Alaska, and not the reverse. The residents of Chukotka (primarily the Eskimos) were allies of the

---

[520] Report of N. I. Daurkin about the trip: Pallas 1781: 246; trade on Ratmanova Island: Efimov 1948: 138; Alekseev 1961: 33.
[521] KPTs: 186; cf.: Belov 1954: 184.

residents of Big Diomede Island.

In this same year the "Chukchi girl Itteni" told about herself:

> By origin she is from the Great Land, to which the toothed Chukchi residing on a Large island in the sea arrived for a fight based on their disagreement, and she was taken prisoner by them when she was ten years old and taken away; and went from that land to the left, and lived among those Chukchi in service for two years and was sold by those toothed Chukchi to the Reindeer Chukchi here by the name Khargypenu, who gave an iron spear and two white fawns for her; she lived with him one year, and she was brought by him to the lower Anadyr' River to the Chekaevo tract and given to the Anadyrsk Cossack Aleksei Shipunov, for which he, Shipunov, gave to that Chukchi a copper kettle, and was brought here by him, Shupunov, in 761 and christened by baptism and was given the name Tat'yana.

Thus, the Alaskan Eskimo woman Itteni was taken captive by the residents of Ratmanova Island, then sold to the Reindeer Chukchi, who then gave her to a Cossack. Judging by the text, the raid of the Ratmanova people on Alaska was in 1758, when they seized the ten-year-old girl. Thus, the "Chukchi" crossed some "Kuka" River half a verst wide that flows into the sea. This could be the Kobuk River about 0.5 km wide.[522]

In 1778 the situation in the strait is described by Captain T. I. Shmalev, from the words of Chukchi toyons who arrived in Gizhiga:

> In previous years the Chukchi went to them on campaigns repeatedly in baidary. And they tell that they fight naked from bows and with spears. The Chukchi always observe where they take off their clothes, then these were first repelled, and because of this many of them are also killed and taken captive. But the Americans were not left to pay, because the Chukchi were often taken based on the weather from Chukotskii Nose to this land in a baidara or two, then the Americans did not release them alive, and took the wives and children captive and kept as servants.[523]

In general, we see that the Asians emerged as the aggressive side, whereas the Alaskans only defended and killed by chance hunters thrown up by a storm.

---

[522] KPTs: 186–187. Based on the assumption of L. S. Berg (1946: 114), Kuka is the Koyuk River which falls into Norton Sound (pro: Markov 1948: 34).

[523] RGADA. F. 199. No. 539. Ch. 2. D. 6. L. 24 ob.

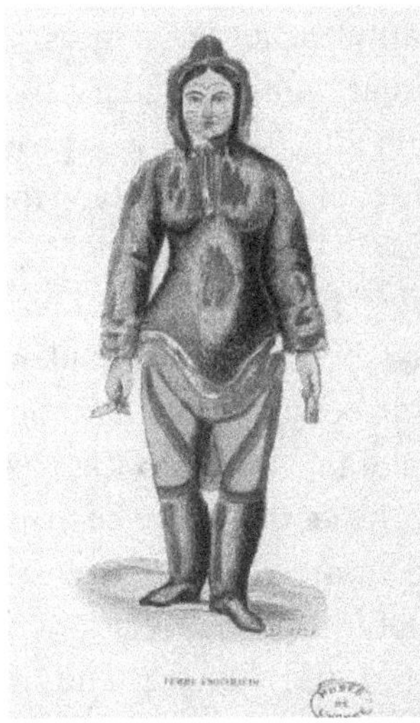

Figure 34. Eskimo woman (first third of the 19th century). (Reproduced based on Wahlen 1844: 341).

Raids to Alaska can also be attested by information from sotnik Ivan Kobelev, who in his diary in 1779 wrote about the "foot Chukchi" Ekhipka Opukhin from the "Kangunskii Evunminskii stockade"—Ekhipka "had been on American land for a campaign and for trade up to five times and had a friend on Ukipan Island. The mentioned friend arrived from that to Imaglin Island."[524] Consequently, the coastal resident Ekhipka had a companion on King Island, with whom he met for trade on Ratmanova Island, whereas Ekhipka himself went to America about five times—probably primarily for military purposes, for seizing valuable goods.

Telling about his trip to the islands in Bering Strait in 1789, I. Kobelev noted that "formerly the Chukchi went repeatedly to this land on campaigns in a hundred baidary." Also agreeing with this was Captain Shmalev, who somewhat earlier, in 1778, wrote about raids to Alaska: "Formerly the Chukchi went to this land on campaigns in a hundred baidary, but now they have ceased and have only trade with the island residents."[525]

---

[524] Efimov 1948: 231; cf.: Fedorova 2011: 17.

[525] I. Kobelev's quotation: Vdovin 1944: 263; cf.: Vdovin 1950: 96; 1965: 93; Malaurie 1974: 140; Nefёdkin 2003: 197–198; T. Shmalev's quotation: RGADA. F. 199. No. 539. Ch. 2. D. 6. L. 24 ob., note.

These were probably earlier large raids, in the same 18th century. If we consider that the usual hunting baidary were involved in the raids, then there were 800 people attacking. If it is proposed that large baidary, which held 20 to 30 people each, were involved then those attacking would be many more.

During I. Kobelev's first trip to the Gvozdev Islands in 1779 no active hostility was noted between the residents of the two Gvozdev Islands. Even the coastal tributary "Chukchi" only required the Eskimos of Ratmanova Island to take the sotnik to the next island, Kruzenshterna, but not to Alaska, where he could have been retained or killed, which would bring a negative reaction from the Tsarist government. In general, the Chukchi had rather stable friendly relations with the residents of the western Diomede Island. Later, at the end of August–beginning of September 1791, two Ratmanova Island residents, who had been in Alaska, fearlessly joined the Reindeer Chukchi from the J. Billings expedition. Attesting to the non-hostile relations of the Eskimos of Kruzenshterna Island with Alaska is the fact that in 1779 a chief on the island was a settler from America.[526]

Captain G. A. Sarychev thus wrote about the relations in Beringia at the end of the 18th century: the Chukchi "sometimes, by their inherent custom toward pillage, sail across Bering Strait and attack the settlements of the savage Americans with arms; robbing them and taking prisoners, they depart with the booty." Another description by another member of Billings' expedition, Dr. C. Merck agrees with the preceding: "For the same purpose, that is, robbery, the Settled Chukchi go in their baidary to America, attack the camps, kill the men, and take the women and children as captives; as a result of the attack on the Americans they also obtain furs, which they trade to the Russians." Further Merck describes events contemporary to him:

> In the present year the Settled Chukchi went in several baidary to America for a raid, but found empty dwellings, and the resident were all gone, fearing the revenge of the Chukchi, since in the previous year during a storm three Chukchi baidary of four were thrown up on the American shore, and the Americans

[526] Circumstances of 1779: Efimov 1948: 230; Eskimos in J. Billings' expedition: Sarychev 1952: 244; EM: 55.

> killed all the men. The Chukchi of the fourth baidara, since the old cover of their boat burst and their prospect only remained to drown, they stabbed each other with knives.[527]

As we see, generally the situation in the strait was not changed: the "Chukchi" attacked the Alaska coast to seize furs and captives. Merck clearly attests that raids were conducted by the Settled Chukchi, that is, primarily Asian Eskimos. Thus, Merck mentions two conflicts that occurred on the shores of Alaska. In 1790 four baidary arrived on the shore from Chukotka by a storm. These baidary were probably not the large campaign type but rather hunting baidary with an eight-man crew. The male part of the crew of the three baidary thrown onto the shore were killed by the Alaskans. There were also probably a few women in the boats who could have been taken captive. The fourth baidara was still afloat, but its leather shell was damaged and it was leaking. The mariners killed each other with knives—preferring to accept an honorable death and go to the upper people than to drown or surrender, followed by inevitable death after the usual, in such situations, torture.

Based on the law of blood revenge, in the summer of 1791 a responding raid by coastal residents of Chukotka in several baidary followed. The Alaskans, realizing the inevitability of revenge and, learning from the islanders about the intentions of the enemy to sail to them, hid. As a result, this raid came to nothing, though evidently revenge had to follow later.

Merck clearly received his information from interpreters of the expedition of I. Kobelev and/or N. Daurkin, who had sailed to Alaska in the summer of 1791. Kobelev tells about the beginning of the expedition in his log book on 26 May: "Being at the Uvelensk [Uelen] stockade, I found the best foot man Opreya, with whom through the straits to the American land (he always had friendship with the Americans) I would see the local peoples." Ten baidary set off from Uelen, then on the Diomede Islands their number was increased to seventeen, as much as can be judged based on a letter by N. Daurkin made for J. Billings on a walrus tusk. There is no report about the trip by Daurkin himself, but Kobelev kept a log in which under 11 June is the following entry:

---

[527] Sarychev 1952: 186, 188; Merck 1978: 121, cf.: 112.

On this day we went across to the American land, to a village that they call Kigigmen, in which I myself was. The residency is not large; there are up to fifty yurts . . . only there are no resident people there. It is said, allegedly from an immense famine, leaving their villages, they left to northern and southern countries. . . . Besides they found out in winter (they arrived to them on dogsleds from the local land of the Uvelensk [Uelen] stockade, when the sea was covered with ice), that those of the local shore intend to go to them on a campaign, who used to go on them fairly often before on campaigns, killed many, took wives and children captive; in large part, it seems they left from fear. However, as was apparent to me, and seemed the most correct, they gathered baidary up to twenty to go on a campaign, a total of up to a hundred and fifty people, to the northern country to the locality by the name Tapkhan.[528]

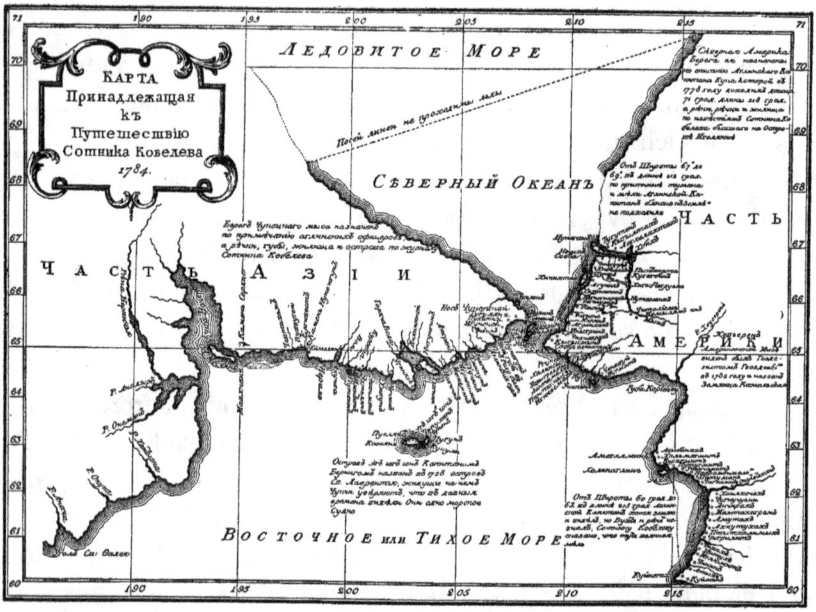

Figure 35. Map of Chukotka and Alaska composed by sotnik I. Kobelev in 1784. (Reproduced after *Letopis' Severa* [Chronicle of the North]. 1957. Vol. 2: 125).

528 Quotations of I. Kobelev, see: Chernenko 1957: 131–132; cf.: Slovtsov 1856: 20–21 (Alaskan Eskimos learned about the raid from the islanders—message to the Senate of the Irkutsk governor general A. Pil' from the report of J. Billings based on information from I. Kobelev); Alekseev 1961: 73–74; Burch 1998: 119; Sheppard 2009: 52. About the concurrence of C. Merck's notes with the log of I. Kobelev: Chernenko 1957: 135.

Thus, according to the description of the sotnik in the first half of 1791, between the residents of Uelen (then still Eskimos) and the residents of the large village of Kygmik on the Seward Peninsula there existed for some time friendly connections. During a winter trip on sleds, the Uelen people warned the Kygmik people about an impending raid, which had become known earlier, at least half a year; indeed, the attackers had been getting ready for the campaign, assembling forces. If this information is compared with Merck's account, then Merck is talking about the past year's murder of those thrown up by the storm, as a result of which the conflict emerged and the raid had been prepared. Since the Uelen people themselves did not participate in any of the warfare, then the Alaskans probably did not kill Uelen residents but rather those of another village. Having learned of the impending raid, the residents of the coast went into the depths of the mainland, probably warned by observers who saw a flotilla of 17 baidary in the sea, in which I. Kobelev sailed. And not only the Kygmit people fled. During the voyage south from Cape Prince of Wales to Port Clarence Bay, Kobelev noted: "Along the seashore all the villages are often small, only empty."[529] Consequently, the information spread along the coast and the residents fled. And they ran not only to the south and to the north of the mainland, but also to King Island, where the sotnik met about ten people, who had arrived in three baidary from the Kheveren River as early as the summer of 1790. The Alaskans apparently took Kobelev's expedition for an anticipated military campaign, which can be judged by Billings' report to the Irkutsk governor general I. A. Pil' of 31 July 1792, written on the basis of Kobelev's report: "That same summer, when I visited with the above-mentioned Americans, the Chukchi intended, as was their custom, to go to America on a campaign, and the Americans learned of their intentions through the islanders, and having left all their villages, located nearby, dispersed from them to different places in the north and south, and others to Ukipen Island."[530]

Kobelev's expedition was not planned as a military one, but the Alaskans took it for another one—the one that was told by the sotnik himself. No Uelen people were in the expedition, but rather residents, as they said, of "the local shore," perhaps Naukan people; indeed, as

[529] Chernenko 1957: 132.
[530] Fedorova 1971: 167; Divin 1979: 400.

indicated in the log, and before this, they had carried out raids many times. This expedition actually occurred: two dozen small hunting baidary went, in which should have been on the order of 160 people. They set off to the north from Kygmik to the region of Shishmaref on the Seward Peninsula against the Tapqaġmiut Eskimos. Merck described the end of the expedition: there also the residents hid.

Another member of J. Billings' Northeast Expedition left a story about a similar subject: M. Sauer, the secretary to the captain commander, concluded on the basis of his observations in August 1791 in Lavrentiya Bay, that in trade among the local residents were women "taken captive from their American neighbors, with whom they often waged war. The reason for the last event between them was this: both sides, meeting on the hunt for sea mammals, argued; a skirmish ensued, in which the Americans seized one baidara and took the crew captive; the second, returning, brought reinforcements, made a raid on the American coast, seized several women, but then peace was restored." At this time Billings stopped in the northern part of Lavrentiya Bay at a village called by members of the expedition Unyagma—evidently Nunyamo.[531] Sauer tells about a specific case that served as the reason for the emergence of hostile relations—a conflict in hunting in the strait. Probably one village in Chukotka was involved in the conflict, the residents of which carried out a responding raid, after which the conflict ran its course and peace was restored, which was being maintained in 1791. This was a small hunting conflict, but at the same time typical for the period.

[531] Sauer 1802: 252. About the site of the expedition at the beginning of August 1791: Sarychev 1952: 182, 276.

Figure 36. Chukchi yarangas in the bay of Svyatogo Lavrentiya (1791). (Reproduced based on *Puteshestvie kapitana Sarycheva po severo-vostochnoi chasti Sibiri, Ledovitomu moryu i Vostochnomu okeanu* [Voyage of Captain Sarychev along the Northeast Part of Siberia, the Arctic Sea, and the Eastern Ocean]. St. Petersburg: Tipografiya Shnora, 1802. Pl. 4.

In 1791 the Alaskan Eskimos complained to Billings that the foot [Settled] and Reindeer Chukchi "go almost annually in large numbers on a campaign across the strait in baidary to the American land, decimate by murder the American people, plunder their property, divide it among themselves, and take captive the wives and children." The captain even ransomed an "American woman" from the Chukchi and took her to Yakutsk, but the Chukchi did not give up her two children,[532] considering them theirs. The Eskimos who lived at Cape Rodney (west of Nome on the Seward Peninsula), where the expedition of J. Billings stopped on 28–29 July 1791, complained about the Siberians.[533] Subsequently, in the second half of the 1780s, after the termination of war with the Russians and Koryak, the primary front of warfare for the residents of Chukotka was Bering Strait. Raids became practically annual—the thirst for booty pushed toward this.

[532] Divin 1979: 400; cf.: Sauer 1802: 251–252; Slovtsov 1856: 20; Vdovin 1950: 96.
[533] Sauer 1802: 243–247; Sarychev 1952: 179–181; "Zapiska o chukotskom narode." 1858: 103; Fedorova 1971: 167.

On 17 January 1793, the State Council discussed J. Billings' report, in which the captain noted that "the northeastern Americans, expressing the desire to have friendly treatment with the Russians, ask for protection from attack and robbery by the Chukchi." The council, according to the recommendation of the Irkutsk and Kolyvansk governor general A. I. Piell, decided to leave the request without a direct answer, only giving hope for the future so that the Russians did not bear high material costs and not aggravate peaceful relations with the Chukchi.[534]

On 31 July 1794, in the instructions of the merchants G. I. Shelikhov and A. E. Polevoi to I. F. Popov, the director of their company, it was pointed out: "It is highly desirable to us that You more firmly establish a trade fair on the American cape lying directly [= opposite] the Chukotka land. On this cape the Chukchi receive martens and other good animals, and the Chukchi receive more from the Americans by taking than by trading. And if You locate yourself well in this place and coddle the Americans, then, we think, the benefit will be good."[535] Consequently, in 1794 the Russian merchants believed that the residents of Chukotka constantly sailed to the Seward Peninsula lying opposite the Chukchi Peninsula for furs, but did not receive them as much by trade as by force.

The exiled Polish brigadier Józef Kopeć described the position in the strait this way (1796): "Between the latter [the New World.—A. N.] and the mainland exist three islands. The Chukchi attack these islands and take from the residents all the property." Evidently the brigadier had in mind attacks of the coastal residents of Chukotka on the Eskimos of the islands of St. Lawrence, King, and Sledge, or he simply incorrectly generalized information about the three islands of the strait (the Diomedes and St. Lawrence Island), and Chukchi raids. Even at the beginning of the 19th century the situation with trading and raiding in the strait, it was thought, did not especially change.[536]

As we see, the Russian documents show us that the hostile relations between the residents of Chukotka and Alaska still continued in

---

[534] *Arkhiv Gosudarstvennogo soveta*. Vol. I. Pt. 2. 1869: 261–262; cf.: Slovtsov 1856: 20; Divin 1979: 397–401.
[535] Pavlov 1957: 55; cf.: Vdovin 1964a: 120.
[536] Quotation: Kopets 1896: 997; beginning of the 19th century: Gedenshtrom 1830: 100.

the first half of the 1790s, that is, up to the beginning of active exploration of Alaska by the Russians. Let us now examine data from the folklore of the Alaskan Eskimos. On the basis of these data it is possible, following the American anthropologist W. Sheppard, to distinguish nine regions where warfare of the Alaskans against the Siberians occurred:

(I) Along the northwest coast of the Seward Peninsula, from modern Wales to Shishmaref in the north and farther: 1) the village of Kinygan (Kiŋigan, modern Wales)—the place of many battles with the Naukan people and/or Uelen Chukchi; 2) Milytavuk (Miqłliqtaġvik, Miłłitaavik), where two Siberian baidary arrived and began to scout the area, but the crews were killed; 3) Pingu (Piŋu): a father and son repelled an attack by Siberians; also the local Eskimos here got in an ambush; 4) Ikpik; two baidary from Chukotka were met first amiably and set up camp, but then the Eskimos (Tapqaġmiut) attacked and killed them all; 5) Yakpatakhak (Yakpatakhaq): the Siberians slaughtered the whole village, except one person, who escaped to Kinygan; 6) Kotlik Lagoon, north of Cape Krusenstern, north of Kotzebue Sound: annihilation of a party of Siberians.

(II) The region of Port Clarence: (7) Sin'yugamiut (modern Teller): two-day battle of Kinygan and Sin'yugamiut warriors against Nykvak Sibiryans, who arrived in two baidary, carried out trade, and then extermination of the remaining Siberians as well—the last battle near the village.

(III) The region of Kawerak (Qawiaraq) in the interior of the Seward Peninsula. According to the folklore of the Kawerak Eskimos, their war with the Siberians began after one arrived for the fur trade in Port Clarence and kidnaped a woman, after which an expedition to Cape Dezhneva followed and the kidnaped woman was returned; 8) Tuksuk Channel between Grantley Harbor and Imuruk Basin: there were several attacks by Siberians there, met on the eastern end of the channel by Kawerak warriors; before the attack the Siberians had already destroyed several small villages; battling with them continued for two days. Based on E. Burch's information, numerous battles against the Naukan Eskimos and Uelen Chukchi occurred in this locality (Tapqaq); the women hid in caves, sometimes they threw stones from the steep bank of the channel onto the baidary, or even diving, ripped their bottoms; the severed heads of the Siberians were left on the place of victory. According to one of the legends, three baidary arrived, of which the first was captured by ambush, the second also, while the third simply retreated; according to another legend, the warriors of the Eskimos set an ambush in a willow thicket, while the women hid in a cave; in a third legend the Kawerak warriors, having discovered the approach of the Siberians, began to dance and jump in order to show the enemy that they knew about their approach; 9) the village of Kawerak itself at the eastern part of Imuruk Lake—the place of many (according to W. Sheppard—a minimum of two, but more probably four or more) attacks by the Siberians, during the last of which the local Eskimos, notified by advance scouting, constructed a fortification of poles and coverings of baidary for protection of the women and children, as well as reinforced the approaches to the fortification with hidden pits with spikes and stakes; in battle the Eskimos, jumping from ambush in the willows, killed all the Siberians, leaving only messengers about the battle; and in the second half of the 19th century the remains of these Siberian baidary were still seen on the shore.[537]

Figure 37. Village of Alaskan Eskimos in the fall (1866). W. Dall calls this village Kegiktowruk—it is the village of Kiktaguk on the southern shore of Norton Sound 22 miles (35.4 km) east of Saint Michael Redoubt. (Reproduced after Dall 1870: 128–129).

[537] Burch 1988: 231 (point no. 6 was taken from the text); 1998: 223–229; Sheppard 2009: 51–55 (excluding point no. 6, derived from information of I. Kobelev); Malaurie 1974: 147. About the beginning of conflict between the residents of Chukotka and Kawerak: Sheppard 2009: 57, 59, n. 12; cf.: Shvaittser and Golovko 2001: 30.

Of course, folklore does not give us a complete picture of the raids, outlining this or that subject or episode, tying it to a specific geographical point. And one and the same episode can be outlined from different points of view with varying details and thus be perceived as different events. In folklore there is usually either the most significant events or latest, in our case—warfare, in which the victors, according to the genre, were the Alaskans. Information from folklore does not give clear chronological boundaries, but it is clear that the events belong to the pre-firearms period, that is, earlier than the middle of the 19th century. The attackers in the Alaskan folklore are residents of Chukotka who arrive in several baidary, probably large campaign baidary. In such case, those who attacked were no more than hundreds. A raid of such number of arrivals was designed primarily as a surprise attack, and not for a frontal battle, since having gathered together the Alaskans could assemble a much larger number of warriors. The object of the Siberians' attack was the western coast of the Seward Peninsula, primarily Wales. On the whole, the folklore of the Alaskan Eskimos gives us a good idea of the place and course of the raid, but not about the chronology of the events.

Information obtained from the local residents at the boundary of the 1870s and 1880s by the American zoologist and ethnologist E. Nelson states:

> In ancient times the Eskimo of Bering strait were constantly at war with one another, the people of the Diomede Islands being leagued with the Eskimo of the Siberian shore against the combined forces of those on King Island and the American shore from near the head of Kotzebue Sound to Cape Prince of Wales and Port Clarence. . . . at one time the inhabitants of the lesser Diomede Island became angry with those of the greater Diomede Island and united with the people of Cape Prince of Wales against them, but were defeated.[538]

Thus, Nelson's informants believed that the residents of both Gvozdev Islands, in union with the residents of the Chukchi coast, once fought against the Eskimos of King Island, whose allies were the Alaskans of the northwest coast of the Seward Peninsula; the residents of Kruzenshterna Island were still in a union with the Eskimos of Wales (Kinygan) and Port Clarence in the first half of the 19th century.

[538] Nelson 1899: 330; cf.: Vdovin 1965: 57; Burch 1998: 118; Sheppard 2009: 47.

According to the vague memories of the modern residents of Little Diomede (Kruzenshterna), they were now on the side of the Siberians, now the Alaskans, depending on the circumstances. According to Alaskan folklore it is also known that the residents of Kruzenshterna Island were sometimes in union with the Eskimos of King Island, the villages of the western Seward Peninsula—Kinygan, Tapkak, and Kawerak. According to C. Merck, the residents of the Diomede Islands and King Island "depend on the Americans, just as the Settled Chukchi on the Reindeer, and fear them. But the residents of Imaglin Island are more closely connected with the Chukchi."[539] Consequently, the islanders on the whole gravitated toward the Eskimos of Alaska who were close to them in language, but the residents of Ratmanova Island, by virtue of geographic location, were more closely connected with Chukotka.

On a map of Siberia, composed in 1715 in Tobol'sk by the captured Swedish officer Philip Johan von Strahlenberg (1676–1747), on a large island in the strait depicting Alaska, there is the inscription: "Those people living on this island differ from the inhabitants of the mainland by language and customs; the mainland residents call them Pukhokhotski or Pogukotski (Puchochotski, Pogukotzki); the latter conduct war with the Gyukhegy (Giuchiegy)," where as the Gyukhegy/Gyukhetsi (Giuchegy, Guichetzi) themselves appear on the map as living on small islands between Chukotka and the Great Island. On a map from the materials of the researcher of Siberia D. G. Messerschmidt, the Pogukotski (Pogukotzcky) are those living on a large island, denoting Alaska, while the Gyutegy (Giuthiegy) are shown on an island opposite Chukotka (1720–1726). Though L. S. Berg proposes comparing the latter with the Alaskan Kigi, and the Pukhokhodtsy with the Eskimos of the Gvozdev Islands, who wear clothing of bird skins, judging by the maps, the ethnic attribution indicates otherwise: the Gyukhegi appear to be Eskimos of the Diomede Islands, primarily large Ratmanova Island, who conducted war with the Alaskans.[540] It is possible that the author confused ethnonyms,

---

[539] Information from Little Diomede: Shvaittser and Golovko 2001: 31, 35, note 9; data from Alaska: Sheppard 2009: 47; alliance of the residents of Kruzenstern Island with the Alaskans: Hickey 1979: 419; information of C. Merck: Merck 1978: 100.

[540] Berg 1945: 174; 1946: 55–56; map of D. G. Messerschmidt: AAN. F. 98. Op. 2. D. 13. L. 1; Tunkina 2013: 241; Fig. 12. Comparison of the Pukhokhodtsy with Alaskans: Markov 1948: 22.

but it can also be supposed that these names are Russian renditions of some local ethnonyms. However, most important for us is the fact of the constant conflicts between the residents of the islands of the strait and the Alaskans during the middle of the second decade of the 18th century, which was also noted by the Russian informants of the Swedish captain.

If the above-stated data is placed with the information of written sources, it is possible to say that the residents of Ratmanova Island were in a state of peace with the Siberians in 1718; they helped the Chukchi in 1731. Warfare between the islanders and the Alaskans was conducted earlier than 1718, that is, at the very beginning of the 18th century or even earlier, according to W. H. Dall—approximately in the middle of the 16th century.[541] The date of war between the residents of the two Gvozdev Islands is unclear: This could be both a long conflict (in an earlier period), and also a brief confrontation that could have taken place also in the course of the 18th century. In 1732 the residents of both islands opposed the Russian mariners led by M. S. Gvozdev, which would suggest their interaction.

At the end of the 18th century trade in Bering Strait is developed; the residents of Chukotka, for trade with the Russians, needed furs, which they acquired in Alaska and exchanged for wares obtained from the Russians. In "Sobranie izvestii o raznykh plemenakh inovertsev, obitavshikh v Irkutskoi gubernii" [Collection of News on the Different Tribes of Infidels Who Live in the Irkutsk Province] (1789) we read: "In summer in calm weather in baidary and in winter over the ice on reindeer it is possible to cross in one day to Imoglin Island, lying almost in the middle of the strait, and from there to America; those Chukchi often visit that part of the world and trade there, and sometimes also carry out robbery there."[542] Consequently, in the opinion of the composer of the document, the main thing in the strait now was not war, but trade.

---

[541] Dall 1870: 375.
[542] Vdovin 1964: 4; 1964a: 121; 1987: 42–43.

Figure 38. Trade of a resident of Chukotka and an Alaskan Malamute Eskimo for a bag of fat. (Reproduced after Dall 1870: 378–379).

C. Merck tells specifically about methods of trade:

> Only a few of the Settled Chukchi go themselves to America to trade. From March to April only individual Chukchi dare to go there from the northeastern cape on foot or on dogsleds across the ice with some risk for themselves. More often they send there married captive women from those places or arrange trade on the islands in the strait, offering iron artifacts and beads and receiving in exchange parkas of marten and mouse furs, wolf, lynx, wolverine, fox, and otter skins. The residents of Okipen Island meet, based on their custom, the baidary of the Chukchi in armor, with bows, arrows, and knives, as well as accompany them upon their departure, and the Chukchi respond in the same way. The islanders assign such guests singly among themselves and go with them, and sit with them in their dwellings in large part with exposed knives. The islanders have up to eight wives each and offer them to visitors.[543]

As we see, at the beginning of the 1790s trade in the strait continued to be dangerous. Therefore, the residents of Chukotka sent captured American women, taken as wives, to America to do the trading.

---

543 Merck 1978: 121–122.

The usual trade between the Asian and American Eskimos occurred on Ratmanova Island. Merck described a trip of the Chukchi to King Island clearly based on materials of I. Kobelev's journey. The latter told in his journal that on 12 June 1791 he sailed with Uelen people in baidary to the island and both sides wore armor, took spears, stretched bowstrings, and began to trade. For furs, the Uelen people gave Russian wares: spear points, knives, glaives, axes, kettles, small corals, and beads; also, upon the expedition's departure both sides were armed. As we see, the partners did not trust each other and were openly armed while trading and even carried knives openly when accompanying the new arrivals on the island.

One of the basic objects of trade was iron weapons. Members of J. Billings' expedition in July 1791, upon landing at Cape Rodney (southwestern Seward Peninsula), noted that the spears of the Eskimos had points of walrus tusk and only among a few—of iron from Gizhiga, while the arrowheads were only of stone, bone, or tusk, but not of iron; tobacco in general did not interest the local residents. Evidently, the volume of trade at this time was insignificant. Later, in the first third of the 19th century, the residents of Ayak (Sledge) Island emerged as intermediaries in trade of the Siberians with the Americans. They went to Ratmanova Island, traded Russian wares there, and distributed them along the coast.[544]

The description of the commander of the brig *Ryurik*, Lieutenant O. von Kotzebue, attests to the strained relations of the Siberians and Alaskans during the first quarter of the 19th century. He, meeting in 1816 a group of Reindeer Chukchi, led by an elder, who were summering on the coast of Lavrentiya Bay, described intertribal relations in the following way:

> The Chukchi are eternal enemies of the Americans. My guest called them simply villains. He told that the Americans adopt the guise of friendliness only when they feel themselves weaker, but as soon as they recognize themselves the strongest and do not perceive any danger for themselves, then without the slightest twinge they rob and kill the stranger; in this case, he said, they carry knives in their sleeves, and use their wives as

---

[544] I. Kobelev's trip in 1791: Chernenko 1957: 133–135; EM: 163; Eskimos at Cape Rodney: Sauer 1802: 243; Sarychev 1952: 180; EM: 90; trade in the first third of the 19th century: Nechto o Sibiri. . . 1806. No. 8: 96; Wrangell 1835: 610; O torgovykh snosheniyakh. . . 1851: 108–109.

> bait for strangers. Seeing several images of residents of the shores of America painted by our artist, they immediately recognized them by the bones in the lower lip, and one of the guests, drawing a knife, fiercely exclaimed: "Wherever I were to meet a man with two bones, I would stab him!"[545]

In general, the matter in this passage is not about large raids, but rather about trade operations, during which the Alaskan Eskimos, given a chance, rob and kill the Chukchi. The Alaskans for the sake of safety carry knives in the sleeves of their parkas, and they offer their women, in the opinion of the Chukchi, to foreigners not based on the general laws of hospitality, but rather only as bait for lulling the vigilance of partners. Because of the mutual distrust, hostility also occurred. Such unfriendly atmosphere in Beringia is also confirmed by the communication of an old Eskimo from Sledge Island, who told that "formerly it was a usual custom among the residents of the Siberian coast to immediately kill any Eskimo from the American shore who might be brought across the strait in a baidara or on an ice floe by a storm."[546]

In the middle of the 19th century Siberian-Alaskan trade was still a dangerous undertaking. The Chukchi described their trade with the Eskimos of Alaska entirely the same way: "They say that their neighbors adopt the guise of friendship only when they are weak; but when they recognize themselves stronger, they rob and kill the arriving merchants, and therefore trade with them is conducted with weapons in hand."[547] In general, it seems that trade even in the middle of the 19th century could at any moment turn into a conflict with the use of weapons.

The fears of the local residents concerning trade were not groundless. Trading at any moment could turn into conflict. W. G. Bogoras told of one expedition from Naukan three generations ago (approximately in the second quarter of the 19th century). The Naukan people sailed in large baidary, with thirty paddlers in each; they were joined by Chukchi in small baidary. The members of the expedition, depending on the circumstances, engaged now in trade, now in robbery. In fact, according

---

545 Kotsebu [Kotzebue] 1948: 103; cf.: Ibid: 82; Semivskii 1817: 77; Antropova 1957: 178.
546 Nelson 1899: 330; cf.: Sheppard 2009: 48.
547 Mushkin 1853: 1422; cf.: Mitchel' 1865: 324; Neiman 1877: 87; Avgustinovich 1878: 52; 1880: 717; E. M. 1892: 233.

to a Prince of Wales legend, after a battle and capitulation the Siberians began to trade with the Alaskans, but then killed the latter.[548]

3.2. The Hypothesis of M. Krauss about Alaskan Eskimos in Siberia

The American linguist M. Krauss advanced the suggestion according to which the Alaskan Eskimos colonized the north coast of Chukotka as early as the 17th century for the purpose of conducting trade with the Russians, and Wrangel Island was their intermediate base, memory of which was preserved in the legend about the Khrakhai. The reports of G. S. Shishmarev and A. P. Lazarev (1821) about the presence of people with labrets in the region of Cape Shelagskii served as the basis for such supposition. This information was obtained from the elder Paigdau (Pil'gaigym) on the south shore of Lavrentiya Bay, who understood quite well the Koryak interpreter of the sloop *Blagonamerennyi*. The elder told that not far from Cape Shelagskii live the Chavaki, who have lips with slits; they also have draft reindeer and baidary. They trade with the Chukchi on a certain Amylik River.[549] People with labrets—these are clearly Eskimos of Alaska or the Gvozdev Islands. In 1816 O. von Kotzebue, through an interpreter, tried to find out from a Reindeer Chukchi in the village of Nunyamo where the Alaskans obtained iron; it turned out that either from the Kolyma, or from the Chukchi. L. A. Zagoskin asserts directly that the residents of Norton Sound obtained Kolyma iron artifacts through the Chukchi and through relay trade. Only an insignificant number of Alaskan traders went to the Anyui fair. According to the report of the Irkutsk governor general Lieutenant I. A. Piell to the Senate, in February 1794 two Alaskans with their wives arrived at the fair, who promised to bring here their countrymen. The wives of the Americans were probably Asians, therefore their husbands took them with them. Later, in March 1821, two more Eskimos from Alaska arrived at the fair, bringing furs.[550] Consequently, a certain number of Russian wares fell

[548] Bogoraz-Tan 1934: 79; cf.: Antropova 1957: 178; Prince of Wales legend: Garber 1940: 58.

[549] Krauss 2005: 171–180; see: Shishmarev 1852: 187, 191–192; Lazarev 1950: 299, 308.

[550] Kotsebu 1948: 103; Zagoskin 1956: 76; 1794: Slovtsov 1856: 21; 1821: Cochrane 1824: 198; cf.: Choris 1822: 7.

to the Alaskans directly from the fair. In general, the description of Paigdau was obtained through an interpreter and could contain some inaccuracies. If the American Eskimos in fact lived in the north of Chukotka, this circumstance would be clearly reflected in Russian sources. The legend of the Khrakhai pictures them as reindeer people, which excludes comparing them directly with natives of Alaska. The Chavaki—in this description these are Chaun Chukchi, who possessed both reindeer herds and baidary. In general, the whole supposition of M. Krauss looks rather an artificial construction and does not deserve support.

3.3. Establishment of Peace in the Strait

One version about the establishment of peace is cited by N. A. Galkin, who visited Chukotka in 1924–1925 as a commercial agent:

> Based on the stories of the old people, a hundred years ago between the Chukchi and the Eskimos of Alaska who came to the peninsula to hunt, there was constant misunderstanding and even slaughter. Finally, they resolved the situation peacefully. On the Diomede Islands they organized a council, in which they agreed to delineate hunting areas and to establish trade and friendly relations. And in fact, from that time discord stopped and the sides faithfully complied with the terms.[551]

This legend probably represents the Uelen point of view. It clearly points to the reason for conflict—dispute over hunting territories. And, based on this version, it was the Eskimos sailing in Asian territory. An encounter occurred in neutral territory, probably on Ratmanova Island, where the conditions for maintenance of peace were worked out. This pacification in the neutral territory was probably the first stage of establishment of peace, after which followed a visit of the Asians to the Americans.

Informants of E. Nelson reported the following about the last raid of the Siberians on the American Eskimos: "The last war party in this district came in a fleet of umiaks from East cape, Siberia, and the Diomede islands, and sailed up Port Clarence, but meeting a large force of the American Eskimo, both sides agreed upon a peace, which has not since been broken."[552] Subsequently, a flotilla from Cape Dezhneva (from Uelen and/or Naukan), in union with the Eskimos from the Ratmanova

[551] Galkin 1929: 72–73.

[552] Nelson 1899: 330; cf.: Sheppard 2009: 48.

Islands, who evidently joined along the way, arrived on the western coast of the Seward Peninsula, but encountered numerous militant Eskimos, who managed, having learned of the approach of the enemy, to assemble to repel the enemy. The Eskimo informant probably did not know whether the Asians initially had peaceful intention.

In 1924 the Kamchadal Petr Kosygin (P. Cosigan) from Uelen told the Danish traveler Knud Rasmussen about the relationship of the Chukchi and Eskimos in the following way: "Once a whole party of Chukchi was attacked and all annihilated, to the last man. After this incident the Chukchi set off in substantial number to the foreign people, to the Eskimos, and proposed that they should live in peace and trade with each other."[553] It was probably a matter of the same event as in the preceding report, to which was added the reason for the expedition of the residents of Chukotka to the Eskimos—the annihilation of a certain group of "Chukchi." The circumstances of the incident are not specified.

According to Naukan legend, the residents of Uelen and Naukan, because of human losses and lack of prospects in wars, decided to stop hostility with the Alaskan Kygmit people and sent to Alaska baidary with young people led by Ekëtamyn; during a meeting with competitions, in which a Naukan won at jumping, pacification finally occurred.[554] In principle, the foregoing legends (American and Siberian) are not contradictory and can completely be telling about the same event from two different points of view, Siberian and Alaskan. On the one hand, the establishment of peace could also have occurred in several stages—between different communities of Chukotka and Alaska.

---

553 Rasmussen 1952: 145. About the trader and interpreter P. Kosygin (Cosigan) it is known that he was a Kamchadal, studied at an American university, by 1925, despite the desire to go to the mainland, he lived for 12 years in a yaranga in Uelen with his Chukchi wife and daughter (Galkin 1929: 63).

554 Voskoboinikov and Menovshchikov 1959: 540–541; Menovshchikov 1958: 168–169; 1969: 226–227; 1987: 214–215; 1988: 308–309.

Figure 39. Eskimo from Sarichef Island, which is located at Shishmaref Inlet in Alaska (first quarter of the 19th century). (Reproduced after Wahlen 1844: 357).

The time of cessation of warfare in Bering Strait, based on folkloric data, is unclear. The story of Ekëtamyn, written down in 1948 from the non-literate Naukan woman Khul'khena, in distinction from other Eskimo legends, contained the names of the heroes on the Chukotka side, which clearly attests to the relatively brief antiquity of the legend. In 1975, old people from Teller (probably providing information once obtained from Siberians during trade) told that when the Siberians started to use guns, they began to think about a new war with the Alaskans, but one old man, who had been wounded in the last conflict, dissuaded them, pointing out that they always lost in clashes with the residents of Kawerak.[555] That is, it is the matter of events of the middle of the 19th century, when the coastal resident began to receive firearms from whalers. The idea of using firearms against the Alaskans was not new: as early as 1778 Captain T. I. Shmalev, transmitting the arguments of the Chukchi, noted that the latter wanted for successful action against the Americans to invite a hundred Russians with firearms to help them.[556]

[555] About the legend from Teller, see: Sheppard 2009: 54.
[556] RGADA. F. 199. No. 539. Ch. 2. D. 6. L. 24 ob., note.

Figure 40. Alaskan Chugach Eskimos in a camp of upturned baidary in Prince William Sound (1789–1790). An engraving. (Reproduced after Sarychev [1802] 1952. Pl. 33). In the 1952 edition the numbers of the pictures are different.

For refinement of the dating it pays to turn to the written sources. On 31 July 1816, when Lieutenant O. von Kotzebue landed at the shore with the brig *Ryurik*, the residents of Cape Prince of Wales ran away in fear, suspecting possible attack on them. Later, in 1835 F. P. Wrangell, governor of Russian America in 1829–1835, noted that in Bering Strait at that time there were no wars, only trade. Lieutenant L. A. Zagoskin, who investigated western Alaska in 1842–1844, reported only about trade of the "Chukchi" in Kotzebue Sound, but not about conflicts in the strait, whereas he mentions warfare within Alaska itself in 1839. The British lieutenant W. H. Hooper, who wintered over on the coast of Chukotka in 1848–1849, noted that the residents of Chukotka traded with the Alaskan Eskimos through the residents of the Gvozdev Islands, whereas open hostility existed with the islanders of St. Lawrence, which was expressed in the capture of women. The missionary A. I. Argentov, in describing his trip to the Anyui fair in March 1848, noted: "It has been ten years since we have heard anything about open raids or large robberies, of which the Chukchi were previously great pursuers," that is, according to Argentov, military clashes in Chukotka were earlier than 1840—the missionary probably had the sea raid of that year in mind. The geologist K. von Ditmar, who served in Kamchatka, wrote in 1853 that the Chukchi "formerly . . . conducted bloody wars and only recently made peace" with the Americans. The English traveler F. Whymper only noted that in the 1860s the Chukchi traded actively in the strait, sailing to the region of Port Clarence; in general, he does not mention wars. Doctor F. M. Augustynowicz noted in the mid-1870s that Chukchi traded with the Kargaul of the islands in the strait, keeping knives bared in their hands, since trade was dangerous.[557]

We will now examine the opinions expressed on this issue. N. A. Galkin, citing stories of Uelen (?) old people in 1924–1925, says that peace with the Alaskans was concluded 100 years ago. In 1899, W. G. Bogoras thought that the attacks of the Chukchi on the Aivan on the other shore of the sea ceased 50 years ago, that is, in the mid-19th century. Modern American anthropologists agree with this opinion as well. W. Sheppard, on the basis of recorded oral reports of the Eskimos, suggests that conflicts in the strait stopped earlier than 1850. These conclusions, in general, do not contradict the observations of E. Burch, who believes that traditional military affairs of the Eskimos of northwestern Alaska declined in the 1850s, only one or two raids being carried out in the 1860s, and warfare completely ceasing in the 1870s. According to J. Malaurie, the state of constant wars in western Alaska was still preserved in 1860–1880s.[558]

At the end of the 19th century relations in the strait were quite peaceful: in 1895 nine baidary arrived in Uelen from Alaska, and ten more from the islands, though primary trade was conducted on Ratmanova Island, where Uelen people and some of the Reindeer Chukchi, and of course Alaskans, gathered in June.[559]

---

[557] Kotsebu 1948: 79; Vrangel' 1835: 608; Wrangell 1839: 57–136 (also see: O torgovykh snosheniyakh . . . 1851: 108–109); Zagoskin 1956: 48, 76, 259; Hooper 1853: 40, 210; Ditmar 1856: 36; Whymper 1869: 161–163; Argentov 1887: 21; Avgustinovich 1880: 715; cf.: Vdovin 1965: 60–61.

[558] Bogoras 1900: XXVI; Weinstein 2004: XIV; Galkin 1929: 72; Malaurie 1974: 151; Burch 1998: 216; Sheppard 2009: 50, 59, note 3; cf.: Shvaittser and Golovko 2001: 28. I. S. Vdovin (1964: 1–2, 5; 1964a: 123; 1987: 42–43) dates the establishment of peace in the strait to the beginning of the 19th century; however, he did not take into account all the information.

[559] Gondatti 1898: VI, XI; cf.: Vdovin 1964a: 126; cf.: Smith 1970: 236–238. About a trade-music meeting at the end of summer in Uelen, see: Rytkheu 2004: 392.

3.4. The Relations of the Chukchi and the Eskimos of St. Lawrence Island

The relationship of the residents of the Chukotka coast developed in a predominantly hostile vein with the residents of St. Lawrence Island. In historical times eleven clans lived on the island, four of which had an Asian origin. Up to the 1920s there were several waves of resettlement to the island, the primary of these immigrations being in the 18th century. Based on legend, once 40 hunters from Uelen, being carried by a storm to the island, tried to attack a village but were killed. In a legend, told in 2013 by the Eskimologist Ludmila I. Ainana (b. 1934), the islanders killed one of the men from a Chaplino baidara carried to the island. In revenge the Asians gathered forces, sailed in several baidary to the village at night, and killed all the local men, leaving alive the Chaplino people, who knew the Chukchi language and who lived on the island. In another Chaplino story, the residents of Unazik sailed to the island, where they kidnaped the girl Ukamanan and her brother, then sold the girl to reindeer herders for 30 reindeer. A historical legend, told by the Napakutak Eskimo Yatylin in 1974, tells of clashes: the islanders made slaves of hunters brought to the island by a storm, and in spring sacrificed them; this happened to a Chaplino person, whom they were going to kill, and to Lorino Chukchi, brought to sacrifice. In the following year, after this became known, a flotilla of Asian Eskimos and Chukchi arrived at the island and killed the local residents. W. G. Bogoras heard a variant of this legend as well: two men from Unazik (Staroe [Old] Chaplino) were carried on an ice floe to the village of Sivukak on the northwest of St. Lawrence Island; one of them was killed, while the second fled and told about the incident. In the following summer, a coalition flotilla set off to the island, killed the islanders in battle on the shore, and took many women captive. Four years later the islanders carried out a reciprocal raid at night. The next year the islanders concluded peace with those who sailed to them.[560] This in fact

[560] Ethnic composition of the island and migrations: Krupnik and Chlenov 1979: 24; Krupnik 1981: 111–113; legend of St. Lawrence : Krupnik and Krutak 2002: 220–221; story about Ukamanan: Rubtsova 1954: 225–231; Menovshchikov 1958: 121–124; 1969: 164–167; 1988: 121–124; 1988a: 188–190; legend of Yatylin: Menovshchikov 1988: 317–320; Nefёdkin 2003: 292–296; legend given by W. G. Bogoras: Bogoraz-Tan 1934: 174–175; Nefёdkin 2003: 291–292; information from L. I. Ainana: Nefёdkin 2016: 226–228.

was possibly the last raid, after which peace was established. However, even during warfare the residents of Chukotka had comrades on the island—*proxenia*, with whom the guests stayed.[561] These were primarily descendants of immigrants from Siberia.

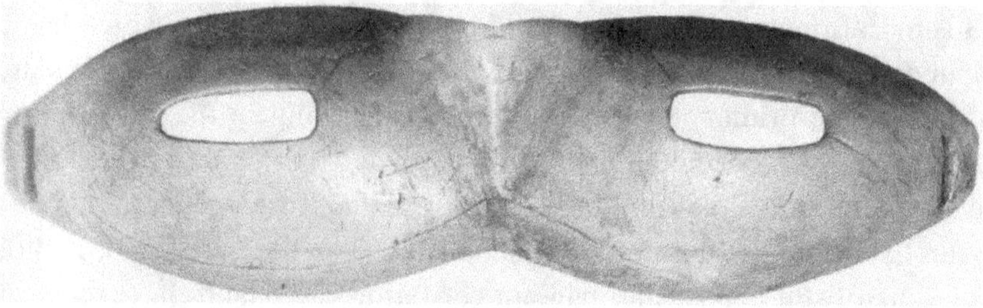

Figure 41. Snow goggles from St. Lawrence Island. It is presumed that they are older than the Punuk culture (300–1300). (Reproduced after Bandi 1977: 44).

If we turn to the written records we see that in 1817 the residents of the island, hiding knives in their sleeves, asked the Russians of the brig *Ryurik* if they intended to kill them. In the middle of the 19th century the Chukchi traders from Lavrentiya Bay described their trade on St. Lawrence Island to a Russian missionary (possibly A. I. Argentov) in the following way: "We trade with them, always keeping a knife in hand, because if they're not happy with our wares they kill us." In the 1870s the Chaplino people brought (probably kidnaped) men from the island, whom they sold to the Reindeer. In 1880 Captain E. P. Herendeen learned that the Chukchi trade with the islanders, wearing armor because of the danger of trade, but in the 1890s friendly relations were formed between the Eskimos of Unazik and the islanders: the Chaplino people went freely to the island not only for trade but also to hunt walruses, and even spent the winter there; the following fact attests to the intensity of the contacts: in the summer of 1895 16 whaleboats sailed from Chaplino to the island, and nine arrived in Chaplino from the island. And Chaplino people themselves settled on the island even in the period of hostility.[562]

---

[561] Rubtsova 1954: 230.

[562] Quotation: Mushkin 1853: 1422; about 1816: Kotsebu 1948: 208; relations in 1870–1890s: Gondatti 1898: 14, 30; E. P. Herendeen: Hough 1895: 634–635; Chaplino people on the island: Bogoras 1934: 106; story of L. I. Ainana from Chaplino (October 2013), see: Nefëdkin 2016: 226–228.

Figure 42. Eskimo breastplate of bone plates from St. Lawrence Island (beginning of the 19th century). (MAE. No. 571-99).

The course of one of the last sea raids is described by the missionary A. I. Argentov:

> Driven by a sense of revenge for some old grudge and feeling greedy, the Chukchi thought about making a raid. This was in 1840. Numbering 70 men they approached some island on a foggy night; their first business was to puncture the natives' baidary; then they attacked unawares the sleeping inhabitants and stabbed many. Frightened by the uproar of the attackers and the desperate cries of their own people, the survivors rushed to escaped in their baidary: they pushed off into the sea and sank. All the fit women and children were taken captive, while the sickly, ailing, afflicted, and aged were all killed. "Why leave the miserable in torment," added the narrator.[563]

The story was told to Argentov by the Chaun Chukchi Gallya through a Chuvan interpreter,[564] the latter not giving the name of the island on which the raid was carried out. This was most probably St. Lawrence Island, less probably Kruzenshterna, King, or Sledge Island. The reason

563 Argentov 1857a: 37; 1886: 30–31.
564 Argentov 1886: 27, 30.

for the raid was probably blood revenge, combined with the desire to seize booty. The attackers, in order not to be noticed, approached the shore at night when the residents slept; the Eskimos evidently had not set patrols—there was no direct threat of attack—and, as a consequence, the enemy was not noticed. The assailants attacked unexpectedly in order not to allow the opponent to prepare for battle and not to incur in it significant losses. In particular, for the surprise attack they skillfully used the fog to conceal their approach.

* * *

Figure 43. St. Michael Redoubt in Norton Sound. Illustration by I. G. Voznesenskii (1843). On the right—the manager's house and barracks with a watchtower; on the front left—a garden; and on the right—footbridge of the pier (description: Blomkvist 1951: 266, 268). (Reproduced based on Blomkvist 1951: 267, Fig. 22).

In general, it is possible to speak of raids to Alaska at least from the 17th century, and usually the advancing side were the Naukan and Uelen Eskimos, to whom were joined the Chukchi, both the Settled and, in smaller degree, the Reindeer. In the 18th century (with breaks in the 1740s) residents of the Gvozdev Islands, in particular Ratmanova Island, became allies of the Siberians. In the 17th and 18th centuries attacks by residents of Chukotka and Alaska were mutual, and they probably attacked in small groups. The reason for such raids was blood revenge for slain relatives, to which mercenary motives could be attached. In the opinion of Eskimos from Wales, the Siberians wanted to seize their hunting grounds.[565] In the second–third quarter of the 18th century (earlier than 1778) large raids were conducted by the Siberians, in which hundreds of baidary participated in each, that is, at least 800 men in each raid. With

[565] Garber 1940: 169.

such a number of warriors they did not count on a surprise raid, but rather on frontal battle with the enemy. At the end of the 18th century, when war with the Koryak had terminated, the center of military action moved to Bering Strait: raids on Alaska began to be conducted almost annually. These raids bore a piratical character: a small flotilla of baidary arrived and plundered the residents. In such raids they now counted on the suddenness of the attack, which permitted the attackers to avoid unnecessary losses. Since our information about these military actions originate primarily from Alaskan folklore, which attributes victory to their own people, it is impossible to say precisely who won in the campaign. But it seems that small parties of Siberians that were spotted were in fact usually repelled and/or annihilated. Indeed, even T. I. Shmalev remarked in 1778 on the success of these raids: "It is noted that in these raids luck came to the Chukchi at great expense."[566] When several baidary moved inside the Seward Peninsula toward Kawerak, they were usually noticed and won by some trickery. If the Alaskans won, all those who arrived were killed; and if the attackers obtained victory, they killed the men and took the women and children captive, also taking the valuable property. In some degree, the Chukchi need for furs for trade with the Russians contributed to escalation of the conflict. Indeed, during this same period the role of trade increased, which earlier was not as significant; in trade between the Russians and Eskimos of Alaska the residents of Chukotka were the middlemen. On the other hand, the need for trade and the increasing need of the Siberians for furs also contributed to the establishment of peaceful economic relations, which in addition was favored by the institution of hospitality between the residents of Chukotka and the islands in Bering Strait. The founding of Russian forts in western Alaska in the 1830s and the appearance of numerous whaling ships in the strait, beginning in 1848, also contributed to pacification. Trade became the primary form of interrelations of the residents of Chukotka and Alaska, with conflicts occurring only sporadically in the first third of the 19th century. Above all, conflicts with the Alaskans end at the beginning of the 1830s, and later (probably in the 1870s), with the residents of St. Lawrence Island. Trade and friendly connections with Alaska were maintained up to 1948, after

[566] RGADA. F. 199. No. 539. Ch. 2. D. 6. L. 24 ob., note.

which Chukotka was separated from the United States by the "Iron Curtain."[567]

[567] Leonova 2008: 157.

CHAPTER 6
CONFLICTS OF THE 19TH CENTURY

1. Clashes between Local Groups of Chukchi and Neighboring Peoples

For the 19th century, large-scale military actions, even by local standards, were not characteristic; however, small clashes occurred. Between the 1980s and 1990s a septuagenarian reindeer herder from the Ust'-Belaya tundra, Ainakvyrgin, told this about past intertribal relations in the region:

> The Chuvan people and Koryak were enemies of the Chukchi. They attacked unexpectedly. And the Chukchi people were frequently in discord, but the common danger quickly united them. The discord usually occurred between nomad camps. . . . And these were the Vaezhskie—those the Tel'kepyl'yt [Tel'kep]. They were aggressive, attacked both the Chukchi and the Russians, but they were always defeated. They strove toward independence, and usually supported the Chuvan and Koryak more than the Chukchi.

Information about hostility with the Koryak and Chuvan people and consolidation of the Chukchi ethnos that followed from it clearly belongs to the 18th century, as does communication about the southern Tel'kep Chukchi. However, information about hostility between the nomad camps can be possibly assigned to a later period. According to the legends, the Tumanskie (Tel'kep) Chukchi went to the Ust'-Bel'skie for the purpose of driving off the reindeer.[568] The Koryak ethnographer L. S. Ivtagina explains the negative attitude toward the Tel'kep Chukchi by their northern fellow tribesmen by the fact that the Tel'kep were not purebred Chukchi, and since they were mixed with the Koryak, the others

[568] Ainakvyrgin: Radov 2008: 74; hostility with the Ust'-Bel'skie: information from V. N. Nuvano (Anadyr', September 2013).

considered them not fully their own. In fact, as early as the third quarter of the 18th century, according to Chukchi legends, the reindeer herder Motlynto kept moving from a Chukchi camp to the Koryak and back. In 1825 navigator G. F. Nerodov noted that the southern ("Tumanskie") Chukchi migrate together with the Ammatovskie Koryak, with whom they are intermarried; at the boundary of the 19th and 20th centuries W. G. Bogoras met at the border of Koryak-Chukchi contact the second generation of people from mixed marriages, who spoke a mixed Chukchi-Koryak dialect and were not able to decide whether they were Chukchi or Koryak.[569]

As early as the boundary of the 19th and 20th centuries the bulk of the Chukchi treated with apprehension some other groups of their fellow tribesmen who lived in outlying regions. The geologist I. P. Tolmachev (1872–1950), member of the Northeast Expedition of G. Ya. Sedov, noted: "The Shelagskie Chukchi are not in favor among the residents of the east coast: with many of the latter they had debt and even bloody scores." The Chaun people enjoyed the same reputation among the Kolyma Chukchi.[570] It was evidently primarily a matter of trade contacts degenerating into disputes, during which murder could happen, which led to blood feud.

Hostility existed not only between separate families and clans, but was spread to whole villages. In this hostility, blood feud could correspondingly be a subsequent secondary phenomenon. Based on Bogoras' observations, between the neighboring villages of L'uren (modern Lorino) and Yanranai (Yandagai) the hostility lasted ten generations (250–300 years) and was still in effect two generations before him, that is, up to the middle of the 19th century. On 19 April 1901 he noted in his diary: "They arrive even from Kuluchin [Kolyuchin Island.—A. N.] with hides to trade in Uŋyin [Uniin = Unazik.—A. N.]. But they go directly over the field, fearing the residents of the village of Lûren [Lorino.—A. N.] who kill those who pass and take from them captives and

[569] Opinion: Ivtagina 2008: 301; Mushkin 1853: 1421; Bogoras 1921: 52; Bogoraz-Tan 1934: 6; Gurvich 1966: 185, 189; Motlynto: Baboshina 1958: 242–243; Kostenetskaya 1973: 172–185; Menovshchikov 1974: 208–306; Leont'ev 1976: 242–243; Belikov 1982: 183–197; Lebedev and Simchenko 1983: 98–100; Nutekeu 2005: 12–14.

[570] Quotation: Tolmachev 1911: 57, cf.: 101; Mushkin 1853a: 844; Kolyma Chukchi: Dionisii 1884: 151–152; Bogoraz-Tan 1934: 175.

goods, sleds. They were killed, but they do not stop." In general, open hostility of neighboring villages lasted approximately three hundred years (from the second half of the 16th century), ending toward the middle of the 19th century; however, robberies periodically continued and at the beginning of the 20th century the British lieutenant W. H. Hooper, who spent the winter of 1848–1849 on the east coast of Chukotka, noted that the residents of neighboring villages refused to go to Yanranai, where they would simply be killed, since hostility existed between Lûren and Yanranai, which incidentally had not passed into an active phase at that time.[571]

At the boundary of the 19th and 20th centuries there were rumors about the settled Chukchi Papaika who lived at Cape Kytrytkyn in Lavrentiya Bay, that he robbed trading caravans, as his grandfather Troochgyn had also done, and killed for this by a spear on sea ice.[572] The rumors about such robberies reached the Markovo people. In 1895 the latter were afraid to lead the head of the Anadyrsk stockade, N. L. Gondatti, to Provideniya Bay; indeed, "part of the Sitting Chukchi enjoys here [in Markovo] a poor reputation, partly as robbers and brigands, and partly as wizards, able to completely ruin a person."[573]

Hostility also existed between other communities. The Naukan local historian V. G. Leonova (b. 1961) reports that earlier the Naukan people were friendly with the Uelen Chukchi, but were at odds with the residents of Inchoun and Chegitun. In 1830 the midshipman A. K. Etolin, commander of the brig *Chichagov*, visited St. Lawrence Island, noting that there were two large villages—Kukulyak (in the north) and Kialegak (in the east), but "the residents of both villages are in eternal hostility among themselves."[574]

---

[571] Quotation from diary of W. G. Bogoras: AAN. F. 250. Op. 1. No. 118. L. 26 ob.; information: Bogoraz-Tan 1934: 29, 176; Hooper 1853: 153, 161, 167. It can be noted that at the end of the 18th century in Yanranai there were two winter pit houses and sixteen yarangas, whereas in L'uren there were four pit houses and fourteen yarangas (Sarychev 1952: 237; Vdovin 1950: 88–89).

[572] Papaika: Tevlyankau 2008: 16; Rytkheu 2010: 172; grandfather: Rytkheu 2004: 368–369.

[573] Gondatti 1898: 2.

[574] Opinion: Leonova 2008: 157; A. K. Etolin: Tikhmenev 1861: 284–285.

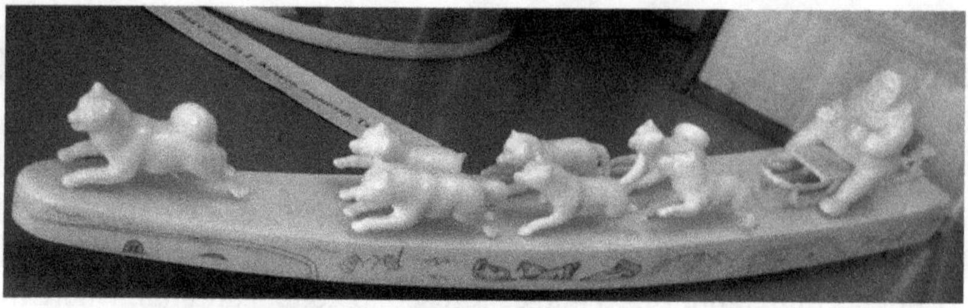

Figure 44. Sleds hitched to dogs. Eskimo bone carving. Archaeology Museum, Sozopol, Bulgaria. A gift to Bulgaria from the USSR in 1967. (Author's photo on location, 29 July 2009).

There also existed concealed ethnic tension between the Chukchi and the Eskimos. In 1899 W. G. Bogoras, analyzing the legend of Elendi, writes: ". . . and at the present time, between them [Aivany] and the Chukchi there is tribal hostility, being repeatedly resolved by individual bloody clashes." Of course, it is not a matter of open conflict, but rather simply about some kind of nationalism and snobbery, particularly among the Reindeer Chukchi, which could spill out in the typical way of clarifying relations in the form of a duel. Bogoras had in mind a duel with weapons, though they were also able to sort things out without them. For the Eskimos, in the first third of the 20th century fighting was "the only form of public trial, as proof on one's rightness"; it lasted until one of the sides admitted defeat.[575]

G. Maydell thus described the methods of resolving conflicts among the Nose Chukchi in the third quarter of the 19th century: "It is surprising that with such kind of life bloody clashes are relatively rarely encountered among them. Disputes happen and are resolved by a spear duel or robbery, if one of the offended can gather several comrades, who help him beat on the opponent and run off part of his herd. But still this kind of violence must be rather rare. . . ."[576] Baron Maydell described the standard methods of resolving everyday conflicts, probably not directly concerning a blood feud.

Conflicts with the Koryak and Even occurred at reindeer river crossings on the lower Anadyr'. A. S. Sgibnev thus described relations contemporary to him (middle of the 19th century) during a hunt:

---

[575] First quotation: Bogoras 1899a: 350; cf.: Bogoraz-Tan 1934: 186; second quotation: Ushakov 1972: 154.

[576] Maidel' 1925: 30–31; cf.: Olsuf'ev 1896: 108.

> One of these crossings is on the Anadyr' River near the Krasnaya River, and the Koryak and Tungus usually wait for the reindeer on the south side of the Anadyr' River, and on the north, the Chukchi. Upon following the reindeer to the north the Koryak try as long as possible to keep the herd on the south side of the Anadyr', in order to be able to kill more reindeer. For this they arrange fires along the whole shore, which cause the reindeer confusion. The Chukchi, awaiting the herds in boats on the Anadyr and on the other shore, of course, have a reason to quarrel with the Koryak.[577]

At that time, one of the largest reindeer crossings was still on the lower Anadyr, over which the herds of animals moved north in summer. The Chukchi hunted in light boats with special long spears the reindeer swimming across the river. And since the Koryak on the south shore frightened the approaching reindeer herds with fires, this caused conflicts to emerge.

In the 19th century conflicts sometimes occurred between the Chukchi and the settled Kerek, whose territory was *de facto* not controlled by Russian authority, and life there flowed in the traditional way. On 14 March 1901 Bogoras recorded in his diary a story of the southern Kerek living between the Opuka and Khatyrka Rivers about their past: "The Chukchi formerly made raids: they come, slaughter the residents, burn the huts, take the food, and lead the women away. But we've heard nothing for 50 years." Kerek legends still told about such activities of the Chukchi in the first half of the 20th century, but without clear dating of the events, which could be assigned to a much earlier time (to the 18th century). But in the beginning of the 20th century memory of Chukchi raids was still fresh—raids stopped only at the turn of the 1840s and 1850s. The last raid on a neighboring Kerek village was dated to ten years later—the turn of the 1850s and 1860s: because of hunger two men on dogsleds from Mólan (Mallen Lagoon) made a raid on a village located 20 verst away; they killed five people there and took food. "After 5 years a nephew of one of those murdered killed one of the assassins. This was the last raid."[578]

---

577 Sgibnev 1869: 28, note 2.

578 Diary of W. G. Bogoras: AAN. F. 250. Op. 1. No. 118. L. 7 ob.–8; attacks of the Chukchi in Kerek Folklore: Leont'ev 1983: 96; Tyneru 2005b: 88; raid of the Kerek themselves (quotation from the diary of W. G. Bogoras): AAN. F. 250. Op. 1. No. 118. L. 7 ob.

However, even at the turn of the 19th and 20th centuries the Kerek complained that the southern Tel'kep Chukchi took furs from them and forced the young people to work as herders.[579]

One of the reasons for the conflicts was theft. It was not widespread within a village, and this pattern of behavior did not work well beyond the boundaries of the latter. However, among the Chukchi, even at the end of the 19th century, there was an extreme kind of sport—stealing reindeer. Indeed, Bogoras noted that "among the Reindeer Chukchi theft between camps and herds, it can be said, was a custom." In addition, in the pasture the herds of different camps could simply get mixed, which also could become one of the reasons for the emergence of disputes, conflicts, and, in case of murder, revenge. Bogoras also says that ". . . for each family it is important in the highest degree to preserve intact the pasture closest to home. Violation of this right is considered a grave offence and often ends in bloody reprisal leading to long-term violence and bloody vengeance."[580]

In the territories bordering the Reindeer Koryak the situation was similar. In 1808 the Kamchatka commandant General Major I. G. Petrovskii stated that the Chukchi "almost constantly fight with their neighbors, the Reindeer Koryak, due to irreconcilable ancient hostility." In 1867 G. Maydell noted: "Bloody clashes have now not occurred for a long time; however, there is every kind of robbery and theft in such places where the camps of the Chukchi are located near Koryak camps, and therefore they try to live far away from each other."[581] In general, in the middle of the 19th century small stealing raids on the neighbors still occurred in the bordering territories, because of which the Reindeer Chukchi and Koryak preferred to have a neutral zone between them.

---

[579] Bogoraz-Tan 1934: 29; Orekhov 1987: 159.

[580] Pattern of behavior: Russkaya missiya. 1890: 100–101; Sverdrup 1930: 323; stealing reindeer: Bogoras 1899b: 19; Beretti 1929: 20; first quotation: Bogoraz-Tan 1934: 28; cf.: Bogoras 1899b: 19; 1902a: 84; Beretti 1929: 20; Menovshchikov 1974: 330–331; mixing of herds: Bogoras 1921: 85; second quotation: Bogoras 1921: 86.

[581] About I. G. Petrovskii: Semivskii 1817: 77, note (second pagination); cf.: Lesseps 1801. Pt. II: 155; quotation: Maidel' 1925: 25; cf.: Ditmar 1855: 52; Jochelson 1892: 12–13; Bogoraz-Tan 1939: 132.

2. Vendetta

In the 19th century the most frequent reason for clashes within an ethnic group, and outside it, continued to be blood revenge. W. G. Bogoras noted that "blood vengeance occurs only when murder is carried out by a member of a different clan. On the other hand, murder committed within a clan or in the closer framework of the family, is left unpunished"—the Chukchi preferred not to get involved in the affairs of another family. In 1895 old people of the Reindeer Chukchi in a camp near the Anadyr' explained to N. L. Gondatti the reason for disputes among them by the fact that "murders often occur either from alcohol, which the Sedentary deliver, or from mushrooms sold by the Russians." Of course, alcohol contributed to the manifestation of aggressiveness and uncontrolled actions. A characteristic case occurred in the 1860s in the village in Svyatoi Lavrentii [St. Lawrence] Bay, when one of the residents "went drunk through the settlement with a gun and shot into every hut as he passed it. In one of the nearest he killed a young girl, but her twelve-year-old brother grabbed a gun and shot him to death." In 1895 Bogoras noted that the Nose Chukchi Kulinenok had carved "on the back . . . of a knife eight notches. This meant that Kulinenok had killed 8 people."[582] In general, murder was a rather typical phenomenon in Chukotka at the end of the19th century.

Dr. F. M. Augustynowicz, who worked in 1875–1876 in the Kolyma district, described the typical behavior of the Chukchi thus:

> They never pardon an offence and silently wait for a chance for vengeance on their enemy, absolutely trying to kill him. If a Chukchi does not manage to do this in his life, the revenge falls to his son, who in any manner possible should carry out the will of his father. Therefore, the vengeance of the Chukchi is very long-lasting; it sometimes passes from generation to generation until there is a chance to end the feud with murder.[583]

[582] First quotation: Bogoras 1899b: 41; cf.: Zavadskii-Krasnopol'skii 1873: 34; Tan 1898. No. 116: 2; Bogoraz-Tan 1934: 179–184; Sverdrup 1930: 324; second quotation: Gondatti 1898: 5; cf.: Krause 1881: 17; third quotation: De Windt 1898: 262; Kulinenok: AAN. F. 250. Op. 1. No. 112. L. 3 ob.

[583] Avgustinovich 1878: 54; 1880: 725; cf.: Kolymskii okrug 1879. No. 6: 46; Sil'nitskii 1897: 52.

As the missionary A. I. Argentov related, in 1842 several Chukchi, who spoke with some kind of accent, arrived on the island of Kolyuchin as a consequence of hunger. They were killed, since they were considered the descendants of the rebellious Erruva who a long time ago fled beyond the sea.[584] Consequently, the Chukchi, who spoke a different dialect, were killed according to the law of blood revenge, in spite of the fact that centuries had passed since Erruva's flight.

Figure 45. A Sitting (Settled) Chukchi (mid-19th century). (Reproduced after Bulychev 1856: Pl. 61).

Among the coastal residents, in spite of the development of trade relations, during the second half of the 19th century the custom of blood revenge was also preserved. So, it is known that, when one Eskimo in a

---

[584] Argentov 1857: 103; 1861: 612; 1861a: 21.

drunken state killed another, he was pursued by all the men of the village which was located in Provideniya Bay. Having caught the murderer, they delivered him to the relatives of the one killed, who cut the meat from his cheeks, chest, arms, and thighs and left to die (an event earlier than 1865).[585]

Later, in 1895, Naukan old people complained with regret that "murders have now become perhaps more than before, since in former times there was still fear of revenge, which now occurs more rarely." These relations are described in more detail by the adjutant of the troop commander of the Priamur'e Military District, Staff Captain A. A. Resin (1857–1933), who was sent for inspection of the region in 1885:

> The Chukchi differ from the Koryak in militancy, judging by the armed clashes that often occur between them. The reason for such clashes is commonly a dispute between members of two clans because of some nonsense, stolen reindeer, and the like. The dispute, usually under the influence of vodka, turns into a fight and often ends in murder. The relatives of the one killed believe it their sacred duty to avenge him and thus, one clan rises and opposes the other, the affair reaching armed conflict; they fight with spears, but sometimes in the affair they also use Winchesters, though in general it is noted that where there are these guns (especially among the Sitting Chukchi) fighting occurs more rarely. Often a clash ends in peace: an advancing clan, aggrieved, demands satisfaction, and indemnity and, if an agreement occurs, the opposing side pays each of those concerned a reindeer, or two. Even in a peaceful time the Chukchi do not forget to fight with spears, since they have duels, which continue until one breaks his spear or the spear is not knocked from the hands of the other.[586]

As we see, the author described two kinds of conflict: clan, resulting in a blood feud, and personal, the resolution of which was in duel combat with spears.

3. Chukchi Clashes with the Even

If at the turn of the 19th and 20th centuries as W. G. Bogoras suggested, on the basis of folkloric information, that the clashes of the Chukchi with the Tungus had occurred in deep antiquity, later he began to

---

[585] Dall 1870: 382.
[586] 1895: Gondatti 1898: 18; Resin 1888: 171–172.

believe that these conflicts were in not such remote times. In any case, these events are almost not reflected in Chukchi folklore. And even the researcher himself, in his monograph on the Chukchi, dedicated only 4.5 lines to this subject, and I. S. Vdovin omitted this subject in his monograph all together.[587]

However, there nevertheless is information on this topic in Chukchi folklore. The writer Yu. S. Rytkheu (1930–2008) tells, on the basis of family legends, that his ancestor eight times removed, Mlakoran went on a campaign to the Even to seize reindeer. Though the author himself assigned this event to the pre-Dezhnev period, if a generation is calculated as 30 years, then we get the middle of the 18th century as the time of this raid. This last calculation is supported by the fact that the son of Mlarokoran, Tynemlen, lived at the same time as Kunlelyu—the hero who was active in the second half of the 18th century.[588]

In 1781 Corporal G. Sheikin, on the basis of observations he made during his service in Anadyrsk in 1756–1764, stated that the Lamut who lived on the Anyui, never underwent attacks by the Chukchi. At the turn of the 1980s and 1990s Ainakvyrgin thus described the relationship of the Chukchi with the Even in antiquity: "The Lamut—Koraramkyt—in the tundra were never beaten or killed. They could go peacefully among irreconcilable enemies when and where they wanted."[589] This situation probably existed on the southern border of Chukchi settlement during the big wars between the Koryak and Chukchi.

Judging by I. S. Gurvich's map, at the end of the 18th century the Chukchi entered into contact with the Even in the region between the Penzhina and Omolon. In 1763 the merchant N. P. Shalaurov stated in a report to the senate that the Chukchi stole reindeer from the Koryak, Yukaghir, and Lamut, who lived between the Anadyr' and Penzhina. In February 1759 the Koryak attacked the Lamut at the place of confluence of the Oloi and the Omolon. According to another document of the 18th century, the Chukchi on their campaigns came to the Ola River, falling

---

[587] Tan-Bogoraz 1930: 68 (clashes with the Tungus in the recent past); Bogoraz-Tan 1934: 175; Vdovin 1965: cf.: Antropova 1957: 182.
[588] Rytkheu 2004: 9, 33–40, 79–91.
[589] G. Sheikin's information: AII. F. 36. Op. 1. No. 643. L. 583; Nefёdkin 2016a: 97; Ainakvyrgin's opinion: Radov 2008: 75.

into Tauisk Bay, and took Lamut captive. W. G. Bogoras supposed that the settling of the Chukchi to the west and south began later, from the 1820s, when they started to push the Lamut into the forest or settle near them. It is evidently the matter of massive resettlement. G. Maydell noted that the Lamut began to actively appear on the international arena after the abolition of Anadyrsk.[590] It is exactly from this time that it is possible to speak of permanent mutual relations between the Chukchi and the Lamut. In the legend of the Kolyma Chukchi, recorded in 1896, it is told that the first encounter of the Reindeer Chukchi, who arrived in the land of the Lamut, occurred much later than the opening of the Anyui fair (1789) and that this encounter happened at the slopes of the Stanovoi Range, almost turning into an armed conflict.[591]

The historical tradition, told to W. G. Bogoras by the Lamut Khirkan from the Chaun tundra in 1896, describes a raid of the Chukchi on the Lamut on the upper reaches of the Korkodon River for the purpose of seizing their reindeer. Another Even legend about two brothers, recorded in 1870, tells of how the two brothers—accurate archers—repulsed an attack on them by a much superior number of Chukchi. Later legends point to the reason for the hostility: the Chukchi stealing reindeer (and then children as well) from the Even.[592] This is quite probable, considering that the Chukchi used the larger Lamut reindeer for breeding their herds. And since the Even did not always want to sell their animals, the relationship could move in the direction of confrontation. About the time of the establishment of peaceful relations A. I. Argentov noted: ". . . friendly intercourse between the Chukchi and the Lamut is still new, and the exchange of reindeer between them began no earlier than 1844."[593]

---

[590] 1759: RGADA. F. 199. No. 539. Ch. 2. D. 6. L. 32 ob.–33; Bogoras 1899b: 5; 1902a: 73; Nikol'skii 1900: 23; Maidel' 1894: 64; 1763: Belov 1954: 181; Ola River: Vdovin 1973: 257–258. For the diagram of resettlement of the Even, see: Gurvich 1966: 268, cf.: 269, 271–272. B. V. Zonov (1931: 61) suggests that the Chukchi appeared on the Omolon from the Anyui at the beginning of the 19th century.

[591] Bogoras 1900: 391–392.

[592] Story of Khirkan: Bogoras 1918: 28–29; legend of brothers: Schiefner 1874: 353–354; Burykin 2007: 381; Robbek 2005: 272; stealing reindeer: Novikova 1987: 108; Robbek 2005: 270–271.

[593] Chukchi trading reindeer from the Lamut: Argentov 1860: 25; 1889: 85; quotation: Argentov 1860: 25; 1889: 85. A. G. Vol'fson (1991: 18), incidentally, noted that Chukchi reindeer were usually bred with wild ones.

From this time peaceful relations had just begun to be established, though old conflicts still made themselves felt.

The most important source for the study of Even-Chukchi relations of the second half of the 19th century are Tungus texts collected by G. Maydell on the Anadyr' in January 1870. He scrupulously recorded stories, using Latin transcription, following several or even a single narrator, who spoke in the east-central dialect of the Even language. This is close to the modern language among the Even of Ulakhan-Chistaisk settlement of the Momsk district in northeastern Yakutia, though in the 19th century distribution of this dialect might have been broader. Interesting us in this case are stories Nos. 31, 32, and 35, describing clashes with the Chukchi. Though Alexei A. Burykin believes that these stories belong to some kind of folklore of the heroic tale type and it is impossible to view them as a historical source, he does not provide weighty arguments in favor of his point of view. In addition, the texts tell about the way of life of the Lamut (including about conflicts with the Chukchi) in the first person, on behalf of the narrator, while other historical texts point directly to a time long passed (Nos. 53 and 58). These stories tell about clashes in Angara region, where the Even entered in order to hunt. In particular, the Lamut told that when they set up camp, the Chukchi attacked them. When the men went on the hunt, the camp was devastated by the Chukchi, who destroyed the tents and killed the women and children. Even on the eve of the meeting with G. Maydell there was a shootout in which the narrator was wounded. The latter planned to take revenge and attack the Chukchi.[594]

Maydell did not translate the folklore texts himself; they and their content most probably remained not really known to him, including reports about Chukchi-Lamut conflicts. And events in the Chukotka zone did not fall within the area of responsibility of Russian authorities, though it is possible to find hints of these conflicts in the texts of Maydell. Thus, he noted about the Anyui fair, that "when certain Lamut or Chukchi, who were known as frequent visitors to the fair, disappear, you sometimes hear that somewhere in the wilderness there was a life and death battle and that

[594] Opinion: Burykin 2007: 368–369; texts, see: Schiefner 1874: 341–343; Burykin 2007: 376–377.

the one who disappeared remained in that place."[595] In such cases, not only duels but also larger clashes could occur.

The Kolyma district police officer Baron Gerhard von Maydell cited another reason for hostility in the second half of the 19th century:

> The Chukchi, as rich herd owners, especially detest the Lamut. The Chukchi blame them for hunting reindeer and not always troubling themselves to distinguish the wild animals from those belonging to the Chukchi, or in other words, they kill everything that turns up, even if know very well that before them is a tame herd; if they are caught at the place of the crime and try to take the killed animal, they allegedly grab a gun to defend themselves . . . the Chukchi are almost helpless against the agile and deft Lamut; with their spear, no matter how well he wields it, the Chukchi cannot approach the Lamut armed with firearms.[596]

Such actions of the Lamut can be largely explained by the fact that the Chukchi began to graze their herds in pastures that earlier were occupied by wild reindeer, as a result the latter changed the territory of their habitation and routes of spring and fall migration. The Lamut and Yukaghir complained of impoverishment of their traditional hunting grounds because of this. It is known that in 1866 ten Chukchi families (about 50 people) were permitted to go across the Kolyma into the Great Tundra, where the Lamut began to rob them of reindeer; this brought on the complaint of erem Amvraugin to the Russian administration with a request for protection.[597] To the demand of the Tungus and Lamut that they get paid for use of their pastures, the western Chukchi answered that they already pay tribute to the tsar. There was nothing left for the Even and Yukaghir than for them to go as herders for the Chukchi, in order to receive payment of reindeer.[598]

---

[595] Maidel' 1894: 68.

[596] Maidel' 1894: 67–68; cf.: Olsuf'ev 1896: 143; Maidel' 1925: 25 (the western Chukchi are not militant and are not able to resist Lamut and Yukaghir armed with firearms).

[597] Neiman 1871. No. 4–5: 10; Maidel' 1925:21–22, 28; Bogoraz-Tan 1933: 243.

[598] Bogoras 1921: 50–51; Bogoraz-Tan 1933: 242.

Figure 46. The last Chukchi erem Tynapo (1930). Dressed in the customary *kukhlyanka* and *kamus* [footwear], and on top—a long red cloth cloak lined with ermine, on his head—a hat of cherry-colored satin. The toion has symbols of authority: medals on his neck, a dagger on his belt, and a saber on a shoulder belt (Polyakov 1983: 76–77). We note that in 1875 the grandfather of Tynapo, erem Amvraurgin, wore as symbols of authority a red kaftan with braids and a standing collar, red trousers with gold stripes, a saber, and three medals on his neck, but with this, he wore a Chukchi cap and boots (Avgustinovich 1880: 719; cf.: Argentov 1887: 22). (Reproduced after Polyakov 1983: illustration between pp. 80 and 81).

Conflicts with the Lamut also occurred at river crossings of reindeer, which in June were moving north and in August–September, south. The Lamut shot reindeer at the crossings, which frightened the whole herd, causing it to change direction of movement,[599] and leaving the Chukchi, who came to hunt, without taking any reindeer.

[599] Wrangell 1846: 100–101; Sgibnev 1869: 28, note 2; Neiman 1871. No. 4–5: 27; 1872. No. 2: 58.

In the mid-1870s F. M. Augustynowicz characterized the relationship of the Lamut with the Chukchi in the following way:

> To the Russian government and in general to all Russians they are sincerely devoted; and this sense of loyalty is equivalent to their hatred of the Chukchi, so that in case of any hostile intentions against the Russian government, or in case of offences caused to a Russian population by the Chukchi, the Lamut are always prepared to protect the interest of the Russians, purely from the desire to fight the Chukchi, relying on their agility and well-aimed shots from rifles. The Lamut wait annually at the Anyui fort during the Chukchi fair for such a favorable case for a clash with the Chukchi, offering their services free of charge to the district police for keeping guard and making night patrols on the outside of the fortified enclosure. Their expectations, however, almost never come true, since the Chukchi at present are much more peaceful and do not express obstinacy and innate impudence.[600]

Still in 1895 the chronic hostility of the Chukchi and Lamut was alive: erem Eigeli refused to cooperate with W. G. Bogoras, having learned that the latter was friendly with the Lamut. At the beginning of the 1900s the Chukchi reindeer herder Ivan Valergin, together with his herd of 7,000 head, went from the tundra between the Kolyma and the Indigirka to Chaun because of harassment from the Tungus leader Afanasii Kurilov, nicknamed "king of the tundra."[601]

Since in Chukchi folklore conflicts with the Lamut are not clearly represented, it is evident that for the Chukchi these were insignificant peripheral clashes, whereas Even legends, by contrast, show the importance of these conflicts for the Lamut. If earlier, at the end of the 18th century, and especially in the first half of the 19th century, the Chukchi, resettling to the southwest, entered the territory of the Lamut, then in the second half of this century the process went in the reverse direction: the Lamut in search of hunting grounds entered the territory of the Chukchi. From the mid-19th century it is possible to speak of some stabilization of relations, which developed primarily in a peaceful course.

---

[600] Avgustinovich 1878: 50; 1880: 714; about Lamut guard, see: Argentov 1886: 13; Dioneo 1895: 7.

[601] About Eigeli: Bogoras 1899b: 7; "Sredi chukchei." 1899. No. 38: 3; Polyakov 1983: 66; cf. about hostility in general: Dioneo 1895: 7; Olsuf'ev 1996: 141; about I. Valergin: Krotov 1927: 53, notes 8, 10.

If initially confrontation was because of territory, then subsequently clashes began to occur because of hunting and stock-breeding: on the one hand—stealing reindeer from the Even for breeding, and on the other—hunting domestic reindeer of the Chukchi and conflicts at crossings. These conflicts led to murder, blood revenge, and conservation of common hostile relations.

4. Relations of the Chukchi and the Russians in the Anadyr' Area and at the Anyui Fair

After abolition of the Anadyrsk stockade and formal acceptance of the Chukchi into Russian citizenship, trade relations with them on the Anadyr' resumed, but now as a private initiative, not directly generated by the government. Trade was necessary both for the Chukchi, who needed Russian goods, and for Russian merchants, who received good profits. And since the traders, wishing to obtained maximum earnings, did not avoid deceiving the Chukchi, this sometimes led to conflicts.

The history of trade on the Anadyr' at the turn of the 18th and 19th centuries is the following: the Irkutsk governor general (1783–1789) I. V. Jacobi permitted the Kargopol' merchant Aleksandr Andreevich Baranov (1746–1819) who lived in Gizhiga and his stewards to go for trade to the Chukchi. In January 1788 Baranov and his companion, the Okhotsk merchant Ivan Krichevtsov, set off to the Anadyr' to hunt furs, collect walrus and mammoth tusks, and for trade with the Koryak and Chukchi. The Okhotsk commandant and Gizhiga mayor were supposed to help and supervise the merchants. In this same year Baranov established a trading post 70 verst (74.7 km) below the former stockade on the Anadyr' River, where about 30 men settled, who were occupied with hunting fur-bearing animals, walruses, and trading for furs from the Chukchi. Russian goods were brought from Gizhiga, where the steward himself stayed. In 1790 the post was destroyed by the Chukchi, the steward and workers were killed, and the goods plundered. On 21 February 1792 elders of the eastern and Chaun Chukchi once again filed a petition through Captain J. Billings for restoration of the Anadyrsk stockade for trade, since it was far for them to go to the Anyui and Gizhiga fairs. On 24 November 1792 the Irkutsk governor general I. A. Piell reported to the senate that the Chukchi requested through Captain Billings the resuming the Anadyrsk fort, where they wanted to pay tribute in fox furs at one per man. However, the governor himself suggested recommending to the Chukchi that they trade at the Anyui and in Gizhiga. This idea was supported on 17 January 1793 by the Government Council.[602]

Nevertheless, private trade on the Anadyr' resumed. I. Krichevtsov continued to conduct trade on the river, settling 35 verst (37 km) below the former stockade. In 1795 the acting Gizhiga mayor Adam Laxman sent Ensign Khabarov with twenty Cossacks to observe trading on the Anadyr' and collect tribute from the Chukchi. After the arrival of Khabarov and the report of A. Laxman, the Irkutsk governor ordered the trade transferred to the Main River.[603]

In 1803 the fortified trading post of the Russian-American Company merchant Vorob'ev appeared, the so-called "Maiskaya artel'." It was built on the right bank of the Anadyr' at the mouth of the Main River. The trading post was several structures surrounded by a wooden wall; the steward lived there permanently with the people, whereas Vorob'ev himself visited the trading post in the winter, which was more convenient for travel. In the trading post itself trade was conducted with the Chukchi not only in winter, when the reindeer herders approached the river during the winter migrations, but also in summer, in August, when 20 to 40 baidary of settled "Chukchi" sailed in to trade. In 1805, at a meeting at the Kamensk stockade, the Chukchi elders complained to the Kamchatka governor General Major P. I. Koshelev about the dishonesty of promyshlenniki during trade. And in 1806 the trading post was destroyed by the Chukchi and fourteen promyshlenniki and the steward were killed in an "insidious way." The Chukchi explained this incident to P. I. Koshelev by the fact that the Russians cheated them during trading, and requested the resumption of trade, having restored the Anadyrsk stockade as a means of control over the trade. Incidentally, no permission followed regarding these questions; the affair was limited to

---

602 "Dve zapiski. . ." 1873: 372, 377–378; Khlebnikov 1835: 4–5; Vdovin 1965: 139; Abakumov 1976: 225; Dikov 1989: 99–100; Chukchi petition 1792: KPTs: 189–190; discussion in the senate: Arkhiv Gosudarstvennogo soveta. Vol. 1. Pt. 2. 1869: 261–262; Slovtsov 1856: 19, 22; cf.: Divin 1979: 398.

603 Sgibnev 2008: 44; Lagus 1890: 453, note 851; Vdovin 1965: 139; Safronov 1980: 119.

correspondence between P. I. Koshelev and the Irkutsk office.[604]

The story by the Markovo regional historian A. E. D'yachkov reveals mutual relations on the Anadyr': "They say that the residents of this second Anadyrsk fort, 12 men in number, once arrived for meat to the mouth of the Main River; upon arrival they were kindly accepted by the Chukchi, but in the evening of the same day, when the Russians began to feed the dogs, the Chukchi suddenly unexpectedly attacked them and killed them all."[605] The time of the event is unclear. According to tradition, it is dated to the period of the second Anadyrsk stockade, which corresponds to the time of activity of Vorob'ev and P. Baranov. This is probably the event of 1806, when the Chukchi unexpectedly attacked the Russians.

Russian trading activity on the Anadyr' meanwhile did not stop, nor the deceiving of the Chukchi during trade. Thus, the Chukchi complained to the Gizhiga authorities, that for 104 red foxes and 19 other animals they were given only 1.5 pudy of tobacco. In 1814 the Gizhiga commissar Pshenitsyn arrived with fourteen Cossacks for examination of affairs on the Anadyr'.[606]

From 1815 annual fairs took place in the region of the middle course of the Anadyr' (at the future Fort) in January-February, which lasted three weeks each, and up to 300 Reindeer Chukchi, Koryak, and Lamut visited each of them. In 1819 a member of the Russian-American Company, Gizhiga merchant of the second guild Petr Andreevich Baranov (d. 1844), brother of above-mentioned Aleksandr Andreevich, governor of Russian America, took a concession on the land at the Anadyr' for the gigantic sum of 37,000 rubles. He reestablished a trading post, having built four houses. He fortified it with a wall and set up several cannons, which, however, were never fired. Hence, the name Krepost' [Fort] emerged for the settlement. Hired Gizhiga townsmen and peasants, led by the steward Sobinin (a total of 20 people), began to live in the settlement. Several dozen Russified local residents from Kolyma and Gizhiga (chiefly

---

[604] Kruzenshtern 1950: 172–173; "Lyubov' chukchei. . ." 1811: 18–23; Shakhovskoi 1823: 37 (in 1806 twelve promyshlenniki were killed; quotation); Sgibnev 2008: 43; D'yachkov 1992: 188; Gurvich 1966: 201; Abakumov 1976: 226; Dikov 1989: 100.
[605] D'yachkov 1992: 188.
[606] Sgibnev 2008: 74; Gurvich 1966: 201.

Yukaghir) joined them. Baranov brought his goods from Gizhiga on ten horses. Each August an off-site fair occurred at the mouth of the Anadyr' with settled "Chukchi," who arrived in fifty or more baidary to hunt reindeer. Along with the trade, the Baranov people occupied themselves year-round with hunting for food and furs.[607]

A. E. D'yachkov said this about the relationship of the Chukchi with the workers of P. A. Baranov's artel:

> When the Baranov artel still stood, the Chukchi treated its residents unfriendly, but they committed murder only one time. This case they relate in the following way. The company (that is, Baranov's workers) were annually occupied with the 'ezov'e' [braided dam, weir], that is, catching fish, and after freeze-up they blocked the Main strait approximately one versta from the fort. One morning there were several men at the ezov'e; several dozen Chukchi followed them and at night crept up to the fort itself. The Chukchi decided to kill the Russians by a trick, and three of them approached the fishermen under the pretext of trade and did not arouse any suspicions or precautions; then one of the Chukchi threw himself on a Russian and stabbed him with a knife; the comrades of the victim ran to the fort and safely escaped the pursuing Chukchi, except a young teenager, who fell behind and was captured. The murdered man was the great-grandfather of the current resident Nikolai Ermachkov, and the prisoner, the father of the former headman Nikolai Podymakhin. After a year the Chukchi again arrived at the fort for trade, brought the prisoner, and returned him, but for a certain ransom.[608]

The time of action of the account can be determined only approximately—between 1819 and 1844. The reason for the attack of the Chukchi is not reported—possibly some offence during trade or hunting competition. The reality of the incident is reinforced by family sources of the Markovo people. The attack was carried out during the formation of ice on the Main, that is, in the fall. It was probably done by the Chukchi who migrated to the river in winter. They decided to attack the Russian fishermen near the trading post, and therefore crept up at night and set an

---

[607] Litke 1835: 366–368; 1948: 225–226; Mushkin 1853: 1421; Bush 1872: 384; D'yachkov 1992: 188; Olsuf'ev 1896: 24; Gondatti 1897: 158–159; 1897a: 93–94; Vdovin 1965: 140; Gurvich 1966: 201; Dikov 1989: 100–101.
[608] D'yachkov 1992: 197.

ambush. Three Chukchi approached as if for trade with the Russians, probably having hidden weapons in their clothing. Since there were no recent precedents for an attack, the fishermen did not expect one. The unexpected attack turned them to flight. The Chukchi in the ambush evidently also began to pursue the fleeing. Incidentally, all the fishermen, except the teenager, ran a versta to the fort, where they found refuge. In the following year the Chukchi arrived at Fort for trade as if nothing had happened and gave up the boy for a ransom. As D'yachkov notes, this was the only case of direct hostile action between the Russians and Chukchi, in spite of the general complexity of mutual relations.

After P. A. Baranov's death his artel split up and trade was transferred into the hands of several Gizhiga merchants. In the 1850s the merchant Vasilii Trifonov delivered his goods to the fort from Kolyma on 50 to 80 horses: tobacco, tea, spirits, and iron and copper items. In the 1860s the walls of Fort were washed away by flooding of the river, and the residents gradually resettled to Markovo village, which was founded in the 1840s by a certain Markov. Gizhiga townsmen were settled there, ancestors of whom evidently were born in Anadyrsk. Chuvan, Yukaghir, and Lamut peoples from the Kolyma District (chiefly from Anyui) began to settle with them; in 1862 a church was built there. In the 1890s only two families numbering fifteen people lived at Fort.[609] From the 1820s Settled Chukchi, often baptized, also began to settle on the banks of the Anadyr'.[610]

The primary fair for trade with the Chukchi was the Anyui fair. From 1789 to 1794 trade took place at the small Angara fort, and then the fair was moved to an island in the Bol'shoi [Big] Anyui and nicknamed the "Island." During the flood of 1848 the river washed away part of the island and the fair was moved 10 verst farther down, to the Anyui fort on the high left bank of the Malyi [Little] Anyui.[611] Here, in the fort, protected by wooden walls and a gate tower, were a chapel, barracks, and several huts. The stockade was inhabited only during the fair, in February–March. The district police officer arrived there from Kolyma with lower-Kolyma

---

[609] Bush 1872: 385; D'yachkov 1893: 40–41; 1992: 202; Gondatti 1897a: 83–84; Vakhtin, Golovko, and Shvaittser 2004: 55.
[610] Neiman 1871. No. 4–5: 28; Gondatti 1897a: 100–102; Bogoras 1921: 90.
[611] Vdovin 1965: 140; Safronov 1980: 117–118; Kolesov 1991: 43.

Cossacks and Russian merchants. During the fair the fort was guarded by the Cossacks and Lamut.[612] Trade was held outside the fort after the delivery of tribute and prayer.

Incidents were extremely rare at the fair. In 1895 one Chaun Reindeer Chukchi was arrested for unruly behavior because of the termination of the sale of vodka at the fair. He suddenly died while under arrest. The Chukchi became so indignant that they were prepared to seize the fort. The Russian authorities awarded money to the stepfather of the deceased. Other distant relatives and countrymen of the dead man also wanted to receive blood money. In the following year the brother of the deceased, arriving at the fair, demanded compensation. After receiving gifts—tea, sugar, and tobacco—he announced that he wanted to take the Cossack who, in the opinion of the Chukchi, was the murderer, or another Russian, as replacement for the widow's murdered husband—only in this way could the offence be considered reconciled.[613] In general, the Chukchi were quite satisfied with the redemption, instead of blood revenge. In 1911 I. P. Tolmachev mentioned the following case from the recent past: when a Chukchi was killed at the Anyui fair in a brawl with Russians, the Chukchi set off against the fort and the district police officer himself fled in fear. Perhaps it was a matter of the same incident of 1895.[614]

Before the 1840s communication of the Kolyma people with the Chukchi was limited to encounters once a year at the Anyui fair. From the beginning of the 1820s, after the expedition of F. P. Wrangell, two or three Yukaghir began to go to the Chukchi to trade, and as many smugglers, since official permission from the Russians was necessary for a trip to them. The activity of the missionary A. I. Argentov played a notable role in normalization of relations with the Chukchi. Argentov learned the Chukchi language and founded the Nikolaevskii church in the summer of 1848 on the Rauchua River, where the Chaun parish was created. There was a cautious attitude toward each other on both sides, explained by old stereotypes, which emerged on the basis of poor mutual understanding. Argentov suggests that the Chaun old people used to say this to the young: "The reindeer don't go to wolves—blood will be spilled; don't live with

---

612 Avgustinovich 1878: 50; Argentov 1886: 13, 23; Kostyurin 1893. No. 24: 2.
613 "Stolknovenie. . . ." 1896: 2; Bogoraz-Tan 1934: 26, 177–178, 181.
614 Tolmachev 1911: 102.

the Russians—there will be trouble," whereas the Kolyma people thought: "Is it a reasonable thing to trust animals? Reindeer aren't comrades to wolves." Before 1844 the Chukchi arrived peacefully to Nizhnekolymsk only once, in 1812—at the invitation of the missionary G. Sleptsov. Then in 1844 40 Chukchi yarangas moved to the Kolyma, and the Russians began to visit Chaun and Cape Shelagskii.[615]

During these visits incidents were rare but nevertheless happened. In 1884 an event named the "merchant Druzhinin affair" occurred. The Nizhnekolymsk merchant Semen Druzhinin went to Cape Shelagskii to trade with the coastal Chukchi. He offered too small a price for a seal carcass intended for dog food, which brought on the dissatisfaction of the Chukchi. The angry merchant announced that the Yakutsk governor would come to Kolyma to punish the Chukchi. For this threat, based on a report by another Kolyma person, the merchant was put in jail. And the district police officer sent a courier to the Chukchi, who told them that the Russian authorities had nothing to do with S. Druzhinin's threats. The local Kolyma residents received the order that they should not give the Chukchi the smallest reason to complain.[616]

5. Conflicts of the Chukchi with Crews of Foreign Ships in Bering Strait

According to the agreement of 1824 between Russia and the United States, ships of both countries could visit ports of each other and conduct trade with the natives. However, ships usually did not reach Bering Strait. In 1834 the term of validity of the agreement expired, and passage of American ships in Russian territorial waters was prohibited. However, smuggling trade was conducted: the Americans brought goods necessary for the Chukchi: tobacco, metal items, and guns, as well as rum and whisky, receiving furs, baleen, walrus tusks, and other products from hunting in exchange. Trade with the Americans, who sold higher quality and less expensive goods, contributed to a decline in turnover of Russian-Chukchi trade in the second half of the 19th century, primarily at the Anyui fair.

---

[615] Argentov 1886: 13 (quotations), 34, note 7. About the founding of the church: Mushkin 1853a: 844; Argentov 1887. No. 4: 56; Shklovskii 1892: 100; about trade in 1843: Paryshev 1909: 120.

[616] Bogoraz-Tan 1934: 60; Stepanov 1957: 336.

After the first whaling bark *Superior* of Captain Thomas Roys from New York visited the strait in 1848, ships from the United States, France, Germany, Hawaii, and Australia began to visit here in large numbers for sea mammal hunting. Already in the next year, 1849, 154 American ships were counted, and in 1852 the number of ships reached its maximum—224. For the period from 1849 to 1899 a total of about 2,500 ships visited the Bering Sea, of which 169 were lost to shipwreck, and of them, 67% were the consequence of severe ice conditions. In 1856, 1859, 1868, 1898, and 1899 ships vanished for unclear reasons at the shores of Chukotka.[617]

New reasons for attacks by coastal residents on the non-military ships also appeared—for the sake of booty. The director of the Western Union Telegraph Expedition, William Dall, who spent the winter of 1866–1867 in Provideniya Bay, writes about the coastal residents of Chukotka: "They will steal, and have sometimes attacked small vessels in the Strait. On one occasion they attempted to take a small schooner, commanded by a captain of my acquaintance," but the crew repulsed the attack and in recent years such attempts were not observed.[618]

Therefore, it is possible that some of the missing ships were seized by coastal residents. In particular, such capture could have occurred during a long stop or even the wintering of the crew in Chukotka. Wintering over periodically occurred before 1874, after which patrolling by Russian naval vessels of their territorial waters in the Far East was introduced for the purpose of intercepting illegal trade by foreigners (in particular, Americans) in Chukotka.

The rumor that the residents of Chukotka loved to rob ships circulated among European mariners. It is known that in 1820 the Chukchi stole the lead coating from a cannon and tried to steal a second from the sloop *Blagonamerennyi* in Lavrentiya Bay. In 1851 a clash occurred in Senyavin Strait between the English crew of the ship *Armata*, which sailed under the American flag, and local residents over a piece of anchor chain, which the Chukchi tried to seize. In the skirmish twelve Chukchi were killed and one member of the crew. In September 1852 Reindeer Chukchi

---

617 Dall 1870: 343; Bockstoce 2006: 53–68.

618 Dall 1870: 381.

looted the wrecked American ship *Citizen*, after which they moved away.[619]

The relationships of local residents with crews of ships that arrived were different. Hostility with crews could also emerge because of the mercenary behavior of the captain. Cases are known when an American captain would send a keg of rum or whisky in a dinghy to the residents of a village, then sailed away, waited, watching through the telescope to see when the Chukchi would get completely drunk. When this happened, the ship sailed back to the shore; the crew disembarked and took everything they wanted from the yarangas of the residents.[620]

Figure 47. A baidara with residents of Chukotka paddling up to the American steamship *Rodgers* (1881). (Reproduced after Gil'der 1886: 51).

Of course, when relations followed the course of confrontation, those same principles of blood revenge operated among the Chukchi. Thus, the legend, told in the village of Nunyamo, reports about a clash in the bay of Svyatoi Lavrentiya [St. Lawrence] between Reindeer Chukchi and the captain of an American schooner, in which a Chukchi sailor hired as a wiper was killed. The avengers tracked down the captain while he was hunting on land, surrounded him, and killed him. Having killed the captain

[619] Chukchi—lovers of theft: Mitchel' 1865: 322; 1820: Lazarev 1950: 298; 1851: Vdovin 1965: 148; 1852: Mitchel' 1865: 319; Bockstoce 2006: 58.

[620] E. M. 1892: 233–234; Kostyurin 1893. No. 24: 2. Yu. Rytkheu (2004: 293–294, 297, 320) tells of such a case in Emma Bay, where the American captain left two kegs of whisky.

guilty of the death of a kinsman, the Chukchi no longer sought revenge, being satisfied with a payment of "an eye for an eye." These events most probably are dated to the third quarter of the 19th century. It is known that on 30 March 1863 Captain Brummerhoff from the Hawaiian brig *Kohola* was stabbed by three coastal residents (a father and two brothers from the bay of Svyatoi Lavrentiya) in revenge for the death in November 1862 of their relative who jumped from this ship into the sea and drowned. A few years later B. Cogan, who took the captain's position, arrived in the bay, and the two brothers threw themselves on him with knives. After just a year or two the brothers made peace with the sailors.[621] This is probably the event described in the legend.

* * *

In general, after conclusion by the Chukchi of peace with the Russians and Koryak at the end of the 18th century, no more large military actions were carried out in the region. However, before the end of the 19th century, when the region was not under the control of Russian authorities, intertribal relations followed the traditional course. In the first half of the 19th century the Chukchi not only traded with the residents of Bering Strait, but also continued their raids, the intensity of which however decreased. This was the result of both long-standing hostility and newly emerging hunting conflicts. Although peace of the Chukchi with the Koryak was established under the influence of Gizhiga authorities, hostility did not stop and in their contact zone clashes emerged because of disputes over pastures and theft of reindeer. For these reasons conflicts also arose between the Chukchi camps themselves, which sometimes led to murder and blood revenge. Hunting competition for reindeer and theft of animals brought on conflicts with the Even, which led to the same consequences. Even at the beginning of the 19th century the western Chukchi, who did not conclude peace at Gizhiga, could attack the Kolyma people. Conflicts occurred only rarely during the fairs, and then usually after consuming alcohol. Trading conflicts also arose with crews of foreign ships that were trading in the strait, and were often provoked by the machinations of the traders themselves.

---

[621] Legend: Dikov 1958: 71–72; about the events: De Windt 1898: 260–262; Lebo 2013: 56–57.

CONCLUSION

The archaeological cultures of Chukotka, of course, do not give us a clear idea of the military history of the region in the preliterary period—it is possible to speak only about the evolution of Eskimo cultures, about the changes in types of weapons, and later, about the appearance of fortifications, drawing conjectural conclusions about the reasons for these changes. The evidence from folklore is more informative. According to the folklore, there was early warfare between the Chukchi and the Koryak, conflicts with the Eskimos, and flight of part of the Chukchi to the west beyond the Kolyma because of blood vengeance, which can be dated to the first third of the 17th century. At the moment of appearance of the Russians in the region in the mid-17th century, the Eskimos occupied the coast of Chukotka, while the migrant Chukchi lived in two compact groups: the primary group was the eastern ("Nose")—on the Chukchi Peninsula, and the western—in the Kolyma-Alazeya tundra.

In the middle of the 17th–first half of the 18th century northeastern Siberia was joined to Russia. Most of the explorers were motivated by a banal thirst for profit: the demand for furs in Europe and Asia because of the developing cooling of the climate (Little Ice Age) increased and it was possible to get a good profit. The Cossacks, according to the edicts, were supposed to use, according to the standard scheme, predominantly peaceful means for bringing the local population into citizenship. They proposed to the local residents that they come under the sovereign's hand accompanied by swearing an oath, providing amanaty, and delivering tribute, initially unregulated and small. If the local residents refused to do this, a forced annexation was set in motion. Scattered pockets of resistance flared up practically everywhere along the movement of the servicemen, but the newcomers received especially fierce resistance only in a few places: in Khakasiya from the Yenisei Kyrghiz, in Priamur'e from the Manchu troops of China, and in northeast Siberia from the Chukchi and Koryak, primarily the settled ones. In the northeast the imposition of

Russian rule by force was the prevailing method of annexing territory to the Muscovy State and, later, the Russian Empire in the middle of the 17th–third quarter of the 18th century.

For monitoring the territory and for convenience of collection of tribute the Cossacks built stockades, which served their movement farther into new territories. For Chukotka the closest stockades were Nizhnekolymsk (1644) and Anadyrsk (1649), which were adjacent to the primary territories of Chukchi settlement. The Cossacks collected tribute annually (from the Chukchi—fox pelts and walrus tusks), and with this, tried to impose a tax on the largest number possible of local men. Amanaty, who were kept under guard in a special hut in stockades, served as guarantees for delivery of tribute. However, this system of hostage taking did not work with the Chukchi and Eskimos: they viewed amanaty as members of the group who were already lost and did not pay tribute for them. And tribute was viewed by the Chukchi more probably as a gift for which they in turn also received gifts.

Before the beginning of the 18th century the main theater of contact of the Russians and Yukaghir with the Kolyma-Alazeya group of Chukchi was the region of the lower Kolyma. In 1649–1662 servicemen carried out several campaigns directed at conquest of the Chukchi, the culmination of which was the capture of the Chukchi leader Mita in 1655. Because of this a situation of conflict permanently existed, which was fueled by competition of the Yukaghir and Russians with the Chukchi over hunting grounds in summer; for the years 1646–1680 19 clashes of servicemen with the Chukchi can be counted.[622] In 1670–1690 the servicemen of Nizhnekolymsk, few in number, switched to passive defense. In 1710 the Chukchi carried out the last known, to us, raid on Nizhnekolymsk. And in the second half of the 1710s the local Chukchi were displaced to the west by their eastern countrymen—one of the illustrations of the long civil war that raged probably because of blood revenge that followed in connection with the flight of the Krekhai group in the first quarter of the 17th century. After the disappearance of the Kolyma-Alazeya group of Chukchi, permanent danger of a winter attack on Nizhnekolymsk also disappeared. Following the western Chukchi, the

[622] Cf.: Zuev 2009: 40 (19 clashes before the end of the 17th century, of them eight campaigns were conducted by servicemen and eleven by Chukchi).

Shelagi, losing the struggle with the eastern Chukchi, were also forced to migrate. The Chaun Chukchi, traveling in summer in baidary, began to carry out attacks on the Russians on the lower Kolyma.

In the following period, during the course of the 18th century, the main region of interaction of the Chukchi and Russians was the Anadyr' basin. Clashes usually arose between the Chukchi, on the one hand, and, on the other, the Anadyrsk Cossacks and Yukaghir, who sailed into the region of the lower Anadyr' to hunt the wild reindeer that crossed the river in spring and fall. A hunting confrontation affected economic interests of the settled coastal "Chukchi," who arrived in large number for hunting. Since the latter were the most interested in peaceful relations with the Russians (and it was difficult to flee from their settlements), it was from them that the Russian collectors obtained tribute, though irregularly. From the 1680s to 1756 there was gradual displacement and extermination of the Chuvan and Khodyn peoples from the region by the Chukchi.

Escalation of the Russian-Chukchi conflict occurred from the 1730s to mid-1750s, when Russian authorities relied primarily on military submission of the territory. The half-century Chukchi war (1727–1778) began. This was a large war, by Siberian standards, information about which has been preserved in Chukchi folklore in the form of the historical legends of "Times of War." The reason for the war were the geopolitical interests of Russia, whose leaders feared that these territories might be occupied by other European powers, in particular France, who possessed vast possessions in North America, and with whom Russia had serious opposition in the first half of the 18th century.[623] For accession of northeastern Siberia, the overland military expedition of A. F. Shestakov-D. I. Pavlutskii (1727–1732) was sent, which in fact complemented the non-military maritime expedition of V. Bering (the First Kamchatka Expedition, 1725–1730). The Chukchi themselves provided the specific reasons for escalation of the conflict: from the beginning of the 18th century they, moving to large-herd livestock breeding, began to carry out regular winter raids on the Koryak for the purpose of seizing reindeer, and then from 1756, to take possession of the suitable pastures of the latter, displacing the Koryak to the south. An additional reason for conflicts was

[623] Bagger 2005: 55–57; cf.: Makarova 1982: 32–34.

the Chukchi need for metal items, which could be taken from the Russians, Yukaghir, and Koryak. The Russian authorities had to intercede for its subjects. The campaigns of D. I. Pavlutskii (1731, 1744, and 1746) that followed were in essence a continuation of the taxation policy of the population of Chukotka, but they also bore a reconnaissance character in many ways, preparatory for further establishment of Russian rule in the northeast, inflicting human and material losses on the adversaries. During the first clashes, the Chukchi came out in open frontal battle since they were numerically superior (1702, 1730, 1731, and 1747); however, later their strategy changed: the Chukchi began to avoid such clashes, in which they bore substantial losses from firearms. Perhaps, the explanation for this change lies in part in the Chukchi view of the enemy's number, which they told the Frenchman J.-B. de Lesseps in 1788: first the Chukchi, dealing with the small in number the Siberian population, thought that the Russians are few and that they had all arrived in their country.[624] However, when it turned out that this was not so, both the defeat and destruction of the enemy in battle began to appear not such an effective means of getting rid of the enemy. Strategically, the military action of the Russians against the Chukchi was unsuccessful, which can be explained by the violent opposition, the small number of Russian troops, and the remoteness of the territory—all this did not permit the empire to win in the struggle. Also, the dismantling of the Anadyrsk stockade and the evacuation of the stockade garrison—the Russian settlement closest to Chukchi territory (1764–1771)—in general, can be seen as the defeat of the empire.[625] According to A. S. Zuev's estimates, from 1642 to 1775 there occurred 56 armed clashes between the Russians and Chukchi, in which the Chukchi had 15 victories and suffered 24 defeats; on the order of 1,500 Chukchi warriors and 150–200 servicemen died in these battles, including in the large frontal fighting on the Egacha River (1730)—12 servicemen and on the Orlova River (1747)—41 servicemen.[626]

---

[624] Lesseps 1801: 94–95.

[625] Bagger 2005: 53–54.

[626] Zuev 2009: 297, 299, 313; cf.: Zuev 2001a: 84; 2002: 66. For comparison: the servicemen collided with the not so territorially remote Koryak for this same time 154 times (Zuev 2005: 43).

If initially Chukotka could be viewed by the Russian administration as a springboard for farther movement toward America, this later became Kamchatka. Interest in Chukotka, which did not have at that time rich discovered natural resources, was lost by the Russian authorities, only to emerge again in the last third of the 18th century, when ships of other European powers that could secure these territories for themselves began to arrive in the seas around the region. Only after a century and a half of war and twenty years (from the mid-1750s) of negotiations, which concluded in an agreement in Gizhiga (1778), did the government of Catherine II announce the annexation of Chukotka to Russia and in 1779 ordered boundary pillars with coats of arms be installed.

The tsarist authorities, organizing trade fairs in the last quarter of the 18th century, gradually successfully placed the Chukchi in economic dependence. However, up to the end of the 19th century the Russian administration did not interfere in the affairs of the local residents, limiting itself to small voluntary tribute paid by Chukchi arriving at the fairs. The need for iron items, firearms, and tobacco led to the development of trade relations with the Russians at the annual fairs. Also contributing to the development of trade, including firearms, was trade with American whalers in Bering Strait during the second half of the 19th century. The traditional style of life in Chukotka was preserved, but the past militancy was gradually fading—gone was the enemy factor that held society in tension. However, warfare in the form of isolated raids occurred both because of the desire to obtain booty and because of clashes on grounds of blood revenge and old hostilities. In spite of the establishment of peace with the Koryak in 1781, conflicts periodically arose in the bordering territory over the reindeer; before the 1830s, while the Russian authorities did not yet control the situation, piratical Chukchi raids continued on Alaska, and on St. Lawrence Island up to its annexation to the United States; at the turn of the 1840s and 1850s the last raid on the Kerek was carried out; hostility with the Even, because of reindeer hunting by the latter, was still going on in 1870. The traditional culture, including education and methods of weapon handling, was preserved primarily among the reindeer herders, up to the execution of the policy of complete collectivization in the 1930s–beginning of the 1950s.

BIBLIOGRAPHY

*(For abbreviations see p. update after figures added)

A. 1878 "Merzlyi krai" [Frozen Region]. *Vsemirnyi puteshestvennik: Illyustrirovannyi zhurnal puteshestvii i geograficheskikh otkrytii* [Global Traveler: An Illustrated Travel Journal and Geographic Discoveries]. St. Petersburg. January. Pp. 261–294.

Abakumov, G. S. 1976. "Brat'ya Baranovy—organizatory torgovli na Chukotke (1786–1843 gg.)" [Brothers Baranov—Organizers of Trade in Chukotka (1786–1843)]. *Istoriya i kul'tura narodov Severo-Vostoka SSSR: Materialy 3-i sessii Dal'nevostochnykh istoricheskikh chtenii po problemam istorii Severo-Vostoka SSSR* [The History and Culture of the Peoples of Northeastern USSR: Materials of the 3rd Session of the Far East Historical Readings on Problems of the History of the Northeastern USSR]. Edited by B. I. Mukhachev. Vladivostok: In-t istorii, arkheologii i etnografii narodov Dal'nego Vostok. Pp. 224–227.

Akkuratov, V. I. 1984. Led i pepel: Zapiski shturmana [Ice and Ash: The Notes of a Navigator]. Moscow: Sovremennik.

Akty istoricheskie. 1984. *Akty istoricheskie* [Historical Acts]. Vol. V. St. Petersburg: Tipografiya Vtorogo otd-niya sobstevnnoi E. I. V. kantselyarii, 1842.

Alekseev, A. I. 1958. *Brat'ya Shmalevy: Istoricheskii ocherk* [The Brothers Shmalev: A Historical Essay]. Magadan: Magadanskoe knizhnoe izdatelstvo, 1958.

________. 1961. *Uchenyi chukcha Nikolai Daurkin* [Scholar of the Chukchi Nikolai Daurkin]. Magadan: Magadanskoe knizhnoe izdatelstvo.

Alekseev, A. N. 1996. *Pervye russkie poseleniya XVII–XVIII vv. na Severo-Vostoke Yakutii. (Istoriya i kul'tura Vostoka Azii)* [The First Russian Settlements of the 17th–18th Centuries in Northeastern Yakutia. (The History and Culture of Eastern Asia)]. Novosibirsk: Izd-vo Un-ta arkheologii i etnografii SO RAN.

Alekseeva, T. I., V. P. Alekseev, S. A. Arutyunov, and D. A. Sergeev. 1983. "Nekotorye itogi istoriko-etnologicheskikh i populyatsionno-antropologicheskikh issledovanii na Chukotskom poluostrove" [Some Results of Historical-Ethnological and Populational-Anthropological Research on the Chukchi Peninsula]. *Na styke Chukotki i Alyaski* [At the Juncture of Chukotka and Alaska]. Edited by V. P. Alekseev. Moscow: Nauka. Pp. 3–64.

Andreev, A. I. 1940. "Zametki po istoricheskoi geografii Sibiri XVI–XVIII vv." [Notes on the Historical Geography of Siberia of the 16th–18th Centuries]. *IVGO* 72(2): 152–157.

________. 1948. *Russkie otkrytiya v Tikhom okeane i Severnoi Amerike v XVIII v.* [Russian Discoveries in the Pacific Ocean and North America in the 18th Century]. Edited by A. I. Andreev. Moscow: Gos. Izdatel'stvo geograficheskoi literatury.

Andrievich, V. K. 1889. *Istoriya Sibiri* [The History of Siberia]. Pt. II. St. Petersburg: Tipografiya V. V. Komarova.

Antropova, V. V. 1947. "K voprosu o mezhplemennykh otnosheniyakh i voennykh ukrepleniyaky u narodov severo-vostoka Sibiri" [On the Question of Intertribal Relations and Military Fortifications among the Peoples of Northeastern Siberia]. *KSIE* 3: 64–65.

________. 1957. "Voprosy voennoi organizatsii i voennogo dela u narodov krainego severo-vostoka Sibiri" [Questions of Military Organization and Warfare among the Peoples of Extreme Northeast Siberia]. *SES*. Moscow-Leningrad. Issue 2 (*TIE*. New series. Vol. 35). Pp. 99–245.

________. 1971. *Kul'tura i byt koryakov* [The Culture and Way of Life of the Koryak]. Leningraed: Nauka.

Anuchin, D. N. 1918. "Gornye chuvantsy" [The Mountain Chuvan People]. *Russkii antropologicheskii zhurnal* [Russian Anthropological Journal]. Moscow. Nos. 1–2, pp. 41–48.

Argentov, A. I. 1857. "Opisanie Nikolaevskogo Chaunskogo prikhoda" [Description of the Nikolaevskii Chaun Parish]. *ZSOIRGO*. Bk. 3. Sec. 1. Pp. 79–106.

________. 1857a. "Putevye zapiski svyashchennika missionera A. Argentova v pripolyarnoi mestnosti" [Travel Notes of Priest Missionary A. Argentov in the Polar Zone]. *ZSOIRGO*. Bk. 4. Pp. 1–59.

________. 1857b. "Statisticheskie svedeniya o prikhozhanakh Spasskoi tserkvi v Nizhnekolymske, otnosyashchiesya k 1855 godu i sostavlennye do narodnoi perepisi" [Statistical Information on the Members of the Savior's Church in Nizhnekolymsk, Belonging to the Year 1855 and Compiled for the National Census]. *ZSOIRGO*. Bk. 3, sec. 3, pp. 19–21.

________. 1860. "Oleni Zalenskogo kraya" [Reindeer of the Zalenskii Province]. *Akklimatizatsiya: ezhemesyachnoe izdanie Komiteta akklimatizatsii* [Acclimatization: Monthly Publication of the Committee on Acclimatization]. Vol. 1, issue 1, pp. 20–33. Moscow: Tipografiya L. Stepanovoi.

________. 1860a. "Izvlechenie iz putevykh zapisok" [Extract from Travel Notes]. *Amur. Ezhenedel'noe izdanie* [Weekly Publication]. Irkutsk. No. 14 (5 April). Pp. 185–187; No. 15 (12 April). Pp. 195–199.

________. 1861. "O polyarnoi severnoi zemle" [On the Polar Northern Land]. *Amur*. No. 67 (26 August). Pp. 600–602; No. 68 (29 August). Pp. 612–614.

________. 1861a. "Severnaya zemlya" [Northern Land]. *ZIRGO*. Bk. 2, pp. 1–35.

________. 1879. "Nizhnekolymskii krai [Nizhnekolymsk Province]. *IIRGO*. Vol. 15, issue 6, pp. 433–451.

________. 1886. *Putevye zapiski* [Travel Notes]. 2nd edition. N. Novgorod: Tipografiya Roiskogo i Dushina.

________. 1887. "God na Krainem Severe" [A Year in the Far North]. *Russkii palomnik* [Russian Pilgrim]. St. Petersburg. Vol. III, no. 1, pp. 9–10; no. 2, pp. 20–22; no. 3, pp. 44–45; no. 4, pp. 56–57; no. 5, p. 70; no. 7, pp. 88–91.

________. 1887a. "Pyatnadtsat' let v Nizhnekolymskom krae: 1843–1857 gg." [Fifteen Years in the Nizhnekolymsk Krai: 1843–1857]. *Sibirskii vestnik* [Siberian Bulletin] 58(22 May): 1–3. Tomsk.

________. 1889. "Pyatnadtsat' let v Nizhnekolymskom krae: 1843–1857 gg. (iz vospominanii missionera)" [Fifteen Years in the Nizhnekolymsk Krai: 1843–1857 (From the Memories of a Missionary)]. *Vsemirnyi puteshestvennik* [World Traveler] 5: 65–80; 6: 81–91. St. Petersburg.

Arkhiv Gosudarstvennogo soveta. 1869. *Arkhiv Gosudarstvennogo soveta* [Archive of the State Council]. T. 1, ch. 2. St. Petersburg: Tipografiya 2-go Otd-niya Sobstvennoi E. I. V. kantselyarii.

Arutyunov, S. A., E. A. Glinskii, and D. A. Sergeev. 1977. "K probleme proniknoveniya zheleza v Arktiku k eskimosam" [On the Problem of the Penetration of Iron into the Arctic to the Eskimos]. *SE* 1: 101–106.

Arutyunov, S. A., I. I. Krupnik, and M. A. Chlenov. 1982. *"Kitovaya alleya" (Drevnosti ostrovov proliva Senyavina)* ["Whale Alley" (The Antiquity of Islands in Senyavin Strait)]. Moscow: Nauka.

Arutyunov, S. A., and D. A. Sergeev. 1975. *Problemy etnicheskoi istorii Beringomor'ya (Ekvenskii mogil'nik)*. Moscow: Nauka. [Published in translation as *Problems of Ethnic History in the Bering Sea: The Ekven Cemetery* by S. A. Arutiunov and D. A. Sergeev. Anchorage, Alaska: National Park Service. 2006].

________. 1983. "Nauchnye rezul'taty rabot na Ekvenskom Drevneeskimosskom mogil'nike (1970–1974 gg.)" [Scientific Results of Work at the Ekven Old Eskimo Cemetery (1970–1974)]. *Na styke Chukotki i Alyaski* [At the Crossroads of Chukotka and Alaska]. Edited by V. P. Alekseev. Moscow: Nauka. Pp. 200–229.

Asadova, S. R. 2010. *Skazki babushki Tymnekvynoi* [Stories of the Grandmother of Tymnekvynyi]. Moscow: Izdatel'skii, obrazovatel'nyi i kul'turnyi tsentr "Detstvo. Otrochestvo. Yunost'.

________. 2012. *Sbornik legend i predanii: Po argishnomu puti kanchalanskogo chauchu* [Collection of Legends and Traditions: Based on the Argish Way of the Kanchalan Chauchu]. Moscow: Russkii raritet.

Avgustinovich, F. M. 1878. "O plemenakh, naselyayushchikh Kolymskii okrug" [On the Tribes Inhabiting the Kolyma Region]. *Izv. Imp. o-va lyubitelei estestvoznaniya, antropologii i etnografii pri Moskovskom un-te* [Proceedings of the Imperial Society of Aficionados of Natural Science, Anthropology, and Ethnography at Moscow University]. Vol. 31: *Trudy antropologicheskogo otdela* [Works of the Anthropology Department]; Vol.

4: Prilozhenie [Appendix]. (Antropologicheskaya vystavka Imp. o-va lyubitelei estestvoznaniya, antropologii i etnografii [Anthropological Exhibition of the Imperial Society of Aficionados of Natural Science, Anthropology, and Ethnography]. Vol. 2, issue 2). Moscow: Tipografiya M. N. Vavrova, 1878. Pp. 43–56.

________. 1880. "Tri goda v Severo-Vostochnoi Sibiri za polyarnym krugom" [Three Years in Northeastern Siberia above the Arctic Circle]. *Drevnyaya i novaya Rossiya: Ezhemesyachnyi istoricheskii zhurnal* [Old and New Russia: A Monthly Historical Journal] XVIII(12): 641–733.

Baboshina, O. E. 1958. *Skazki Chukotki* [Chukchi Stories]. Moscow: Khudozhestvennaya literatura.

Bagger, H. 2005. "Tjukotkas indlemmelse i Det russiske imperium" [Chukchi Incorporation into the Russian Empire]. *Tjukotka i fortid og nutid* [Chukotka in the Past and Present]. Ed. by B. Nielsen. (*Eskimologis skrifter*, no 18). København: Afdeling for Eskimologi og arktiske studier, Københavns Universitet. Pp. 39–64.

Bagrin, E. A. 2007. "Zashchitnoe vooruzhenie sluzhilykh lyudei v Sibiri i na Dal'nem Vostoke v XVII–XVIII vv. (po pis'mennym istochnikam)" [Protective Armament of Servicemen in Siberia and the Far East in the 17th–18th Centuries (Based on Written Sources)]. *Russkie pervoprokhodtsy na Dal'nem Vostoke v XVII–XIX vv.: Istorikoarkheologicheskie issledovaniya* [Russian Explorers in the Far East in the 17th–19th Centuries: Historical Archaeological Investigations]. Vol. V, pt. 1, pp. 269–283. Vladivostok: DVO RAN.

Bakai, N. 1895. "Istoriko-geograficheskie materialy, otnosyashchiesya do Yakutskoi oblasti vo vtoroi chetverti XVIII veka" [Historical Geographic Materials Related to the Yakutsk Region in the Second Quarter of the 18th Century]. *IVSOIRGO* 25(4–5): 82–94.

Bandi, H.-G. 1977. *Die Kunst der Eskimos auf der St.-Lorenz-Insel in Alaska* [The Art of the Eskimos on St. Lawrence Island]. (*Orbis pictus*, 65). Bern; Stuttgart: Hallwag Verlag,

Bandi, H.-G. 1995. "Siberian Eskimos as Whaler and Warriors." *Hunting the Largest Animals: Native Whaling in the Western Arctic and Subarctic*. Ed. A. P. McCartney. (*Studies in Whaling*, no 3; *Occasional Publication*, no 36). Edmonton: The Canadian Circumpolar Institute, Pp. 165–183.

Belikov, L. V. 1956. Osnovnye vidy ustnogo narodnogo tvorchestva chukchei [The Basic Kinds of Oral Folk Creation of the Chukchi]. Abstract of Dissertation for Candidacy in Philological Sciences. Leningrad: [B. i.].

________. 1960. "K kharakteristike ustnogo narodnogo tvorchestva" [On the Characteristics of Oral Folk Creation of the Chukchi]. *UZLGPI* 167: 271–293.

________. 1965. "Geroicheskie skazaniya chukchei" [Heroic Stories of the Chukchi]. *Yazyki i fol'klor narodov Krainego Severa* [Language and Folklore of the Peoples of the Extreme North] (*USLGPI*, vol. 269). Leningrad. Pp. 154–169.

________. 1982. *Chukotskie narodnye skazki, mify i predaniya* [Chukchi Folk Tales, Myths, and Traditions]. Magadan: Magadanskoe knizhnoe izdatel'stvo.

Belov, M. I. 1952. *Russkie morekhody v Ledovitom i Tikhom okeanakh: Sbornik dokumentov o velikikh russkikh geograficheskikh otkrytiyakh na severovostoke Azii v XVII veke* [Russian Mariners in the Arctic and Pacific Oceans: Collection of Documents on the Great Russian Geographic Discoveries in Northeast Asia in the 17th Century]. Moscow-Leningrad: Izdatel'stvo Glavsevnorputi.

________. 1954. "Novye materialy o pokhodakh ustyuzhskogo kuptsa Nikity Shalaurova" [New Materials on the Campaigns of the Ustyug Merchant Nikita Shalaurov]. *Geograficheskii sbornik. III: Istoriya geograficheskikh znanii i geograficheskikh otkrytii* [Geographic Collection. III: The History of Geographic Knowledge and Geographic Discoveries]. Moscow-Leningrad: Izdatel'stvo AN SSSR. Pp. 160–184.

________. 1956. *Arkticheskoe moreplavanie s drevneishikh vremen do serediny XIX veka. (Istoriya otkrytiya i osvoeniya Severnogo morskogo puti.* T. I) [Arctic Voyages from the Earliest Times to the Middle of the 19th Century]. Moscow: Morskoi transport.

________. 1957. "Novye dannye o suzhbakh Vladimira Atlasova i pervykh pokhodakh russkikh na Kamchatku" [New Data on the Service of Vladimir Atlasov and the First Campaigns of the Russians in Kamchatka]. *Letopis' Severa* [Chronicle of the North], No. 2. Moscow: Gos. izd-vo geograficheskoi literatury. Pp. 86–106.

________. 1973. *Podvig Semena Dezhneva* [The Exploit of Semen Dezhnev]. Moscow: Mysl'.

Beregovaya, N. A. 1954. "Arkheologicheskie nakhodki na ostrove Chetyrekhstolbovom" [Archaeological Finds on Chetyrekhstolbovyi Island]. *Sovetskaya arkheologiya* [Soviet Archaeology] 20: 288–312. Moscow.

Beretti, N. N. 1929. "Na krainem Severo-Vostoke" [In the Extreme Northeast]. *Zap. Vladivostokskogo otd. Gos. Russkogo geograficheskogo o-va. (O-vo izucheniya Amurskogo kraya)* [Notes of the Vladivostok Division of the State Russian Geographic Society. (Society for the Study of the Amur Province)] IV(XXI): 5–102. Vladivostok.

Berg, L.S. 1945. "Pervye russkie svedeniya ob Alyaske" [The First Russian Information about Alaska]. IVGO 77(3): 172–174.

________. 1946. Otkrytie Kamchatki i ekspeditsiya Beringa, 1725–1742 [The Discovery of Kamchatka and the Expedition of Bering, 1725–1742]. 4th ed. Moscow-Leningrad: Izdatel'stvo AN SSSR.

Berkh, V. N. 1819. "Puteshestvie kazach'ego golovy Afanasiya Shestakova i pokhod maiora Pavlutskogo v 1729 i 1730 godakh (Otryvok iz I chasti 'Istorii rossiiskikh geograficheskikh otkrytii,' sochinyaemoi V. N. Berkhom" [The Trip of Cossack Leader Afanasii Shestakov and the Campaign of Major Pavlutskii in 1729 and 1730 (An Excerpt from Part I of "The History of Russian Geographic Discoveries," Composed by V. N. Berkh)]. *Syn*

Otechestva: Istoricheskii, politicheskii i literaturnyi zhurnal, izdavaemyi N. Grechem [Son of the Fatherland: Historical, Political, and Literary Journal, Published by N. Grech]. St. Petersburg. Pt. 54. No. 20. Pp. 3–17.

________. 1821–1823. *Khronologicheskaya istoriya vsekh puteshestvii v severnye polyarnye strany* [A Chronological History of All Travels in Northern Polar Countries]. Pts. I–II. St. Peterburg: Voennaya tipografiya General'nogo shtaba.

________. 1825. "Raznye izvestiya i pokazaniya o Chukotskoi zemle" [Various News and Readings about Chukchi Land]. *Severnyi arkhiv: Zhurnal drevnostei i novostei po chasti istorii, statistiki, puteshestvii, pravovedeniya i nravov* [Northern Archive: Journal of Things Old and New on the Part of History, Statistics, Travels, Law, and Morals]. St. Petersburg. Pt. 19. No. 22. Pp. 164–201.

Blomkvist, E. E. 1951. "Risunki I. G. Voznesenskogo (Ekspeditsiya 1839–1849 gg.) [Illustrations of I. G. Voznesenskii (Expedition of 1839–1849)]. *SMAE* 13: 230–303.

Bockstoce, J. R. 2006. "Nineteenth Century Commercial Shipping Losses in the Northern Bering Sea, Chukchi Sea, and Beaufort Sea." *The Northern Mariner / La marin du nord* 16(2): 53–68.

Bogoras, W. G. [Bogoraz, V. G.]. 1899. "Russkie na Kolyme" [Russians on the Kolyma]. *Zhizn': Literaturnyi, nauchnyi i politicheskii zhurnal* [Life: Literary, Science, and Political Journal] VI (June): 103 –125. St. Petersburg.

________. 1899a. "Skazanie ob Elendi i ego synov'yakh (perevod s chukotskogo)" [The Story of Elendi and His Brothers (Translation from the Chukchi)]. *Zhivaya starina. God IX* [Living Antiquity. Year 9] 3: 350–370. St. Petersburg.

________. 1899b. " Kratkii otchet ob issledovanii chukoch Kolymskogo kraya" [A Brief Account of Investigations of the Chukchi of the Kolyma Province]. *IVSOIRGO* 30(1): 1–52.

________. 1900. *Materialy po izucheniyu chukotskogo yazyka i fol'klora, sobrannye v Kolymskom okruge* [Materials on the Study of the Chukchi Language and Folklore, Collected in the Kolyma Region]. Pt. I. (*Trudy Yakutskoi ekspeditsii, snaryazhennoi na sredstva I. M. Sibiryakova*. Otd. III. T. XI, ch. 3) [Works of the Yakutsk Expedition, Outfitted on the Means of I. M. Sibiryakov. Sec. III. Vol. XI, pt. 3)]. St. Petersburg: Imp. akademiya nauk.

________. 1901. *Oblastnoi slovar' kolymskogo russkogo narechiya* [Regional Dictionary of Kolyma Russian Dialect]. St. Petersburg: Imperatorskaya akademiya nauk.

________. 1902. "O prinyatii chukoch v russkoe poddanstvo" [On the Adoption by the Chukchi of Russian Citizenship]. *Zhivaya starina. God XII* [Living Antiquity. Year 12] 2: 147–164. St. Petersburg.

________. 1902a. "Chukchi." *Nauchnoe obozrenie: Ezhemesyachnyi nauchno-filosofskii i literaturnyi zhurnal. God IX* [Science Review: Monthly Science-Philosopy and Literary Journal. Year 9] 1: 70–88. St. Petersburg.

________. 1910. *Chuckchee Mythology*. (MAMNH JE. Vol. VIII. Pt. 1). Leiden; New York.

________. 1913. *The Eskimo of Siberia*. (MAMNH JE. Vol. VIII. Pt. 2). Leiden; New York.

________. 1918. *Tales of Yukaghir, Lamut, and Russianized Natives of Eastern Siberia*. (*Anthropological Papers of the American Museum of Natural History*. Vol. XX. Pt. 1). New York: Published by order of the Trustees.

________. 1919. "Narodnaya literatura paleoaziatov" [Folk Literature of the Paleoasiatics]. Literatura Vostoka: Sb. st. [Literature of the East: Collection of Articles]. Issue 1. Petrograd: Vsemirnaya literatura. Pp. 50–68.

________. 1921. *Predely rasseleniya chukotskogo plemeni v drevnie i novye vremena* [The Boundaries of Settlement of the Chukchi Tribe in Antiquity and New Times]. AAN. F. 135. Op. 2. No. 57. L. 1–98.

________. 1927. "Drevnie pereseleniya narodov v severnoi Evrazii i v Amerike" [Ancient Settlements of the Peoples of Northern Eurasia and in America]. *SMAE*. VI. Leningrad: Izdatel'stvo AN SSSR. Pp. 37–62.

________. 1931. "Materialy po lamutskomu yazyku" [Materials on the Lamut Language]. *Tungusskii sbornik* [Tungus Collection]. Vol. I. Leningrad: Izdatel'stvo AN SSR. Pp. 1–106.

________. 1934. "Yuitskii (aziatsko-eskimosskii) yazyk" [The Yuit (Asian Eskimo) Language]. *Yazyki i pis'mennost' narodov Severa* [Languages and Writing of the Peoples of the North]. Edited by Ya. P. Al'kor. Pt. 3. (*Trudy po lingvistike Nauchno-issledovatel'skoi assotsiatsii In-ta narodov Severa TsIK SSSR*) [Works on Linguistics of the Science-Research Association of the Institute of Peoples of the North TsIK SSSR). Vol. III. Moscow; Leningrad: OGIZ. Pp. 105–128.

Bogoraz-Tan, V. G.

1928 "Novye dannye k voprosu o protoaziatakh" [New Data on the Question of Proto-Asians]. Izv. LGU. [News of Leningrad State University]. Vol. 1. Leningrad: [B. i.]. Pp. 235–243.

________. 1933. "Olenevodstvo" [Reindeer Herding]. Problemy proiskhozhdeniya domashnikh zhivotnykh [Problems of the Origin of Domestic Animals]. Issue 1. (Trudy laboratorii genetiki AN SSSR) [(Works of the Genetics Laboratory AN SSSR)]. Leningrad: [B. i.]. Pp. 219–251.

________. 1934a. *Chukotskoe obshchestvo: Kratkii putevoditel' po vystavke* [Chukchi Society: A Brief Guide to the Exhibition]. Leningrad: Izdatel'stvo AN SSSR.

________. 1934–1939. *Chukchi: avtorizovannyi per. s angl.* [The Chukchi: An Authorized Translation from English]. Pts. I–II. Leningrad: Izdatel'stvo Instituta narodov Severa TsIK SSSR.

________. 1936. "Osnovnye tipy fol'klora Severnoi Evrazii i Severnoi Ameriki" [The Primary Types of Folklore of Northern Eurasia and North America]. *Sovetskii folk'lor: sb. statei i materialov* [Soviet Folklore: A Collection of Articles and Materials]. Nos. 4–5. Moscow; Leningrad: Izdatel'stvo AN SSSR. Pp. 29–50.

Boyakova, S. I. 2001. *Osvoenie Arktiki i narody Severo-Vostoka Azii (XIX v.–1917 g.)* [The Developing of the Arctic and Peoples of Northeast Asia (19th Century–1917)]. Novosibirsk: Nauka.

Breitfuss, L. 1939. "Early Maps of North-Eastern Asia and the Lands around the North Pacific: Controversy between G. F. Müller and N. Delisle." *Imago mundi: A Periodical Review of Early Cartography*. No 3. London, Pp. 87–99.

Bronshtein, M. M. 2007. "Ekven i Paipel'gak glazami etnologov" [Ekven and Paipel'gak in the Eyes of Ethnologists]. *MAZ*. Pp. 26–47.

Bronshtein, M. M., and K. A. Dneprovskii. 2001. "Zhilishche morskikh zveroboev drevnei Chukotki" [A Dwelling of Sea Mammal Hunters of Ancient Chukotka]. *Pamyatniki kul'tury. Novye otkrytiya. 2000* [Cultural Sites. New Discoveries. 2000]. Compiled by T. B. Knyazevskaya. Moscow: Nauka. Pp. 587–619.

Bronshtein, M. M., K. A. Dneprovskii, and E. S. Sukhorukova. 2007. *Mir arkticheskikh zveroboev: Shagi v nepoznannoe. Katalog vystavki* [The World of Sea Mammal Hunters. Steps into the Unknown. Catalog for an Exhibition]. Moscow; Anadyr': Departament kul'tury, sporta, turizma i informatsionnoi politiki Chukotskogo avtonomnogo okruga; GMIV; MAZ.

Bulychev, I. D. 1856. *Puteshestvie po Vostochnoi Sibiri. (Al'bom risunkov)* [A Trip through Eastern Siberia. (An Album of Illustrations)]. St. Petersburg: Tipografiya Imp. akademii nauk.

Burch, E. S., Jr. 1988. "War and Trade." *Crossroads of Continents: Cultures of Siberia and Alaska*. Ed. W. Fitzhugh, A. Crowell. Washington: Smithsonian Institution. Pp. 227–240.

________. 1998. *The Cultural and Natural Heritage of Northwest Alaska*. Vol. VII: International Affairs. Kotzebue; Anchorage: NANA Museum of the Arctic and U. S. National Park Service, Alaska Region.

Burykin, A. A. 1993. Chuvantsy: Proiskhozhdenie i etnicheskaya istoriya. Material'naya i dukhovnaya kul'tura (konets XIX–nachalo XX v.) [The Chuvan People: Origin and Ethnic History. Material and Spiritual Culture (End of the 19th–Beginning of the 20th Century)]. Anadyr': [B. i.].

________. 1998. "Neloga, Pogycha, Kovycha—legendarnye reki russkikh dokumentov serediny XVII veka i sovremennaya karta Chukotki" [The Neloga, Pogycha, Kovycha—Legendary Rivers of Russian Documents of the Mid-17th Century and a Contemporary Map of Chukotka]. *EO* 6: 79–88.

________. 2000. "K interpretatsii nekotorykh zagadochnykh etnonimov Arkticheskogo poberezh'ya Chukotki" [On the Interpretation of Some Enigmatic Ethnonyms of the Arctic Coast of Chukotka]. *EO* 4: 45–59.

________. 2004. O praktike obshcheniya russkikh zemleprokhodtsev s korennym naseleniem Severo-Vostoka Azii v seredine i vtoroi polovine XVII veka" [On the Practice of Communication of Russian Explorers with the Native Population of Northeast Asia in the Middle and Second Half of the 17th Century]. *Yakutiya-Forpost osvoeniya Severo-Vostoka Sibiri, Dal'nego Vostoka i Russkoi Ameriki (XVII–XX veka)* [Yakutia—An Outpost in the

Development of Northeastern Siberia, the Far East, and Russian America (17th–20th)]. Edited by V. N. Ivanov. Yakutsk: YaF Izd-va SO RAN. Pp. 35–44.

________. 2006. *Istoriko-etnograficheskie i istorikokul'turnye aspekty izucheniya onomasticheskogo prostranstva regiona (toponimika i etnonimika Vostochnoi Sibiri). (Ocherki po toponikike i etnonikike Vostochnoi Sibiri)* [Historical-Ethnographic and Historical-Cultural Aspects of the Study of the Onomastic Expanse of the Region (Toponymy and Ethnonymics of Eastern Siberia). (Essays on Toponymy and Ethnonyms of Eastern Siberia)]. St. Petersburg: Peterburgskoe Vostokovedenie.

________. 2007. "Pervoe sobranie obrazotsov fol'klora evenov Yakutii (k 135-letiyu izdaniya)" [The First Collection of Samples of Folklore of the Even of Yakutia (On the 135th Year of Publication)]. *Acta Linguistica Petropolitana.* Vol. III, pt. 3, pp. 361–392. St. Petersburg.

________. 2013. *Imena sobstvennye kak istoricheskii istochnik (Po materialam russkikh dokumentov ob otkrytii i osvoenii Sibiri i Dal'nego Vostoka Rossii XVI–XIX vekov)* [Proper Names as a Historical Source (Based on Materials from Russian Documents on the Discovery and Development of Siberia and the Far East of Russia in the 16th–19th Centuries)]. St. Petersburg: Peterburgskoe Vostokovedenie.

Bush, R. J. 1872. *Reindeer, Dogs, and Snow-Shoes: A Journal of Siberian Travel and Explorations Made in the Years 1865, 1866, and 1867.* London: Sampson Low, son, and Marston.

Bushnev, N. 1995. "Konkistador Chukotki" [Conqueror of Chukotka]. *Russkaya Amerika: istoriko-geograficheskii zhurnal* [Russian America: Historical-Geographic Journal] 6: 16–19. Vologda.

________. 1996. "Poteryannaya ekspeditsiya" [Lost Expedition]. *Severnaya Patsifika: Regional'nyi informatsionnyi daidzhest* [North Pacific: Regional Informational Digest] 1: 64–68. Petropavlovsk-Kamchatskii.

Chernenko, M. B. 1957. "Puteshestvie po Chukotskoi zemle i plavanie na Alyasku kazach'ego sotnika Ivana Kobeleva v 1779 i 1789–1791 gg." [A Trip through Chukchi Land and a Voyage to Alaska of Cossack Sotnik Ivan Kobelev in 1779 and 1789–1791]. *Letopis' Severa* [Chronicle of the North]. Issue 2. Moscow: Gos. izd-vo geograficheskoi literatury. Pp. 121–141.

Chesnokov, Yu. V. 1993. "Yaranga olennykh chukchei: adaptatsionnye svoistva" [The Yaranga of the Reindeer Chukchi: Adaptive Features]. *Rossiiskii etnograf* [Russian Ethnographer]. Moscow: [B. i.]. No. 12. Pp. 221–231.

Chlenov, M. A. 1988. "Ekologicheskie faktory etnicheskoi istorii raiona Beringova proliva" [Ecological Factors of the Ethnic History in the Region of Bering Strait]. *Ekologiya amerikanskikh indeitsev i eskimosov: Problemy indeanistiki* [The Ecology of the American Indians and Eskimos: Problems of Indian Studies]. Edited by V. A. Tishkov. Moscow: Nauka. Pp. 64–75.

Choris, L. 1822. *Voyage pittoresque autour du monde* [A Picturesque Voyage around the World]. Paris: F. Didot.

Chukchi. 1835. "Chukchi" [Chukchi]. *Moskovskie vedomosti* [Moscow News]. No. 47 (12 June). Pp. 2337–2340.

Chukchi i zemlya ikh. . . . 1851. "Chukchi i zemlya ikh, s otkrytiya etogo kraya do nastoyashchego vremeni" [The Chukchi and Their Land, from the Discovery of this Region to the Present Time]. *ZhMVD*. Pt. 34. Pp. 456–490.

Chulkov, M. D. 1785. *Istoricheskoe opisanie rossiiskoi kommertsii pri vsekh portakh i granitsakh ot drevnikh vremen do nyne nastoyashchego. . .* [A Historical Description of Russian Commerce at All Ports and Borders from Ancient Times to the Present Day. . .]. Vol. III. Bks. 1–2. Moscow: Universitetskaya tipografiya.

Cochrane, J. D. 1824. *Narrative of a Pedestrian Journey through Russia and Siberian Tartary, from the Frontiers of China to the Frozen Sea and Kamtchatka; Performed during the Years 1820, 1821, 1822 and 1823.* Philadelphia: H. C. Carey and I. Lea; New York: Collins and Hannay.

Collins, H. B., Jr. 1937. *Archaeology of St. Lawrence Island, Alaska.* (*Smithsonian Miscellaneous Collections*. Vol. 96, no 1). Washington: Smithsonian Institution.

Cooper, H. K., and J. B. Gabriel. 2013. "Metal Armor from St. Lawrence Island." *Arctic Anthropology* 50(1): 1–19.

DAI 1867: *Dopolneniya k Aktam istoricheskim* [Supplements to the Historical Acts]. St. Petersburg: Tipografiya E. Pratsa. Vol. II, 1846; Vol. III, 1848; Vol. IV, 1851; Vol. VI, 1857; Vol. VIII, 1862; Vol. X.

Dall, W. H. 1870. *Alaska and Its Resources*. Boston: Lee and Shepard.

De Windt, H. 1898. *Through the Gold-Fields of Alaska to Bering Strait*. London: Chatto and Windus.

[Dikov, N. N.]. 1958. "Fol'klor" [Folklore]. *ZChKM*. Issue 1, pp. 69–72.

Dikov, N. N. 1958a. "Predvaritel'nyi otchet o rabote arkheologicheskoi ekspeditsii Chukotskogo kraevedcheskogo muzeya v 1956 g." [Preliminary Report on the Work of the Archaeological Expedition of the Chukotka Regional Museum in 1956]. *ZChKM*. Issue 1, pp. 32–44.

________. 1971. *Naskal'nye zagadki drevnei Chukotki (Petroglify Pegtymelya)*. Moscow: Nauka. [Published in English as *Mysteries in the Rocks of Ancient Chukotka (Petroglyphs of Pegtymel')*. Anchorage, Alaska: Shared Beringian Heritage Program, National Park Service, 1999.]

________. 1976. "Na ostrov Vrangelya—k drevnim eskimosam" [On Wrangel Island —The Ancient Eskimos]. *Znanie—sila* [Knowledge Is Power]. No. 10(592): 19–22. Moscow.

________. 1977. *Arkheologicheskie pamyatniki Kamchatki, Chukotki i Verkhnei Kolymy: Aziya na styke s Amerikoi v drevnosti*. Moscow: Nauka. [Published in English as *Archaeological Sites of Kamchatka, Chukotka, and the Upper Kolyma*. Anchorage, Alaska: Shared Beringian Heritage Program, National Park Service, 2003.]

________. 1977a. "Otkrytie paleoeskimosskoi kul'tury na ostrove Vrangelya" [Discovery of a Paleo-Eskimo Culture on Wrangel Island]. *KZOKM*. Issue 11, pp. 127–130.

________. 1979. *Drevnie kul'tury Severo-Vostochnoi Azii (Aziya na styke s Amerike v drevnosti)*. Moscow: Nauka. [Published in English as *Early Cultures of Northeast Asia*. Anchorage, Alaska: Shared Beringian Heritage Program, National Park Service, 2004.]

________. 1989. *Istoriya Chukotki s drevneishikh vremen do nashikh dnei* [The History of Chukotka from Earliest Times to Our Days]. 2nd ed. Edited by N. N. Dikov. Moscow: Mysl'.

________. 1996. "Arkheologicheskoe proshloe obitatelei Severo-Vostoka Azii" [The Archaeological Past of the Inhabitants of Northeast Asia]. *Severo-Vostok Rossii s drevneishikh vremen do nashikh dnei: Novye ekskursy v istoriyu* [Northeastern Russia from the Earliest Times to Our Days]. Edited by N. N. Dikov. Magadan: SVKNII DVO RAN. Pp. 13–31.

Dioneo. 1895. *Na krainem Severo-Vostoke Sibiri* [In Extreme Northeastern Siberia]. St. Petersburg: Izdanie L. F. Panteleeva.

Dionisii. 1884. "Poezdka Preosvyashchennogo Dionisiya, episkopa Yakutskogo, v Chukotskuyu missiyu v 1868/69 g." [Visit of His Grace Dionysius, Bishop of Yakutsk, at the Yakutsk Mission 1868/69]. *Trudy pravoslavnykh missii Vostochnoi Sibiri: Izdanie Irkutskogo komiteta Pravoslavnogo missionerskogo obshchestva* [Works of the Orthodox Mission of Eastern Siberia: Publication of the Irkutsk Committee of the Orthodox Mission Society]. Vol. II, pp. 141–163. Irkutsk: Tipografiya N. N. Sinitsyna.

Ditmar, K. M. von. 1855. "O koryakakh i ves'ma blizkikh k nim po proiskhozhdeniyu chukchakh" [On the Koryak and the Very Close to Them in Origin Chukchi]. *VIRGO*. Pt. 15, pp. 51–63.

Ditmar, K. M. von. 1856. "O koryakakh i ves'ma blizkikh k nim po proiskhozhdeniyu chukchakh" [On the Koryak and the Very Close to Them in Origin Chukchi]. *VIRGO*. Pt. 16, pp. 19–39 (fifth pagination).

Divin, V. A. 1956. *K beregam Ameriki: Plavanie i issledovaniya M. S. Gvozdeva, pervootkryvatelya Severo-Zapadnoi Ameriki* [To the Shores of America: Voyage and Investigations of M. S. Gvozdev, Initial Discovery of Northwestern America]. Moscow: Gosudarstvennoe izdatel'stvo geograficheskoi literatury.

________. 1979. *Russkaya tikhookeanskaya epopeya* [The Russian Pacific Ocean Epic]. Khabarovsk: Khabarovsk knizhnoe izdatel'stvo.

Dneprovskii, K. A. 2001. Dinamika drevneeskimosskoi kul'tury Chukotki v epokhu birnirka i rannego punuka (po materialam arkheologicheskogo kompleksa Ekven) [The Dynamics of Old Eskimo Culture of Chukotka in the Period of Birnirk and Early Punuk (Based on Materials of the Archaeological Complex from Ekven)]. Abstract of dissertation for candidate in Historical Science. Moscow: [B. i.].

________. 2007. "Drevneeskimosskie pamyatniki, issledovannye Chukotskoi arkheologicheskoi ekspeditsiei" [Old Eskimo Sites Investigated by the Chukchi Archaeological Expedition]. *MAZ*. Pp. 15–25.

Dolgikh, B. O. 1938. *Legendy i skazki nganasan* [Legends and Stories of the Nganasan]. Krasnoyarsk: Krasnoyarskoe kraevoe gosudarstvennoe izdatel'stvo.

________. 1960. *Rodovoi i plemennoi sostav narodov Sibiri v XVII veke* [Clan and Tribal Composition of the Peoples of Siberia in the 17th Century]. (*TIE*. Vol. 55). Moscow: Izdatel'stvo AN SSSR.

________. 1961. *Mifologicheskie skazki i istoricheskie predaniya entsev* [Mythological Stories and Historical Traditions of the Enets]. (*TIE*. Vol. 66). Moscow: Izdatel'stvo AN SSSR.

________. 1976. *Skazki i predaniya nganasan* [Stories and Traditions of the Nganasan]. (*SMNV*). Moscow: Nauka.

"Dve zapiski. . ." 1873. "Dve zapiski o snosheniyakh s chukchami v 1774–1776 godakh" [Two Notes on Relations with the Chukchi in 1774–1776]. *Pamyatniki novoi rasskoi istorii: sb. ist. statei i materialov* [Monuments of New Russian History: Collection of Historical Articles and Materials]. Vol. 3, pp. 360–371. St. Petersburg.

D'yachkov, [A. E.]. 1893. *Anadyrskii krai. (Zapiski Obshchestva izucheniya Amurskogo kraya*. T. II) [The Anadyrsk District. (Notes of the Society for the Study of the Amur District. Vol. II)]. Vladivstok: Tipografiya Sibirskogo flotskogo ekipazha.

________. 1992. "Anadyrskii krai" [Anadyr' District]. *Zhukharev N. A. Povest' ob Afanasii D'yachkove. D'yachkov A. E. Anadyrskii krai* [Zhikharev N. A. The Story of Afanasii D'yachkov. D'yachkov A. E. Anadyrskii Districk]. (DIB). Magadan: Magadanskoe knizhnoe izdatel'stvo. Pp. 160–265.

E. M. 1892. "Pribavlenie k rasskazu o chukchakh" [An Addition to the Story of the Chukchi]. *Yakutskie eparkhial'nye vedomosti* [The Yakutsk Diocesan Bulletin]. No. 15 (1 August), pp. 231–235.

Efimov, A. V. 1948. *Iz istorii russkikh ekspeditsii na tikhom okeane (Pervaya polovina XVIII v.)* [From the History of Russian Expeditions in the Pacific Ocean (First Half of the 18th Century)]. Moscow: Voenizdat.

________. 1950. *Iz istorii velikikh russkikh geograficheskikh otkrytii v Severnom Ledovitom i Tikhom okeanakh (XVII–pervaya polovina XVIII v.)* [From the History of Great Russian Geographic Discoveries in the Arctic and Pacific Oceans (17th–First Half of the 18th Century)]. Moscow: Gosudarsvennoe izdatel'stvo geograficheskoi literatury.

________. 1964. *Atlas geograficheskikh otkrytii v Sibiri i Severo-Zapadnoi Amerike XVII–XVIII vv.* [Atlas of Geographic Discoveries in Siberia and Northwestern America in the 17th–18th Centuries]. Moscow: Nauka.

________. 1971. *Iz istorii velikikh russkikh geograficheskikh otkrytii* [From the History of Great Russian Geographic Discoveries]. Moscow: Nauka.

Elert, A. Kh. 1991. "Novye materialy po istorii chukchei v pervoi polovine XVIII v." [New Materials on the History of the Chukchi in the First Half of the 18th Century] *Izv. Sib. otd-niya Akademii nauk SSSR. Istoriya, filosofiya, filologiya* [News of the Siberian Division of the Academy of Sciences of the USSR. History, Philosophy, Philology]. Issue 3. Novosibirsk. Pp. 27–30.

________. 1996. "Novye materialy po historii russko-koryakskikh otnoshenii v pervoi chetverti XVIII v." [New Materials on the History of Russian-Koryak Relations in the First Quarter of the 18th Century]. *Russkoe obshchestvo i literatura pozdnego feodalizma: sb. nauch. trudov* [Russian Society and Literature of Late Feudalism: Collection of Scientific Works]. Edited by N. N. Pokrovskii. (*Arkheologiya i istochnikovedenie Sibiri* [Archaeology and Source Studies for Siberia]). Novosibirsk: Sibirskii khronograf. Pp. 247–264.

________. 1998. "Problemy vkhozhdeniya korennykh narodov Sibiri v sostav Rossii v neopublikovannykh trudakh G. F. Millera" [Problems of the Incorporation of Native Peoples of Siberia in the Composition of Russia in the Unpublished Works of G. F. Müller]. *Istoriya russkoi Dukhovnoi kul'tury v rukopisnom nasledii XVI–XX vv.: sb. nauch. trudov* [The History of Russian Spiritual Culture in the Manuscript Heritage of the 16th–20th Centuries: Collection of Scientific Works]. Edited by E. K. Romodanovskii. Novosibirsk: Nauka. Pp. 121–133.

________. 1999. *Narody Sibiri v trudakh G. F. Millera* [Peoples of Siberia in the Works G. F. Müller]. Novosibirsk: Izd-vo In-ta arkheologii i etnografii SO RAN.

________. 2009. "Vvedenie" [Introduction]. *Miller G. F. Opisanie sibirskikh narodov. (Istochniki po istorii Sibirii Alyaski v rossiiskikh arkhivakh.* T. VIII.1) [Müller, G. F. Description of Siberian Peoples. (Sources on the History of Siberia and Alaska in Russian Archives. Vol. VIII.1). Moscow: Pamyatniki istoricheskoi mysli. Pp. 7–24.

Erunov, I. 1993. "Russkie starozhily nizov'ev Kolymy: istoriya formirovaniya" [Russian Old Dwellers of the Lower Reaches of the Kolyma: History of the Formation]. Rossiiskii etnograf [Russian Ethnographer]. No. 12, pp. 108–125. Moscow: [B. i.].

Fedorova, S. G. 1960. "Pis'mo na morzhovom klyke" [Letter on a Walrus Tusk]. *Ezhegodnik Gos. Ist. muzeya* (za 1958 g.) [Annual of the State Historical Museum (for 1958)]. Moscow: [B. i.]. Pp. 147–161.

________. 1964. "K voprosu o rannikh russkikh poseleniyakh na Alyaske" [On the Question of Early Russian Settlements in Alaska]. *Letopis' Severa* [Chronicle of the North]. Vol. IV. Moscow. Pp. 97–113.

________. 1971. "Issledovatel' Chukotki i Alyaski kazachii sotnik Ivan Kobelev" [Investigator of Chukotka and Alaska Cossack Sotnik Ivan Kobelev]. *Letopis' Severa* [Chronicle of the North]. Vol. V. Moscow: Mysl'. Pp. 156–172.

________. 2011. *Russkaya Amerika: ot pervykh poselenii do prodazhi Alyaski: konets XVIII veka–1867 god* [Russian America: From the First Settlements to the Sale of Alaska: End of the 18th Century–1867]. Moscow: Lomonosov.

Fedorova, T. S. 1984. *Russkie ekspeditsii po izucheniyu severnoi chasti Tikhogo okeana v pervoi polovine XVIII v.: sb. dokumentov* [Russian Expeditions for Study of the Northern Part of the Pacific Ocean in the First Half of the 18th Century: Collection of Documents]. Compiled by T. S. Fedorova. Moscow: Nauka.

Fel', S. E. 1960. *Kartografiya v Rossii XVIII veka* [Cartography in Russia in 18th Century]. Moscow: Izd-vo geodezicheskoi literatury.

Fisinina, A. K. 1995. *Chukotskoe i eskimosskoe isskusstvo XIX–XX vv. (Iz sobraniya muzeya)* [Chukchi and Eskimo Art of the 19th–20th Centuries. (From Collection of the Museum)]. Moscow: Sergievo-Posadskii istoriko-khudozhestvennyi muzei-zapovednik.

Fitzhugh, W. W. 2009. "Notes on Art Styles, Cultures, and Chronology." *Gifts from the Ancestors: Ancient Ivories of Bering Strait*. Ed. by W. W. Fitzhugh, J. Hollowell, and A. L. Crowell. Princeton: Princeton University Art Museum. Pp. 88–93.

Galkin, N. A. 1929. *V zemle polunochnogo solntsa. (Biblioteka ekspeditsii i puteshestvii)* [In the Land of the Midnight Sun. (Library of Expeditions and Voyages)]. Moscow; Leningrad: Molodaya gvardiya.

Garber, C. M. 1940. *Stories and Legends of the Bering Strait Eskimos*. Boston: The Christopher Publishing House.

Gedenshtrom, M. M. 1822. "Puteshestvie Gedenshtroma po Ledovitomu moryu i ostrovam onago, lezhashchim ot ust'ya Leny k vostoku" [The Voyage of Gedenshtrom along the Arctic Ocean and Its Islands, Lying from the Mouth of the Lena to the East]. *Sibirskii vestnik* [Siberian Bulletin]. Pts. 17–19, 1–168 (based on duplication of end-to-end numbering of the pages: Pts. 17–18, pp. 27–38, 99–116, 170–184, 245–258, 291–304, 355–378; Pt. 19, Pp. 1–18, 85–106). St. Petersburg.

________. 1830. *Otryvki o Sibiri* [Excerpts about Siberia]. St. Petersburg: Tipografiya Med. departamenta MVD.

Gel'mersen, L. K. 1876. "K voprosu ob otkrytii Vrangelevoi zemli" [On the Question of the Discovery of Wrangel Land]. *IIRGO*. Vol. 12, issue 6, pp. 455–487.

Glinskii, E. A. 1986. Luk i strely narodov Severo-Vostoka Sibiri kak istoriko-kul'turnoe yavlenie [Bow and Arrows of the Peoples of Northeastern Siberia as a Historical-Cultural Phenomenon]. Abstract of dissertation for candidacy to Historical Science Leningrad: [B. i.].

________. 1987. "Metodika opisaniya ruchnykh lukov v muzeinykh sobraniyakh (po materialam sibirskikh kollektsii GME narodov SSSR i MAE AN SSSR)" [Method of Description of Hand-Held Bows in the Museum Collections (Based on Materials of the Siberian Collections of the GME of People of the SSSR and MAE AN SSSR)]. *Problemy komplektovaniya, nauchnogo opisaniya i atributsii etnograficheskikh pamyatnikov: sb. nauch. trudov* [Problems of Acquisition, Scientific Description, and Attribution of Ethnographic Items: Collection of Scholarly Works]. Edited by I. I. Shangina. Leningrad: Gos. muzei etnografii narodov SSSR. Pp. 123–128.

Gil'der, U. [Gilder, W. H.]. 1886. *V l'dakh i snegakh: Puteshestvie v Sibir' dlya poiskov ekspeditsii kapitana Delonga* [In Ice and Snow: A Mission into Siberia in Search of Captain Delong's Expedition]. Translated by V. N. Mainov. St. Petersburg: Tipografiya A. S. Suvorina.

Gol'denberg, L. A. 1966. *Fedor Ivanovich Soimonov (1692–1780)*. Moscow: Nauka.

________. 1979. *Katorzhanin—sibirskii gubernator: Zhizn' i trudy F. I. Soimonova* [Convict—Siberian Governor: The Life and Works of F. I. Soimonov]. (DIB). Magadan: Magadanskoe knizhnoe izdatel'stvo.

________. 1980. "Karta Shestakova 1724 g." [Shestakov's Map of 1724]. *Ispol'zovanie starykh kart v geograficheskikh i istoricheskikh issledovaniyakh* [The Use of Old Maps in Geographic and Historical Investigations]. Moscow: Moskovskii filial Geograficheskogo o-va SSSR. Pp. 137–160.

________. 1984. *Mezhdu dvumya ekspeditsiyami beringa* [Between Bering's Two Expeditions]. (DIB). Magadan: Magadanskoe knizhnoe izdatel'stvo.

________. 1985. *Mikhail Spiridonovich Gvozdev (nachalo XVIII v.–posle 1759)* [Mikhail Spiridonovich Gvozdev (Beginning of the 18th Century–after 1759)]. Moscow: Nauka.

Golder, F. A. 1914. *Russian Expansion on the Pacific, 1641–1850*. Cleveland: The Arthur H. Clark Company.

Gondatti, N. L. 1897. "Osedloe naselenie reki Anadyr'" [The Settled Population of the Anadyr River]. *ZPOIRGO*. Vol. III, issue 1, pp. 111–165.

________. 1897a: "Svedeniya o poseleniyakh po Anadyru" [Information on the Settlements along the Anadyr]. *ZPOIRGO*. Vol. III, issue 1, pp. 71–111.

________. 1897b. "Sostav naseleniya Anadyrskoi okrugi" [The Composition of the Population of the Anadyr District]. *ZPOIRGO*. Vol. III, issue 1, pp. 166–178.

________. 1898. "Poezdka iz s. Markova na r. Anadyr' v bukhtu Provideniya (Beringov proliv)" [A Trip from Markovo Village on the Anadyr River to Provideniya Bay (Bering Strait)]. *ZPOIRGO*. Vol. IV, issue 1, pp. 1–42 (first pagination). Appendix, pp. I–XXVII.

Gorosov, O. 2011. "Nepokornye chukchi" [The Rebellious Chukchi]. *Tainy XX veka: Ezhenedel'nik* [Secrets of the 20th Century: Weekly]. No. 15, pp. 14–15.

Grekov, V. I. 1960. *Ocherki iz istorii russkikh geograficheskikh issledovanii v 1725–1765 gg.* [Essays from the History of Russian Geographic Investigations in 1725–1765]. Moscow: Izdatel'stvo AN SSSR.

Grenader, M. B. 1975. "Poslednie gody deyatel'nosti T. I. Shmaleva" [T. I. Shmalev's Last Years of Activity]. *Letopis' Severa* [Chronicle of the North]. Vol. VII, pp. 93–106. Moscow.

Gromov, L. V. 1941. "Ostatki drevnego zhilishcha na ostrove Vrangelya" [The Remains of an Ancient Dwelling on Wrangel Island]. *Problemy Arktiki* [Problems of the Arctic]. Vol. 3, pp. 75–77. Leningrad; Moscow: Izdatel'stvo Glavsevmorputi.

________. 1957. "Sledy starinnogo zaseleniya na ostrove Vrangelya" [Traces of an Ancient Settlement on Wrangel Island]. *Letopis' Severa* [Chronicle of the North]. Vol. 2, pp. 155–156. Moscow.

Gurvich, I. S. 1957. "Yukagiry chuvanskogo roda v seredine XVIII veka" [Yukaghir of the Chuvan Clan in the Mid-18th Century]. *SES*. Issue II. (*TIE*. New Series. Vol. 35). Pp. 246–262. Moscow; Leningrad.

________. 1966. "Etnicheskaya istoriya severo-vostoka Sibiri [The Ethnic History of Northeastern Siberia]. (*TIE*. New Series. Vol. 89). Moscow: Nauka.

________. 1975. "Yukagirskaya problema v svete etnograficheskikh dannykh" [The Yukaghir Problem in Light of Ethnographic Data]. *Yukagiry (istoriko-etnograficheskii ocherk)* [Yukaghir (Historical-Ethnographic Essays)]. Edited by A. P. Okladnikov. Novosibirsk: Nauka. Pp. 12–83.

________. 1980. "Problemy proiskhozhdeniya chukchei, koryakov i itel'menov" [Problems of the Origin of the Chukchi, Koryak, and Itel'men]. *Etnogenez narodov Severa* [The Ethnogenesis of Peoples of the North]. Edited by I. S. Gurvich. Moscow: Nauka. Pp. 210–225.

________. 1982. "Severo-vostochnye paleoaziaty i eskimosy" [The Northeastern Paleoasiatics and Eskimos]. *Etnicheskaya istoriya norodov Severa* [The Ethnic History of Peoples of the North]. Edited by I. S. Gurvich. Moscow: Nauka. Pp. 197–222.

________. 1985. "Etnicheskikh sostav i etnicheskaya istoriya narodov Severo-Vostoka Azii" [The Ethnic Composition and Ethnic History of Peoples of Northeast Asia]. *Narody Dal'nego Vostoka SSSR v XVII–XX vv.* [Peoples of the Far East of the USSR in the 17th–20th Centuries]. Edited by I. S. Gurvich. Moscow: Nauka. Pp. 50–66.

________. 1992. "Chuvantsy" [The Chuvan People]. *EO* 5: 76–83.

Gurvich, I. S., and K. G. Kuzakov. *Koryakskii natsional'nyi okrug (ocherki georgrafii, istorii, etnografii, ekonomiki)* [The Koryak National District (Essays on Geography, History, Ethnography, Economics)]. Moscow: Izdatel'stvo AN SSSR.

Hickey, C. G. 1979. "The Historic Beringian Trade Network: Its Nature and Origins." *Thule Eskimo Culture: An Anthropological Retrospective*. Ed. A. P. McCartney. (*Mercury Series. Archaeological Survey of Canada Paper* No 88). Ottawa: National Museum of Man. Pp. 411–434.

Hooper, W. H. 1853. *Ten Months among the Tents of the Tuski, with Incidents of an Arctic Boat Expedition in Search of Sir John Franklin*. London: L. J. Murray.

Hough, W. 1895. "Primitive American Armor." *Annual Report of the Board of Regents of the Smithsonian Institution for the Year Ending June 30, 1893*. Washington. Pp. 625–651.

Ivanov, V. N.1999. *Vkhozhdenie Severo-Vostoka Azii v sostav russkogo gosudarstva* [The Entrance of Northeast Asia into the Composition of the Russian State]. Novosibirsk: Nauka.

Ivanov, V. F. 1979. *Pis'mennye istochniki po istorii Yakutii XVII veka* [Written Sources on the History of Yakutia of the 17th Century]. Novosibirsk: Nauka.

Ivtagina, L. S. 2008. "Protivopostavlenie 'svoi/chuzhoi' v chukhotskikh geroicheskikh skazaniyakh o stolknoveniyakh s koryakami" [The Apposition of "Us/Them" in Chukchi Heroic Stories about Conflicts with the Koryak]. *Vestnik Rossiiskogo gos. gumanitarnogo un-ta* [Bulletin of the Russian State Humanities University]. Moscow. No. 9, pp. 294–302.

Jochelson, W. I. [Iokhel'son, V. I.]. 1892. "Drama na beregu Ledovitogo okeana" [Drama on the Shore of the Arctic Ocean]. *VO*, No. 9 (1 March), pp. 12–14.

________. 1895. "Zametki o naselenii Yakutskoi oblasti v istoriko-etnograficheskom otnoshenii" [Notes on the Population of the Yakutsk Region in Historical-Ethnographic Regard]. Zhivaya starina [Living Antiquity]. Issue 2, year 5, pp. 127–161. St. Petersburg.

________. 1897. "K voprosu ob ischeznuvshikh narodnostyakh Kolymskogo okruga" [On the Question of Disappearing Nationalities of the Kolyma Region]. *IVSOIRGO*. Vol. 28, no. 2, pp. 160–165.

________. 1898. "Po rekam Yasachnoi i Korkodonu: Drevnii i sovremennyi yukagirskii byt i pis'mena" [Along the Yasachnaya and Korkodon Rivers: The Ancient and Modern Yukaghir Way of Life and Writings]. *IIRGO*. Vol. 34, issue 3, pp. 255–290.

________. 1900. *Materialy po izucheniyu yukagirskogo yazyka i fol'klora, sobrannye v Kolymskom okruge. Ch. I. (Trudy Yakutskoi ekspeditsii, snaryazhennoi na sredstva I. M. Sibiryakova. Otd. III. T. IX, ch. 3)* [Materials on the Study of the Yukaghir Language and Folklore Collected in the Kolyma Region. Pt. I. (Works of the Yakutsk Expedition, Equipped by the Means of I. M. Sibiryakov. Sec. III. Vol. IX, pt. 3). St. Petersburg: Imp. akademiya nauk.

________. 1900a. "Brodyachie rody tundry mezhdu rekami Indigirkoi i Kolymoi, ikh etnicheskii sostav, narechiya, byt, brachnye i inye obrachai i vzaimodeistvie razlichnykh plemennykh elementov" [Nomadic Clans of the Tundra between the Indigirka and Kolyma Rivers, Their Ethnic Composition, Dialect, Way of Life, Marriage, and Other Customs and Interaction of Different Tribal Elements]. *Zhivaya starina* [Living Antiquity]. Issue 1–2, pp. 151–193. St. Petersburg.

________. 1905. *The Koryak*. Pt. 1–2. (MAMNH JE. Vol. VI. Pt. 1–2). Leiden; New York.

________. 1928. *Peoples of Asiatic Russia*. New York: The American Museum of Natural History.

________. 1997. *Koryaki: Material'naya kul'tura i sotsial'naya organizatsiya* [The Koryak: Material Culture and Social Organization]. Translation from English. St. Petersburg: Nauka.

KPMGYa 1936: *Kolonial'naya politika moskovskogo gosudarstva v Yakutii v XVII v.: sb. dokumentov* [Colonial Policy of the Moscow Government in Yakutia in the 17th Century: Collection of Documents]. Edited by Ya. P. Al'kor and B. D. Grekov. Leningrad: Izdatel'stvo Instituta narodov Severa TsIK SSSR im. P. G. Smidovicha.

KPTs 1935: *Kolonial'naya politika tsarizma na Kamchatke i Chukotke v XVIII veke: sb. dokumentov* [Colonial Policy of Tsarism in Kamchatka and Chukotka in the 18th Century: Collection of Documents]. Edited by Ya. P.

Al'kor and A. K. Drezen. Leningrad: Izdatel'stvo Instituta narodov Severa TsIK SSSR.

Kavelin, E. 1931. "V strane anakalin and chaucha" [In the Country of the Anakalin and Chaucha]. *Sovetskii Sever: Obshchestvenno-nauchnyi zhurnal* [The Soviet North: Social Science Journal]. No. 7–8, pp. 75–121. Moscow.

Kekketyn, K. 2010. *Evnyto-pastukh: povesti na koryakskom i russkom yazykakh* [Sheepdog: Stories of the Koryak and Russian Languages]. Petropavlovsk-Kamchatskii: Kamchatpress.

Khlebnikov, K. S. 1835. *Zhizneopisanie Aleksandra Andreevicha Baranova, glavnogo pravitelya rossiiskikh kolonii v Amerike* [Biography of Alexander Andreevich Baranov, Governor of the Russian Colonies in America]. St. Petersburg: Morskaya tipografiya.

Kiber, A. 1823. "Zamechaniya o nekotorykh predmetakh estestvennoi istorii, uchinennykh v Nizhhe-Kolymske i okrestnostyakh onogo v 1821 godu" [Remarks on Some Objects of Natural History, Arranged in Nizhne-Kolymsk and the This Vicinity in 1821]. Translated from French by P. Naumov. *Sibirskii vestnik* [Siberian Bulletin]. Pt. 2, pp. 121–164. St. Petersburg.

________. 1824. "Chukchi." Translated from French by E. Pertsov. *Sibirskii vestnik* [Siberian Bulletin]. Pt. 2, bk. 9–10, pp. 87–126. St. Petersburg.

________. 1827. "Izvlechenie iz zhurnala puteshestviya doktora Kibera, s mneniem ob onom akademika Zakharova" [An Extract from the Journal of the Voyage of Doctor Kiber, with the Opinion about This Academician Zakharov]. *Zapiski, izdavaemye Gos. Admiralteiskim departamentom, otnosyashchiesya k moreplavaniyu, naukam i slovesnosti* [Notes Published by the State Admiralty Department, Belonging to Navigation, Science, and Literature]. Ch. 13, pp. 179–216. St. Petersburg.

Kibrik, A. E., S. V. Kodzasov, and I. A. Murav'eva. 2000. *Yazyk i fol'klor alyutortsev* [The Language and Folklore of the Alyutor People]. Moscow: IMLI RAN "Nasledie."

Kir'yak (Dikova), M. A. 2012. *Zagadochnyi mir drevnikh graffiti: po materialam pozdneneoliticheskoi stoyanki Rauchuvagytgyn I* (Chukotka). Magadan: Koris. (Published in English as *The Enigmatic World of Ancient Graffiti: Rock Art in Chukotka: The Chaunskaya Region, Russia*. Oxford, England: Archaeopress Publishing, Ltd. 2015).

Kolesnitskaya I. M. 1971. "V. G. Bogoraz-Tan—fol'klorist" [W. G. Bogoras-Tan—Folklorist]. *Ocherki istorii russkoi etnografii, fol'kloristiki i antropologii* [Essays on the History of Ethnography, Folklore Studies, and Physical Anthropology]. Issue V. (TIE. New Series. Vol. 95). Moscow. Pp. 139–159.

Kolesov, M. I. 1991. *Istoriya Kolymskogo kraya* [History of the Kolyma District]. Pt. I. Yakutsk: Yakutskoe knizhnoe izdatel'stvo.

Koltun, L. 1904. Kratkii ocherk Anadyrskogo uezda i byta naseleniya: Markovo i ego znachenie dlya kochevykh inorodtsev [A Brief Outline of the Anadyr' District and the Way of Life of Its Population: Markovo and Its Significance for Nomadic Non-Russians]. Vladivostok. (Reprint from the newspaper *Vostochnyi vestnik*).

Kolymskii okrug. 1879. "Kolymskii okrug" [The Kolyma District]. *Missioner: Ezhenedel'noe izdanie Pravostavnogo missionerskogo obshchestva* [Missionary: A Weekly Publication of the Orthodox Mission Society]. Moscow. No. 4, pp. 30–32; No. 5, pp. 39–40; No. 6, pp. 46–47; No. 7, pp. 53–55.

Kondratov, A. M. 1983. *Byla zemlya Arktida. (Sever vokrug nas)* [The Land of Arctida. (The North around Us)]. Magadan: Magadanskoe knizhnoe izdatel'stvo.

Kopets, I. 1896. "Zapiski brigadira Iosifa Koptsya" [The Notes of Brigadier Iosif Kopets']. Translated by G. A. Vorob'ev. *Istoricheskii vestnik: Istoriko-literaturnyi zhurnal* [Historical Bulletin: Historical-Literary Journal]. Year 17, no. 10, pp. 226–244; No. 11, 579–601; No. 12, pp. 991–1016. St. Petersburg.

Korotaeva, D. 2008. "Zhizn' po pogode" [Life Based on the Weather]. *Region 87: Ezhekvartal'nyi zhurnal o vozmozhnom chude* [Region 87: Quarterly Journal on Possible Miracle]. Moscow. No. 1 (5), pp. 4–11

Kostenetskaya, M. G. *Luna kholodnogo litsa* [The Moon Is a Cold Person]. Riga: Liesma.

Kostyurin [Kostyurin, V. F.]. 1893. "Ekonomicheskoe polozhenie chukchei" [The Economic Position of the Chukchi]. *Sibirskii listok* [Siberian Leaflet]. Tobol'sk. No. 24 (4 April), p. 2; No. 26 (11 April), p. 2.

Kosven, M. O. 1961. "Etnograficheskie rezul'taty Velikoi Severnoi ekspeditsii" [The Ethnographic Results of the Great Northern Expedition]. *SES*. Issue III. (*TIE*. Vol. LXIV). Pp. 167–212. Moscow; Leningrad.

________. 1962. "Iz istorii etnografii koryakov v XVIII v. (publikatsiya)" [From the History of the Ethnography of the Koryak in the 18th Century (Publication)]. *SES*. Vol. IV. (*TIE*. New series. Vol. 78). Pp. 276–291. Moscow.

Kotsebu, O. E. [Kotsebue, O. E.]. 1948. *Puteshestvie vokrug sveta*. 2nd ed. Moscow: OGIZ. (Published in English as *A New Voyage round the World in the Years 1823, 24, 25, and 26*. London: H. Colburn & R. Bentley, 1830].

Kozlov, N. V. 1956. *Skazki narodov Severo-Vostoka* [Stories of the Peoples of the Northeast]. Edited by N. V. Kozlov. Magadan: Magadanskoe knizhnoe Izdatel'stvo.

Krasovskii, M. 1894. "Russkie v Yakutskoi oblasti v XVII v." [Russians in the Yakutsk Region in the 17th Century]. *Izv. O-va arkheologii, istorii i etnografii pri Imp. kazanskom un-te* [News of the Society of Archaeology, History, and Ethnography of the Imperial Kazan University]. Vol. XII, issue 2, pp. 143–176. Kazan.

Krause, A. 1881. "Die Expedition der Bremer geogr. Gesellschaft nach der Tschuktschen-Halbinsel" [The Expedition of the Bremen Geographic Society to the Chukchi Peninsula]. Sommer 1881. *Deutsche Geographische Blätter* [German Geographic Sheets]. Vol. V. Bk. 1. Bremen. Pp. 1–35.

Krauss, M. E. 1981. "Yazyki korennogo naseleniya Alyaski: proshloe, nastoyashchee i budushchee" [Languages of the Native Population of Alaska: Past, Present, and Fugure]. *Traditsionnye kul'tury Severnoi Sibiri i Severa Ameriki* [Traditional Cultures of Northern Siberia and North America]. Edited by I. S. Gurvich. Moscow: Nauka. Pp. 149–181.

________. 2005. "Eskimo Language in Asia, 1791 on, and Wrangel Island — Point Hope Connection." *Études / Inuit / Studies* 29(1–2): 163–185.

Krasheninnikov, S. P. 1949. *Opisanie zemli Kamchatki* [A Description of the Land of Kamchatka]. Moscow; Leningrad: Izdatel'stvo Glavsevmorputi.

Krotov, M. A. 1927. "K istorii chukchei" [On the History of the Chukchi]. Sb. trudov issledovatel'skogo o-va "Saqa keskile" [Collection of the Works of the Research Society "Saqa Keskile"]. Yakutsk: [B. i.]. Issue 1 (4), pp. 51–57.

Krupnik, I. I. 1981. "Drevnie poselki i demograficheskaya istoriya aziatskikh eskimosov Yugo-Vostochnoi Chukotki (vklyuchaya ostrov Svyatogo Lavrentiya)" [Ancient Village and Demographic History of the Asiatic Eskimos of Southeastern Chukotka (Including St. Lawrence Island)]. *Traditsionnye kul'tury Severnoi Sibiri i Severa Ameriki* [Traditional Cultures of Northern Siberia and North America]. Edited by I. S. Gurvich. Moscow: Nauka. Pp. 97–118.

________. 1983. "Drevnie i traditsionnye poseleniya eskimosov na Yugo-Vostoke Chukotskogo poluostrova" [Ancient and Traditional Settlements of the Eskimos on the Southeastern Chukchi Peninsula]. *Na styke Chukotki i Alyaski* [At the Crossroads of Chukotka and Alaska]. Edited by V. P. Alekseev. Moscow: Nauka. Pp. 65–95.

________. 1989. *Arkticheskaya etnoekologiya: Modeli traditsionnogo prirodopol'zovaniya morskikh okhotnikov i olenevodov Severnoi Azii* [Arctic Ethnoecology: Models of Traditional Nature-Use of Maritime Hunters and Reindeer Herders of Northern Asia]. Moscow: Nauka.

Krupnik, I. I., and M. A. Chlenov. 1979. "Dinamika etnolingvisticheskoi situatsii u aziatskikh eskimosov (konets XIX v.–1970-e gg.)" [The Dynamics of the Ethnolinguistic Situation among Asiatic Eskimos (End of the 19th Century–1970s)]. *SE* 2: 19–29.

Krupnik, I., and L. Krutak. 2002. *Akuzílleput Igaqullghet: Our Words Put to Paper: Sourcebook in St. Lawrence Island Heritage and History*. Compiled by I. Krupnik and L. Krutak; ed. by I. Krupnik, W. Walunga (Kepelgu), V. Metcalf (Qaakaghlleq). (*Contributions to Circumpolar Anthropology* 3). Washington: Arctic Studies Center, National Museum of Natural History, Smithsonian Institution.

Kruzenshtern, I. F. 1950. *Puteshestvie vokrug sveta v 1803, 1804, 1805 i 1806 godakh na korablyakh "Nadezhde" i "Neve"* [Voyage around the World in 1803, 1804, 1805, and 1806 in the Ships *Nadezhda* and *Neva*]. Moscow: Gosudarstvennoe izdatel'stvo geograficheskoi literatury.

Kulikov, M. I. 2002. *Chukotka: Zigzagi istorii malykh narodov Severa* [Chukotka: Zigzags in the History of Natives Groups of the North]. Velikii Novgorod: Novgorodskii gosudarstvennyi universitet.

Kurilov, G. N. 2005. *Fol'klor yukagirov*. (*Pamyatniki fol'klora narodov Sibiri i Dal'nego Vostoka*. T. 25) [Folklore of the Yukaghir. (Monuments of the Folklore of Peoples of Siberia and the Far East. Vol. 25)]. Novosibirsk: Nauka.

Kuttegina, M. P. 2003. "Eivelkei" [Eivalkei]. *Aiverette: Repertuarnometodicheskii sbornik* [Aiverette: A Repertory Methodical Collection]. Issue 10. Anadyr'. Pp. 47–49.

Kuvaev, O. M. *Dnevnik pribrezhnogo plavaniya* [Diary of Coastal Navigation]. Moscow: Fizkul'tura i sport.

Kuz'minykh, V. I. 1994. "Neizvestnye voiny: Chukotka 1649–1755 gg." [The Unknown War: Chukotka 1649–1755]. *Materialy XXXII Mezhdunar. nauch. stud. konf. "Student i nauchnotekhnicheskii progress". Istoriya* [Materials of the 32nd International Science Student Conference "Student and Scientific-Technical Progress." History]. Novosibirsk: NUG. Pp. 78–81.

Kymytval', A. A. 1987. *Lyg`oravetl'en lymnylte*. [Native Education] Magadan: Magadanskoe knizhnoe izdatel'stvo.

Kytgaut, E. I. 2003. "Pirichyven" [Pirichyven]. *Aiverette: Repertuarnometodicheskii sbornik* [Aiverette: A Repertory Methodical Collection]. Issue 10. Anadyr'. P. 34.

L. 1901. "Zhenshchina—nevol'nitsa v Sibiri" [A Woman Is a Slave in Siberia]. *Enisei* [Yenisei]. Krasnoyarsk. No. 48 (22 April). Pp. 2–3; No. 49 (25 April). Pp. 2–3.

Labanauskas, K. I. 2000. "Zagadki istorii nganasan (na osnove fol'klora)" [Enigmas in the History of the Nganasan (On the Basis of Folklore)]. *Polyarnoe siyanie: Literaturno-khudozhestvennyi al'manakh* [Aurora: A Literary Almanac]. Moscow. Pp. 95–96.

Lagus, V. Ya. I. 1890. *Erik Laksman, ego zhizn', puteshestviya, issledovaniya, perepiska* [Erik Laxman, His Life, Travels, Research, and Correspondence]. Translated from Swedish. St. Petersburg: Izdatel'stvo Imperatorskoi akademii nauk.

Lazarev, A. P. 1950. *Plavanie vokrut sveta voennogo shlyupa "Blagonamerennyi"* [Voyage around the World of the Naval Sloop *Blagonamerennyi*]. Moscow: Gosudarstvennoe izdatel'stgo goegraficheskoi literatury.

Lebedev, D. M. 1957. *Ocherki po istorii geografii v Rossii XVIII v. (1725–1800 gg.)* [Essays on the History of Geography in Russia of the 18th Century (1725–1800)]. Moscow: Izdatel'stvo AN SSSR.

Lebedev, V. V., and Yu. B. Simchenko. 1983. *Achaivayamskaya vesna* [Achaivayamskaya Spring]. Moscow: Mysl'.

Lebo, S. A. 2013. "A Hawaiian Perspective on whaling in the North Pacific." *Anthropological Studies of Whaling*. Ed. by N. Kishigami, H. Hamaguchi, and J. M. Savelle. (*Senri Ethnological Studies*, 84). Osaka: National Museum of Ethnology. Pp. 51–78.

Leonova, V G. 2008. "Pamyat' o Naukane" [Memory of Naukan]. *TB*. Pp. 156–161.

Leont'ev, V. V. 1969. "K voprosu o toponimakh Pevek i Peek" [On the Question of the Toponyms Pevek and Peek]. *Izv. Sib. otd-niya AN SSSR* [News of the Siberian Division of the Academy of Sciences of the USSR]. Novosibirsk. *Seriya: Obshchestvennye nauki* [Series: Social Sciences]. Issue 1, no. 1, pp. 102–107.

________. 1972–1975. "Etnicheskaya istoriya poselka Uelen" [The Ethnic History of the Village of Uelen]. *KZOKM*. Issue 9 (1972), pp. 83–94; issue 10 (1975), pp. 77–87.

________. 1976. *Po zemle drevnikh kerekov: Zapiski etnografa* [In the Land of the Ancient Kerek: Notes of an Ethnographer]. Magadan: Magadanskoe knizhnoe izdatel'stvo.

________. 1976a. "Novye dannye o kerekskoi narodnosti" [New Data on the Kerek People]. *Istoriya i kul'tura narodov Severo-Vostoka SSSR: Materaly 3-i sessii Dal'nevostochnykh istoricheskikh chtenii po problemam istorii Severo-Vostoka SSSR* [The History and Culture of Peoples of the Northeastern USSR: Materials of the 3rd Session of the Far-Eastern Historical Readings on Problems of the History of the Northeastern USSR]. Edited by B. I. Mukhachev. Vladivostok: In-t istorii, arkheologii i etnografii narodov Dal'nego Vostoka. Pp. 232–237.

________. 1983. *Etnografiya i fol'klor kerekov* [The Ethnography and Folklore of the Kerek]. Moscow: Nauka.

Leont'ev, V. V., and K. A. Novikova. 1989. *Toponimicheskii slovar' Severo-Vostoka SSSR* [Toponymic Dictionary of the Northeastern USSR]. Magadan: Magadanskoe knizhnoe izdatel'stvo.

Leont'eva, G. A. 1997. *Yakutskii kazak Vladimir Atlasov—pervoprokhodets zemli Kamchatki* [The Yakutsk Cossack Vladimir Atlasov—Explorer of the Land of Kamchatka]. Moscow: Institut etnografii i antropologii im. N. N. Miklukho-Maklaya.

Lesseps, J.-B. B. de. 1801. *Lessepsovo puteshestvie po Kamchatke i po yuzhnoi storone Sibiri* [Lesseps' Journey through Kamchatka and the Southern Part of Siberia]. Translation from French. Pt. II. Moscow: Gubernskaya tipografiya u A. Reshetnikova.

Limonov, E. V. 2013. Apologiya chukchei: Moi knigi, moi voiny, moi zhenshchiny [Apologia of the Chukchi: My Books, My Wars, My Women]. Moscow: AST.

Lindenau, J. I. 1983. *Opisanie narodov Sibiri (pervaya polovina XVIII veka): Istoriko-etnograficheskie materialy o narodakh Sibiri i Severo-Vostoka* [A Description of the Peoples of Siberia (First Half of the 18th Century): Historical-Ethnographic Material on the Peoples of Siberia and the Far East]. Translated from German by Z. D. Titova. (*DIB*). Magadan: Magadanskoe knizhnoe izdatel'stvo.

Litke, F. P. 1835. "Zamechaniya o chukchakh" [Notes on the Chukchi]. *ZhMVD*. Pt. 16, pp. 356–368.

________. 1948. *Puteshestvie vokrug sveta na voennom shlyupe "Sinyavin", 1826–1829*. 2nd ed. Moscow: OGIZ. (Published in English as *A Voyage around the World, 1826–1829*. Kingston, Ontario: The Limestone Press, 1987).

Lyubov' chukchei . . . 1811. "Lyubov' chukchei k spravedlivosti" [The Love of the Chukchi for Justice]. *Russkii vestnik na 1811i god, izdavaemyi S. Glinkoi* [A Russian Bulletin in 1811, Being Published by S. Glinka]. Moscow. Pt. 15, no. 8, pp. 17–26.

Lyutsidarskaya, A. A. 2006. "Amanaty v Sibiri" [Hostages in Siberia]. *Problemy arkheologii, etnografii, antropologii Sibiri i sopredel'nykh territorii* [Problems of Archaeology, Ethnography, and Physical Anthropology of Siberia and Adjacent Territories]. Edited by A. P. Derevyanko and V. I. Molodin. Vol. XII, pt. 2. Novosibirsk: Izdatelstvo Instituta arkheologii i etnografii SO RAN. Pp. 121–124.

MIYa. 1970. *Materialy po istorii Yakutii XVII veka. (Dokumenty yasachnogo sbora)* [Materials on the History of Yakutia of the 17th Century. (Documents of Tribute Collection)]. Ch. 1–3. Moscow: Nauka.

Machinskii, A. V. 1941. "Drevnyaya eskimosskaya kul'tura na Chukoskom poluostrove" [Ancient Eskimo Culture on the Chukchi Peninsula]. *KSIIMK*. Issue 9, pp. 80–89.

Magidovich, I. P., and V. I. Magidovich. 1984. *Ocherki po istorii geograficheskikh otkrytii. T. III: Geograficheskie otkrytiya i issledovaniya novogo vremeni (seredina XVII–XVIII v.)* [Essays on the History of Geographical Discoveries. Vol. III. Geographic Discoveries and Research of Recent Times (Middle of the 17th–18th Century)]. 3rd ed. Moscow: Prosveshchenie.

Makarova, R. V. 1968. *Russkie na Tikhom okeane vo vtoroi polovine XVIII v.* [Russians on the Pacific Ocean in the Second Half of the 18th Century]. Moscow: Nauka.

________. 1982. *Vneshnyaya politika Rossii na Dal'nem Vostoke (vtoraya polovina XVIII v.–70-e gody XIX v.): uchebnoe posobie po spetskursu* [Foreign Affairs of Russia in the Far East (Second Half of the 18th Century to the 1870s): Textbook for a Specialized Course]. Moscow: [B. i.].

Makhovskii, Ya. 1981. *Ognennyi shar: legendy, pedaniya i skazki eskimosov* [Ball of Fire: Legends, Traditions, and Stories of the Eskimos]. Translated by S. V. Tel'nyuk. Magadan: Magadanskoe knizhnoe izdatel'stvo.

Maksimov, S. V. 1900. *Sibir' i katorga* [Siberia and Servitude]. 3rd ed. St. Petersburg: Izdanie V. I. Gubinskogo.

Malaurie, J. 1974. "Raids et esclavage dans les sociétés utochtones de détroit de Behring" [Raids and Slavery in the Aboriginal Societies of Bering Strait]. *Inter-Nord: Revue international d'études arctiques et nordiques*. no. 13–14. Décembre. Mouton; Paris; La Haye. Pp. 129–155.

Mamyshev, N. R. 1809. "Otchayanie (istoricheskii otryvok)" [Despair (A Historical Excerpt)]. *Vestnik Evropy, izdavaemyi V. Zhukovskim*. Moscow. Ch. 44, No. 5, Pp. 20–27.

Markov, S. N. 1948. *Letopis' Alyaski* [Chronicle of Alaska]. Moscow; Leningrad: Izdatel'stvo Glavsevmorputi.

Martynov, M. N. 1957. "Malye narody Yakutii" [Ethnic Groups of Yakutia]. *Istoriya Yakutskoi ASSR* [History of the Yakutsk ASSR]. Vol. II. Moscow: Izd-vo AN SSSR. Pp. 95–110.

Mason, O. W. 1998. "The Contest between the Ipiutak, Old Bering Sea, and Birnirk Polities and the Origin of Whaling during the First Millennium A. D. along Bering Strait." *Journal of Anthropological Archaeology* 17: 240–325.

Maydell, G. L. 1871. "Otvety Chukotskoi ekspeditsii na voprosy akademika Bera" [Answers of the Chukotka Expedition to the Questions of Academician Baer]. *ISOIRGO*. Vol. II, no. 1, pp. 60–70.

________. 1894. *Puteshestvie po severo-vostochnoi chasti Yakutskoi oblasti v 1868–1870 godakh* [A Voyage along the Northeastern Part of the Yakutsk Region in 1868–1870]. Translated from German by V. L. Bianki. (Prilozhenie k LXXIV tomu *Zapisok Imperskoi akademii nauk*. No. 3) [(Appendix to the 74th volume of Notes of the Imperial Academy of Sciences. No. 3)]. St. Petersburg: Imperatorskaya akademiya nauk.

________. 1925. "Zapiska o narodakh, zhivushchikh v severo-vostochnoi chasti Yakutskoi i Primorskoi oblasti" [A Note on the Peoples Who Live in the Northeastern Part of Yakutia and the Primor'e Region]. *Sb. trudov issledovatel'skogo o-va "Saqa Keskile"* [Collected Works of the Research Society of "Saqa Keskile"]. Issue 1 (August). Yakutsk: [B. i.]. Pp. 20–34.

Medushevskaya, O. M. 1954. "Kartograficheskie istochniki po istorii russkikh geograficheskikh otkrytii na Tikhom okeane vo 2-i polovine XVIII veka" [Cartographic Sources on the History of Russian Geographic Discoveries on the Pacific Ocean in the 2nd Half of the 18th Century]. *Trudy Moskovskogo gos. istoriko-arkhivnogo in-ta* [Works of the Moscow State Historical-Archival Institute]. Vol. 7, pp. 95–127. Moscow: Izdatel'stvo MGU.

Menovshchikov, G. A. 1950. *Chukotskie, eskimosskie, koryakskie skazki* [Chukchi, Eskimo, and Koryak Stories]. Khabarovsk: Dal'nevostochnoe gosudarstvennoe izdatel'stvo.

________. 1958. *Eskimosskie skazki* [Eskimo Stories]. Compiled by G. A. Menovshchikov. Magadan: Magadanskoe knizhnoe izdatel'stvo.

________. 1964. *Yazyk sirenikskikh eskimosov* [Language of the Siriniki Eskimos]. Moscow; Leningrad: Nauka.

________. 1964a. *Ustnoe narodnoe tvorchestvo aziatskikh eskimosov kak istoriko-etnograficheskii istochnik* [Oral Folk Creations of the Asiatic Eskimos as a Historical-Ethnographic Source]. Moscow: Nauka.

________. 1969. *Eskimosskie skazki i legendy* [Eskimo Stories and Legends]. Compiled, translated, introduced, and annotated by G. A. Menovshchikov. Magadan: Magadanskoe knizhnoe izdatel'stvo.

________. 1972. *Mestnye nazvaniya na karte Chukotki* [Local Names on the Map of Chukotka]. Magadan: Magadanskoe knizhnoe izdatel'stvo.

________. 1974. *Skazki i mify narodov Chukotki i Kamchatki* [Stories and Myths of the Peoples of Chukotka and Kamchatka]. Compiled by G. A. Menovshchikov. (*SMNV*). Moscow: Nauka.

________. 1980. *Yazyk eskimosov Beringova proliva* [The Language of the Eskimos of Bering Strait]. Leningrad: Nauka.

________. 1985. *Skazki i mify eskimosov Sibiri, Alyaski, Kanady i Grenlandii* [Stories and Myths of the Eskimos of Siberia, Alaska, Canada, and Greenland]. Compiled by G. A. Menovshchikov. (*SMNV*). Moscow: Nauka.

________. 1987. *Materialy i issledovaniya po yazyku i fol'kloru naukanckikh eskimosov* [Materials and Investigations on the Language and Folklore of the Naukan Eskimos]. Leningrad: Nauka.

________. 1988. *Eskimosskie skazki i mify* [Eskimo Stories and Myths]. Compiled, introduced, and annotated by G. A. Menovshchikov. (*SMNV*). Moscow: Nauka.

________. 1988a. *Materialy i issledovaniya po yazyku i fol'kloreu chaplinskikh eskimosov* [Materials and Investigations on the Language and Folklore of the Chaplino Eskimos]. Leningrad: Nauka.

Merck, C. H. [Merk, K. G.]. 1978. "Opisanie obychaev i obraza zhizni chukchei" [Description of the Customs and Way of Life of the Chukchi]. Translated from German by Z. D. Titova. *EM*. Pp. 98–155.

Middendorff, A. F. 1878. *Puteshestvie na Sever i Vostok Sibiri* [Travel in the North and East of Siberia]. Pt. II, sec. 6. St. Petersburg: Izd-vo Imp. akademii nauk.

Mikhailova, E. A. 1996. "K voprosu ob etno- i kul'turogeneze korennogo naseleniya krainego Severo-Vostoka Azii" [On the Question of Ethnogenesis and the Origin of Culture of the Native Population of Extreme Northeast Asia]. *Sibir': drevnie etnosy i ikh kul'tura* [Siberia: Ancient Ethnic Groups and Their Cultures]. Edited by L. R. Pavlinskaya. St. Petersburg: MAE RAN. Pp. 164–195.

Mitchel', G. 1865. "Zima u chukchei" [Winter among the Chukchi]. *Vokrug sveta* [Around the World]. Year 5. St. Petersburg. No. 10, pp. 313–330.

Mitryukova, L. "Plashch iz kishok morzha" [A Cape of Walrus Gut]. *Severnye prostory* [Northern Spaces]. Moscow. No. 3, p. 41.

Moll, T. A., and P. I. Inenlikei. 2005. *Chukotsko-russkii slovar'* [Chukchi-Russian Dictionary]. 2nd ed. St. Petersburg: Prosveshchenie.

Molodykh, I. F. 1931. *Bibliograficheskii ukazatel' po Kolymskomu krayu Yakutskoi ASSR* (*Materialy partii po issledovaniyu reki Kolymy*. Vyp. III) [Bibliographic Index of the Kolyma District of the Yakutsk ASSR (Materials of the Party on Investigation of the Kolyma River. Issue III)]. Edited by I. F. Molodykh. Irkutsk: [B. i.].

Moskvin, I. S. 1864. "Voevody inachal'niki g. Yakutska i ikh deistviya" [Governors and Leaders of the City of Yakutsk and Their Activities]. *Pamyatnaya knizhka Yakutskoi oblasti na 1863 god* [A Commemorative

Book on the Yakutsk Region in 1863]. St. Petersburg: Tipografiya V. Bezobrazova i Komp. Pp. 165–202.

Müller, G. F. 1756. *Opisanie o torgakh sibirskikh* [A Description of Bargaining of the Siberians]. St. Petersburg: Izdatel'stvo Akademii nauk.

________. 1996. "Opisanie morskikh puteshestvii po ledovitomu i Vostochnomu moryu, s rossiiskoi storony uchinennykh" [A Description of Sea Voyages through the Arctic and Eastern Seas, Carried Out from the Russian Side]. *Miller, G. F. Sochineniya po istorii Rossii: Izbrannoe. (Pamyatniki istoricheskoi mysli)* [Müller, G. F. Works on the History of Russia: Selected. (Monuments of Historical Thought)]. Moscow: Nauka. Pp. 19–126.

________. 2009. *Opisanie sibirskikh narodov. (Istochniki po istorii Sibiri i Alyaski v rossiiskikh arkhivakh.* T. VIII.1) [A Description of the Siberian Peoples. (Sources on the History of Siberia and Alaska in the Russian Archives. Vol. VIII.1)]. Moscow: Pamyatniki istoricheskoi mysli.

Mushkin, P. 1853. "Torgovlya chaukchei s russkimi i zhitelyami ostrovov Vostochnogo morya" [Trade of the Chukchi with the Russians and Residents of the Islands of the Eastern Sea]. *Moskovskie vedomosti* [Moscow Gazette]. Moscow. No. 138 (17 November). Pp. 1421–1422.

________. 1853a. "Russkie missionery u chaukchei" [Russian Missionaries among the Chukchi]. *Moskovskie vedomosti* [Moscow Gazette]. Moscow. No. 82 (9 July). Pp. 843–844; No. 38 (11 July). Pp. 853–854.

N-`. 1897. "N-`. Pokhodskaya Viska" [N-`. Pokhodsk Temple]. *Sibirskii sbornik* [Siberian Bulletin]. Year 12. Irkutsk. Issue 1–2, pp. 1–22.

Nechto o Sibiri. . . 1806. "Nechto o Sibiri i narodakh,v nei obitayushchikh" [Something about Siberia and the Peoples Who Live There]. *Sankt-Peterburgskii zhurnal* [St. Petersburg Journal]. No. 7, pp. 92–108; no. 8, pp. 92–106.

Nefëdkin, A. K. 2003. *Voennoe delo chukchei (seredina XVII–nachalo XX v.).* (*Ethnographica Petropolitana.* X) [Military Affairs of the Chukchi (Middle of the 17th–Beginning of the 20th Century). (Ethnographica Petropolitana. X). St. Peterburgskoe Vostokovedenie.

________. 2013. "'Ledyanoi pokhod' kolymsko-alazeiskikh chukchei v 1716–1721 gg. po svidetel'stvu G. F. Millera" [The "Ice Campaign" of the Kolyma-Alazeya Chukchi in 1716–1721 Based on the Information of G. F. Müller]. *Millerovskie chteniya: K 285-letiyu Arkhiva Rossiiskoi akademii nauk: sb. nauch. statei po materialam Mezhdunar. nauch. konf. 23–25 aprelya 2013 g. Sankt-Peterburg* [The Müller Readings: On the 285th Year of the Archive of the Russian Academy of Sciences: Collection of Scientific Articles Based on Materials of the International Science Conference of 23–25 April 2013. St. Petersburg]. Edited by I. V. Tunikin. (Ad fonts: *Materialy i issledovaniya po istorii nauki.* Vyp. 4) [(Ad. fonts: Materials and Research on the History of Science. Issue 4)]. St. Petersburg: Nestor-Istoriya. Pp. 80–84.

________. 2015. "Zapiska otstavnogo kapala Grigoriya Sheikina" [Note of the Retired Corporal Grigorii Sheikin]. *Vestnik Novosibirskogo gos. un-ta. Seriya: Istoriya, filologiya.* T. 14, vyp. 8: *Istoriya* [Bulletin of the

Novosibirsk State University. Series: History and Philology. Vol, 14, issue 8: History]. Novosibirsk. Pp. 183–190.

________. 2015a. "Plavanie M. S. Gvozdeva v Beringov proliv (1732 g.) kak voennaya ekspeditsiya" [The Voyage of M. S. Gvozdev in Bering Strait (1732) as a Military Expedition]. *Gosudarstvo i obshchestvo v Evrope: Problemy stanovleniya i rezvitiya: sb. nauch. rabot* [State and Society in Europe: Problems of Formation and Development: Collection of Scientific Works]. Edited by R. K. Khabibullin. Ufa: RITs BashGU. Pp. 102–109.

________. 2016. *Geroicheskie skazaniya narodov Chukotki* [Heroic Stories of the Peoples of Chukotka]. St. Petersburg: Lemma.

________. 2016a. *Neopublikovannye dokumenty po istorii Chukotki XVIII veka* [Unpublished Documents on the Chukotka History in the 18th Century]. St. Petersburg: Lemma.

Neiman, K. K. 1871. "Istoricheskii obzor deistvii Chukotskoi ekspeditsii" [A Historical Survey of the Activities of the Chukotka Expedition]. *ISOIRGO*. Vol. I, no. 4–5, pp. 6–31; vol. II, no. 3, pp. 7–28.

________. 1872. "Neskol'ko slov o torgovle i promyshlennosti severnykh okrugov Yakutskoi oblasti" [A Few Words on Trade and Business of the Northern Districts of the Yakutsk Region]. *ISOIRGO*. Vol. III, no. 1, pp. 32–44; no. 2, pp. 57–69.

________. 1877. "Plavanie po Vostochnomu okeanu" [A Voyage in the Eastern Ocean]. *ISOIRGO*. Vol. VIII, no. 1–2, pp. 43–56; no. 3–4, pp. 83–94; no. 5–6, pp. 148–155.

Nelson, E. W. 1899. *The Eskimo about Bering Strait*. Eighteenth Annual Report of the Bureau of American Ethnology to the Secretary of the Smithsonian Institution 1896–1897. Pt. I. Washington. Pp. 19–518.

Nikol'skii, D. [P]. 1900. "O chukchakh Kolymskogo okruga" [About the Chukchi of the Kolyma District]. *Russkii antropologicheskii zhurnal* [Russian Anthropological Journal]. Vol. 2, pp. 23–29.

Novikova, K. A. 1987. Evenskie skazki, predaniya i legendy [Even Stories, Traditions, and Legends]. Compilation, postscript, and notation by K. A. Novikov. Magadan: Magadanskoe knizhnoe izdatel'stvo.

Nordenshel'd, A. E. 1879. "Ekspeditsiya professora A. E. Nordenshel'da" [The Expedition of Professor A. E. Nordenskiöld]. *IIRGO*. Vol. 15, issue 3, pp. 210–226.

________. 1881. *Puteshestvie A. E. Nordenshel'da vokrug Azii na parokhode "Vega"* [The Voyage of A. E. Nordenskiöld around Asia on the Steamship *Vega*]. Translated by S. I. Baranovskii. Pt. I. St. Petersburg: I. I. Vallenius.

Nul. . . 1866. "Nul . . . Otdalennyi sever Sibiri: Anadyrskii ostrog" [Nul . . . Remote Northern Siberia: The Anadyrsk Stockade]. *Sibirskii vestnik* [Siberian Bulletin]. No. 16 (15 August), [pp. 3–4]; no. 17 (30 August), [pp. 3–4].

Nutekeu, E. I. 2005. Chukotskie skazki i legendy (v pereskazakh babushki Nutekeu) [Chukchi Stories and Legends (Retold by Grandmother Nutekeu)]. [B. m.]: [B. i.].

________. 2005a. "Al'katvaam." *Aiverette: repertuarno-metodicheskii sbornik* [Aiverette: A Repertory Methodic Digest]. Issue 11. Anadyr'. Pp. 20–23.

Nuvano, V. N. 2003. "Mezhplemennye otnosheniya na yuge Chukotki v XVIII–XX vekakh: Vaezhkie chukchi" [Inter-Tribal Relations in Southern Chukotka in the 18th–20th Centuries: The Vaezhkie Chukchi]. *Chukotskie issledovaniya: Olenevodstvo sredi chukchei* [Chukchi Investigations: Reindeer Herding among the Chukchi]. Edited by K. Ikeya. Issue 1. Osaka: Chukotka Studies Committee. Pp. 59–56.

________. 2008. "Etnoterritorial'naya gruppa 'vaezhskie chukchi'" [The Ethnoterritorial Group "Vaezhskie Chukchi"]. *TB*. Pp. 78–84.

O torgovykh snosheniyakh. . . 1851. "O torgovykh snosheniyakh mezhdu tuzemtsami severo-vostochnogo berega Azii i Severo-Zapadnoi Ameriki" [About Trade Relations between Natives of the Northeastern Shore of Asia and Northwest America]. *ZhMVD*. Pt. 35, pp. 102–112. St. Petersburg.

Obruchev, S. V. 1954. *V neizvedannye kraya: Puteshestviya na Sever, 1917–1930 gg.* [In Uncharted Territory: Travels in the North, 1917–1930]. Moscow: Molodaya gvardiya.

Ogloblin, N. 1903. "Vostochno-Sibirskie polyarnye morekhody XVII veka" [East Siberian Polar Navigators of the 17th Century]. *ZhMNP*. Ch. 347. May. Pp. 38–62.

Ogorodnikov, V. I. 1922. *Iz istorii pokoreniya Sibiri: Pokorenie Yukagirskoi zemli*. (*Trudy Gos. in-ta narodnogo obrazovaniya v Chita*. Kn. 1) [From the History of the Conquest of Yukaghir Land. (Works of the State Institute of Public Education in Chita. Bk. 1)]. Chita: Tipografiya Dal'profsoveta.

Okhotina-Lind, N., and P. U. Mëller. 2001. *Vtoraya kamchatskaya ekspeditsiya. Dokumenty 1730–1733*. Ch. 1: *Morskie otryady*. (*Istochniki po istorii Sibiri i Alyaski iz rossiiskikh arkhivov*. T. IV.1) [The Second Kamchatka Expedition. Documents of 1730–1733. Pt. 1: Maritime Detachments. (Sources on the History of Siberia and Alaska from the Russian Archives. Vol. IV.1)]. Moscow: Pamyatniki istoricheskoi mysli.

________. 2013. *Vtoraya kamchatskaya ekspeditsiya. Dokumenty 1737–1738*: *Morskie otryady*. (*Istochniki po istorii Sibiri i Alyaski iz rossiiskikh arkhivov*. T. IX) [The Second Kamchatka Expedition. Documents of 1737–1738: Maritime Detachments. (Sources on the History of Siberia and Alaska from the Russian Archives. Vol. IX)]. St. Petersburg: Nestor-Istoriya.

________. 2015. *Vtoraya kamchatskaya ekspeditsiya. Dokumenty 1739–1740*: *Morskie otryady*. (*Istochniki po istorii Sibiri i Alyaski iz rossiiskikh arkhivov*. T. X) [The Second Kamchatka Expedition. Documents of 1739–1740: Maritime Detachments. (Sources on the History of Siberia and Alaska from the Russian Archives. Vol. X)]. St. Petersburg: Nestor-Istoriya.

Okladnikov, A. P. 1947. "Drevnie kul'tury severo-vostochnoi Azii po dannym arkheologicheskikh issledovanii 1946 g. v Kolymskom krae" [Early Cultures of Northeast Asia Based on Data from Archaeological Investigations of 1946 in the Kolyma District]. Vestnik drevnei istorii [Bulletin of Ancient History]. Moscow. No. 1, pp. 176–182.

________. 1948. "K istorii etnograficheskogo izucheniya Yakutii" [On the History of Ethnographic Investigation of Yakutia]. Sb. materialov po etnografii yakutov [Collection of Materials on the Ethnography of the Yakut]. Yakutsk: Yakutskoe knizhnoe izdatel'stvo. Pp. 17–48.

________. 1975. *Yukagiry (istoriko-etnograficheskii ocherk)* [The Yukaghir (A Historical-Ethnographic Outline)]. Edited by A. P. Okladnikov. Novosibirsk: Nauka.

Okladnikov, A. P., and N. A. Beregovaya. 1971. *Drevnie poseleniya Baranova mysa*. Novosibirsk: Nauka. (Published in English as *The Early Sites of Cape Baranov*. Anchorage, Alaska: Shared Beringian Heritage Program, National Park Service, 2008).

Okun', S. B. 1935. *Ocherki po istorii kolonial'noi politiki v Kamchatskom krae* [Essays on the History of Colonial Policies in the Kamchatka District]. Leningrad: OGIZ, Sotsekgiz.

Olsuf'ev, A. V. 1896. *Obshchii ocherk Anadyrskoi okrugi, ee ekonomicheskogo sostoyaniya i byta naseleniya* [A General Survey of the Anadyr' District, Its Economic State, and Way of Life of the Population]. (*ZPOIRGO*. Vol. II, issue 1). St. Petersburg: Tipografiya Imperatorskoi AN.

Olyksandrovich. 1884. "Ugolok dalekogo severa" [The Little Corner of the Far North]. *Sibirskaya gazeta* [The Siberian Gazette]. Tomsk. No. 7 (12 February). Col. 185–188; No. 9 (26 February). Col. 234–237; No. 11 (11 March). Col. 295–297.

Omruv'e, I. V. 2004. "Alya—potomok voina" [Alya—Descendant of a Warrior]. *Krainii Sever* [The Extreme North]. Anadyr'. No. 48 (1444).

Orekhov, A. A. 1987. *Drevnyaya kul'tura Severo-Zapadnogo Beringomor'ya*. Moscow: Nauka. (Published in English as *Early Culture of the Northwestern Bering Sea*. Anchorage, Alaska: Shared Beringian Heritage Program, National Park Service, 1999).

Orlova, E. P. 1957. "Chertezhi Chukotki Yakova Lindenau i Timofeya Perevalova" [The Drawings of Chukotka by Yakov Lindenau and Timofei Perevalov]. *Voprosy geografii Dal'nego Vostoka*. Sb. 3 [Questions of Geography of the Far East. Collection 3]. Khabarovsk. Pp. 120–129.

Orlova, N. S. 1951. *Otkrytiya russkikh zemleprokhodtsev i polyarnykh morekhodov XVII veka na Severo-Vostoke Azii: sbornik dokumentov* [Discoveries of Russian Explorers and Polar Navigators of the 17th Century in Northeast Asia: Collection of Documents]. Moscow: Geografgiz.

PSI 1882–1885: *Pamyatniki sibirskoi istorii XVIII veka*. Kn. 1–2 [Monuments of Siberian History of the 18th Century. Bks. 1–2]. St. Petersburg: Tipografiya MVD.

Pallas, P. 1781. "Besondere Nachrichten über die tschuktschische Landspitze und benachbarte Inseln" [Special Reports on the Turkish Headland and Neighboring Islands]. *Neue nordische Beyträge zur physikalischen und geographischen Erd- und Völkerbeschreibung, Naturgeschichte und Oekonomie* [New Northern Contributions to the Physical and Geographic

Descriptions of Lands and Peoples, Natural History, and Economy]. Vol. I. Sec. 2. St. Petersburg; Leipzig. Pp. 245–248.

Paryshev, S. E. 1909. "Uchrezhdenie Chukotskoi missii i obrazovanie Chaunskogo prikhoda" [Establishment of the Chukotka Mission and Formation of the Chaun Parish]. *Pravoslavnyi blagovestnik* [The Orthodox Evangelist]. Moscow: Russkaya pechatnya. No. 15. Pp. 119–126; No. 21. Pp. 396–400; No. 24. Pp. 546–548.

Pavlov, P. N. 1957. *K istorii Rossiisko-Amerikanskoi kompanii: sb. dokumental'nykh materialov* [On the History of the Russian-American Company: Collection of Documentary Materials]. Edited by P. N. Pavlov. Krasnoyarsk: [B. i.].

Pokrovskii, A. A. 1941. *Ekspeditsiya Beringa: sb. dokumentov* [Bering's Expedition: Collection of Documents]. Edited by A. A. Pokrovskii. Moscow: Glavnoe arkhivnoe upravlenie NKVD SSSR.

Polevoi, B. P. 1965. "Nakhodka chelobit'ya pervootkryvatelei Kolymy" [The Find of a Petition of Pioneers of the Kolyma]. *Sibir' perioda feodalizma.* Vyp. 2: *Ekonomika, upravlenie i kul'tura Sibiri XVI–XIX vv.* [Siberian in the Period of Feudalism. Issue 2: The Economics, Management, and Culture of Siberia in the 16th–19th Centuries]. Novosibirsk: Nauka. Pp. 285–291.

________. 1967. "Iz istorii otkrytiya severo-zapadnoi chasti Ameriki" [From the History of Discovery of the Northwestern Part of America]. *Ot Alyaski do Ogonennoi Zemli: Istoriya i etnografiya stran Ameriki* [From Alaska to Tierra del Fuego: History and Ethnography of the Land of America]. Edited by I. R. Grigulevich. Moscow: Nauka. Pp. 107–120.

________. 1972. "Delo o 'bunte' v Yakutske (Po dokumentam Yakutskoi prikaznoi izby)" [The Case of the "Riot" in Yakutsk (Based on Documents of the Yakutsk Mandative Office)]. *Yakutskii arkhiv: sb. statei i dokumentov* [Yakutsk Archive: Collection of Articles and Documents]. Issue 4. Yakutsk: Yakutskoe kn. izd-vo. Pp. 149–157.

________. 1976. "Vkhozhdenie v sostav Rossii chasti vostochnykh chukchei" [Incorporation into the Composition of Russia Parts of the Eastern Chukchi]. *Istoriya i kul'tura narodov Severo-Vostoka SSSR: Materialy 3-i sessii Dal'nevostochnykh istoricheskikh chtenii po problemam istorii Severo-Vostoka SSSR* [The History and Culture of the Peoples of Northeastern USSR: Materials of the 3rd Session of the Far-Eastern Historical Reading on Problems of the History of Northeastern USSR]. Edited by B. I. Mukhachev. Vladivostok: In-t istorii, arkheologii i etnografii narodov Dal'nego Vostoka. Pp. 219–223.

________. 2000. *Novoe ob otkrytii Kamchatki* [New Information on the Discovery of Kamchatka]. Moscow: IPTs "Russkii raritet."

Polomoshnov, I. V. 1977. Khrabraya Toryt: Chukotskie skazki [Brave Toryt: Chukchi Tales]. Edited by I. V. Polomoshnov. Magadan: Magadanskoe knizhnoe izdatel'stvo.

Polonskii, A. S. 1850. "Pokhod geogezista Mikhaila Gvozdeva v Beringov proliv, 1732 goda" [The Expedition of Geodesist Mikhail Gvozdev in Bering Strait,

1732]. *Morskoi sbornik, izdavaemyi ot Morskogo uchenogo komiteta* [Maritime Collection, Issued by the Maritime Scholarly Committee]. St. Petersburg. Vol. IV, No. 11. Pp. 389–402.

Polunin, F. A. 1773. *Geograficheskii leksikon Rossiiskogo gosudarstva. . .* [Geographic Lexicon of the Russian State. . .]. Moscow: Universitetskoe izd-vo.

Polyakov, A. A. 1933. "Chukotskie prazdniki" [Chukchi Holidays]. *Sovetskii Sever* [Soviet North]. Moscow. No. 3. Pp. 102–106.

________. 1983. *Po neprotorennym tropam* [On Unbeaten Paths]. Moscow: Molodaya gvardiya.

Popov, A. A. 1934. "Zatundrenskie krest'yane" [Peasants of the Lower Yenisei]. *SE*. No. 3. Pp. 77–86.

________. 1934a. "Materialy po rodovomu stroyu dolgan" [Materials on the Kin System of the Dolgan]. *SE*. No. 6. Pp. 116–139.

________. 1937. *Dolganskii fol'klor* [Dolgan Folklore]. Leningrad: Sovetskii pisatel'.

________. 1948. *Nganasany* [The Nganasan]. Issue 1. (*TIE*. New series. Vol. 3). Moscow; Leningrad: Izdatel'stvo AN SSSR.

Popov, I. 1893. "Kolymskii krai" [The Kolyma District]. *Sibirskii vestnik* [Siberian Bulletin]. No. 78 (9 July). Pp. 1–2.

Priklonskii, V. L. 1893. *Materialy dlya bibliografii Yakutskoi oblasti*. (Prilozhenie k gaz. *Vostochnoe obozrenie*: *Sibirskii sbornik*. Vyp. 1) [Materials for a Bibliography of the Yakutsk Province. (Appendix to the Newspaper Eastern Review: Siberian Collection. Issue 1). Irkutsk: Tipografiya K. I. Vitkovskogo.

Puzyrev, V. P. 1994. "Kolymskii zatvornik" [The Kolyma Hermit]. *Voprosy istorii* [Questions of History]. Moscow. No. 9. Pp. 157–160.

Radov, I. A. "Nazvaniya chukotskikh rodov Ust'-Bel'skoi tundry" [Names of the Chukchi Clans of the Ust'-Belaya Tundra]. *TB*. Pp. 72–77.

Rasmussen, K. 1952. *The Alaskan Eskimos as Described in the Posthumous Notes of Dr. Knud Rasmussen by H. Ostermann*. Translated from the Danish by W. E. Calvert. (*Report of the Fifth Thule Expedition, 1921–1924*. Vol. X, no 3). Copenhagen: Gyldendalske Boghandel, Nordisk Forlag.

Raushenbakh, V. M. 1969. *Novye nakhodki na Chetyrekhstolbovom ostrove*. (Trudy Gos. Inst. muzeya: Pamyatniki kul'tury. Vyp. 35) [New Finds on Chetyrekhstolbovoe Island. (Works of the State Historical Museum: Monuments of Culture. Issue 35)]. Moscow: Sovetskaya Rossiya.

Ray, D. J. 1964. "Kauwerak, Lost Village of Alaska." *The Beaver* 295(2): 4–13.

Resin, A. A. 1888. "Ocherk inorodtsev russkogo poberezh'ya Tikhogo okeana" [An Essay on Foreigners of the Russian Coast of the Pacific Ocean]. *IIRGO*. Vol. XXIV, issue 3, pp. 121–198.

Reuse W. J., de. 1994. *Siberian Yupik Eskimo: The Language and Its Contacts with Chukchi. (Studies in Indigenous Languages of the Americas)*. Salt Lake City: University of Utah Press.

Robbek, V. A. 2005. *Fol'klor evenkov Berezovki: Obraztsy shedevrov* [Folklore of the Evenk of Berezovka: Examples of Masterpieces]. Yakutsk: In-t problem malochislennykh narodov Severa SO RAN.

Rozhkov, E. F. 2006. "Zapolyarnyi Sindbal-morekhod: Kniga skazanii chukotskogo okhotnika i moreplavatelya oboshla mnogie strany mira" [The Polar Sinbad the Sailor: A Book of Stories of the Chukchi Hunter and Mariner Bypassing Many Countries of the World]. *Nezavisimaya gazeta-Regiony* [The Independent Newspaper—Regions]. Moscow, 4 September. URL: *http://www.ng.ru/ng_region/2006-09-04/20_sindbad.html* (date of access: 8-25-2016).

________. 2008. "Bronezhilet iz mamontovoi kosti" [Armor of Mammoth Ivory]. *Nezavisimoe voennoe obozrenie* [Independent Military Review]. Moscow. 14 March. P. 8.

Rubtsova, E. S. 1954. *Materialy po yazyku i fol'kloreu eskimosov (chaplinskii dialekt)* [Materials on Language and Folklore of the Eskimos (The Chaplino Dialect)]. Pt. I. Moscow; Leningrad: Izdatelstvo AN SSSR.

Rudenko, S. I. 1947. *Drevnyaya kul'tura Beringova morya i eskimosskaya problema*. Moscow; Leningrad: Izd-vo Glavsevmorputi. (Published in English as *The Ancient Culture of the Bering Sea and the Eskimo problem*. Toronto: Arctic Institute of North America, University of Toronto Press, 1961).

________. 1949. "Tatuirovka aziatskikh eskimosov" [Tattooing of the Asiatic Eskimos]. *SE*. No. 1. Pp. 149–154.

Russkaya missiya. 1890. "Russkaya missiya: chukchi" [The Russian Mission: The Chukchi]. *Semeinye vechera: Illyustrirovannyi zhurnal* [Family Evening: An Illustrated Journal]. St. Petersburg. No. 1. Pp. 95–106.

Rytkheu, Yu. S. 2004. *Poslednii shaman* [The Last Shaman]. St. Petersburg: Zhurnal "Zvezda."

________. 2010. *Dorozhnyi leksikon* [Road Lexicon]. St. Petersburg: Zhurnal "Zvezda."

Safronov, F. G. 1975. "Ssylka v Vostochnuyu Sibir' v pervoi polovine XVIII v." [An Exile in Eastern Siberia in the First Half of the 18th Century]. *Ssylka i katorga v Sibiri (XVIII–nachalo XX v.)* [Exile and Hard Labor in Siberia (18th–Beginning 20th Century)]. Novosibirsk: Nauka. Pp. 15–37.

________. 1980. *Russkie promysly i torgi na severovostoke Azii v XVII–seredine XIX v.* [Russian Businesses and Trade in Northeast Asia in the 17th–Middle of the 19th Century]. Moscow: Nauka.

Samoilov, V. A. 1945. *Semen Dezhnev i ego vremya* [Semen Dezhnev and His Time]. Moscow: Izd-vo Glavsevmorputi.

Sarychev, G. A. 1952. *Puteshestvie po severo-vostochnoi chasti Sibiri, Ledovitomu moryu i Vostochnomu okeanu* [Voyage along the Northeast Part of Siberia, Arctic Sea, and the Eastern Ocean]. Moscow: Gosudarstvennoe izdatel'stvo geograficheskoi literatury.

Sauer, M. 1802. *An Account of a Geographical and Astronomical Expedition to the Northern Parts of Russia*. London: A. Strahan.

________. 1802a. *Voyage fait par ordre de l'impératrice de Russie Catherine II, dans le nord de la Russie Asiatique, dans la Mer glaciale, dans la Mer d'Anadyr, et sur les côtes de l'Amérique depuis 1785 jusqu'en 1794 par le commondore Billings* [Voyage Made by Order of the Empress of Russia, Catherine II. . . .] Translated from the English by J. Castéra. Vol. II. Paris: F. Buisson.

Savel'ev, I. V. 2012. "Rannie russkie poseleniya na Alyaske: mif ili real'nost'" [Early Russian Settlements in Alaska: Myth or Reality]. *Vestnik Severnogo (Arkticheskogo) federal'nogo un-ta.* Seriya: *Gumanitarnye i sotsial'nye nauki* [Bulletin of the Northern (Arctic) Federal University. Series: Humanities and Social Sciences]. Arkhangel'sk. No. 5. Pp. 24–30.

Sbornik Imperatorskogo. . . . 1888. *Sbornik Imperatorskogo rossiiskogo istoricheskogo obshchestva* [Collection of the Imperial Russian Historical Society]. Vol. LXIII. *Protokoly, zhurnaly i ukazy Verkhovnogo tainogo soveta, 1726–1730* [Protocols, Journals, and Orders of the Supreme Privy Council, 1726–1730]. St. Petersburg: Tipografiya I. N. Skorokhodova.

Sel'skii, I. S. 1861. "Ssylka v Vostochnuyu Sibir' zamechatel'nykh lits (1645–1762)" [Exile in Eastern Siberia of Notable Personnel]. *Russkoe slovo: Literaturno-khudozhestvennyi zhurnal* [The Russian Word: Literary-Art Journal]. St. Petersburg. (August). Pp. 1–35.

Sementovskii, N. 1852. "Novye cvedeniya o chukchakh" [New Information on the Chukchi]. *VIRGO*. Pt. 5. Bk. 1. Sec. 5. P. 143.

Semivskii, N. G. 1817. *Noveishie lyubopytnye i dostovernye povestvovaniya o Vostochnoi Sibiri, iz chego mnogoe donyne ne bylo vsem izvestno* [The Most Recent Interesting and Authentic Narratives of Eastern Siberia, Much of Which Was Not Known to All up to Now]. St. Petersburg: Voennaya tipografiya.

Senatskii arkhiv. 1889–1904. St. Petersburg: Tipografiya Pravitel'stvuyushchego senata. [translation?] Vol. II, 1889; Vol. V, 1892; Vol. VI, 1893; Vol. VII, 1895; Vol. VIII, 1897; Vol. IX, 1901; Vol. XI, 1904.

Sergeev, O. I., and V. N. Chernavskaya. 1998. "Kazak M. V. Stadukhin—issledovatel' Severo-Vostoka Azii" [Cossack M. V. Stadukhin—Investigator of Northeast Asia]. *Russkie pervoprokhodtsy na Dal'nem Vostoke v XVII–XIX vv. (Istoriko-arkheologicheskie issledovaniya)* [Russian Explorers in the Far East in the 17th–19th Centuries (Historical-Archaeological Investigations)]. Vol. III. Vladivostok: DVO RAN. Pp. 56–69.

Sergeeva, K. S. 1968. *Skazochnik Kivagme* [Storyteller Kivagme]. 2nd edition. Magadan: Magadanskoe knizhnoe izdatel'stvo.

Sgibnev, A. S. 1868. "Bol'shoi kamchatskii naryad" [The Big Kamchatka Outfit]. *Morskoi sbornik* [Maritime Collection]. Vol. 98, No. 12. St. Petersburg. Pp. 131–139.

________. 1869. "Materialy dlya istorii Kamchatki: Ekspeditsiya Shestakova" [Materials for the History of Kamchatka: The Shestakov Expedition]. *Morskoi sbornik* [Maritime Collection]. Vol. 100, No. 2 (February). St. Petersburg. Pp. 1–34 (second pagination).

________. 2005–2008. “Istoricheskii ocherk glavneishikh sobytii v Kamchatke” [A Historical Essay of the Main Events in Kamchatka]. *Voprosy istorii Kamchatki* [Questions of History of Kamchatka]. Petropavlovsk-Kamchatskii: KamchatGTU. Issue 1. 2005. Pp. 5–103; Issue 2. 2008. Pp. 5–123.

Shakhovskoi, A. 1823. “Vzglyad na torgovlyu, proizvodimuyu chrez Okhotskii port” [A Look at Trade Carried Out through the Okhotsk Port]. *Severnyi arkhiv, zhurnal istorii, statistiki i puteshestvii* [Northern Archive, Journal of History, Statistics and Travel]. St. Petersburg. Pt. VII. No. 13. Pp. 28–45.

Shashkov, S. 1864. “Materialy dlya istorii Severo-Vostochnoi Sibiri v XVIII veke” [Materials for the History of Northeastern Siberia in the 18th Century]. *Chteniya v Imp. o-ve istorii i drevnostei rossiiskikh pri moskovskom un-te* [Readings in the Imperial Society of Russian History and Antiquity at Moscow University]. Moscow. Bk. 3. Pp. 62–93 (fifth pagination).

Shchekatov, A., and L. M. Maksimovich. 1809. Slovar’ geograficheskii Rossiiskogo gosudarstva [Geographic Dictionary of the Russian State]. Pt. VII. Moscow: Tipografiya F. Lyubiya.

Shchukin, N. 1852. “Pokhod kapitana Pavlutskogo v Chukotskuyu zemlyu” [The Campaign of Captain Pavlutskii in Chukchi Land]. Sovremennik [Contemporary]. Vol. 34. No. 7. Pp. 1–16 (sixth pagination).

________. 1854. “Pokhod kapitana Pavlutskogo v Chukotskuyu zemlyu” [The Campaign of Captain Pavlutskii in Chukchi Land]. *Zhurnal dlya chteniya vospitannikam voenno-uchebnykh zavedenii*. God XVII [Journal for the Reading to Pupils of Military Schools. Year XVII]. St. Petersburg. Vol. 107. No. 427. Pp. 409–437.

Shentalinskaya, T. S. 2003. “Poslednii konkistador Chukotki: Fol’klornye ipostasi maiora Pavlutskogo, ili Novye fakty po materialam amerikanskoi Severo-Tikhookeanskoi ekspeditsii” [The Last Conquistador of Chukotka: The Folkloric Incarnations of Major Pavlutskii, or New Facts Based on Materials of the American North Pacific Expedition]. *Mir Severa* [World of the North]. Moscow. Nos. 2–3. Pp. 60–63.

________. 2008. “Fol’klornye ipostasi maiora Pavlutskogo (Novye fakty po materialam Severo-Tikhookeanskoi ekspeditsii)” [The Folkloric Incarnations of Major Pavlutskii (New Facts Based on Materials of the North Pacific Expedition)]. *TB*. Pp. 262–274.

________. 2009. “Fol’klor chuvantsev” [Folklore of the Chuvan People]. *EO*. No. 3. Pp. 60–78.

________. 2012. “Sof’ya Bogoraz—avtor zapisei russkogo fol’klora na Chukotke” [Sofya Bogoraz—Author of Notations of Russian Folklore in Chukotka]. *EO*. No. 1. Pp. 110–120.

Shentalinskii, V. 1978. “Zapovednaya osen’ na Vrangele” [Reserved Autumn on Wrangel]. *Vokrug sveta* [Around the World]. Moscow. No. 9. Pp. 42–50.

Sheppard, W. 2009. “Assault on Qawiaraq: Siberian-Alaskan Warfare in Historical and Cultural Context.” *Chasing the Dark: Perspectives on Place, History, and Alaska Native Land Claims*. Ed. by K. L. Pratt. Vol. I.

Anchorage: U. S. Department of the Interior, Bureau of Indian Affairs, Division of Environmental and Cultural Resources Management, ANCSA Office. Pp. 46–63.

Schiefner, F. A. 1874. "Baron Gerhard von Maydell's Tungusische Sprachproben" [Baron Gerhard von Maydell's Tungus Speech Samples]. *Mélanges asiatiques* [Asian Blends]. Vol. VII. Livraisons 2–3. St. Pétersburg. Pp. 323–377.

Shirokov, Yu. A. 1968. "Gravirovannye klyki v sobranii Gos. muzeya revolyutsii" [Engraved Tusks in the Collection of the State Museum of Revolution]. *ZChKM*. Issue 5. Pp. 87–90.

Shishmarev, G. S. 1852. "Svedeniya o chukchakh kapitana Shishmareva 1821 goda" [Information about the Chukchi of Captain Shishmarev of 1821]. *ZGDMM*. Pt. 10. Pp. 178–200.

Shklovskii, I. V. 1892. *Ocherki krainego Severo-Vostoka*. (*ZVSOIRGO po obshchei geografii*. T. II, vyp. 2) [Essays of the Extreme Northeast. (ZVSOIRGO on General Geography. Vol. II, issue 2)]. Irkutsk: Tipografiya shtaba Irkutskogo voennogo okruga.

Shurkhalo, D. 2004. "Narod-anekdot" [People-Anecdote]. *Tribuna* [Tribune]. 20 November. URL: *http://archive.is/hYAxi* (date accessed: 4 September 2016).

Shvaittser [Schweitzer], P., and E. V. Golovko. 2001. "Pamyat' o voine: konstruirovanie vneshnego konflikta v kul'ture etnicheskikh obshchnostei Beringova proliva" [Memory of War: Construction of External Conflict in the Culture of Ethic Communities of Bering Strait]. *Trudy fakul'teta etnologii Evropeiskogo un-ta v Sankt Peterburge* [Works of the Department of Ethnology of the European University in St. Petersburg]. Issue 1. St. Petersburg. Pp. 26–37.

Simchenko, Yu. B. 1970. "Osobennosti sotsial'noi organizatsii paleoaziatov krainego Severo-Vostoka Sibiri (koryaki, chukchi, itel'meny, eskimosy)" [Features of Social Organization of the Paleoasiatics of Extreme Northeast Siberia (Koryak, Chukchi, Itel'men, and Eskimos)]. *Obshchestvennyi stroi u narodov Severnoi Sibiri (XVII–nachalo XX v.)* [Social System of the Peoples of Northern Siberia (17th–Beginning of the 20th Century)]. Edited by I. S. Gurvich and B. O. Dolgikh. Moscow: Nauka. Pp. 313–331.

________. 1976. *Kul'tura okhotnikov na olenya Severnoi Evrazii* [The Culture of Reindeer Hunters of Northern Eurasia. Moscow: Nauka.

Slëzkin, Yu. 2008. *Arkticheskie zerkala: Rossiya i malye narody Severa*. Moscow: Novoe literaturnoe obozrenie. Authorized translation from English by O. Leont'eva. (Historia Rossica). (Published in English as *Arctic Mirrors: Russia and the Small Peoples of the North*. Cornell University Press, 1996.)

Slovtsov, P. A. 1854. "Svedeniya o chukotskoi zemle v XVIII stoletii" [Information on the Land of the Chukchi in the 18th Century]. *VIRGO*. Pt. 9 (1853), sec. VIII: "Smes'". Pp. 12–14.

________. 1856. "Istoricheskaya zapiska o chukotskom narode, obitayushchem okolo beregov Ledovitogo morya" [A Historical Note on the Chukchi People

Living near the Shores of the Arctic Ocean]. *VIRGO*. Pt. 18, vol. 5. sec. 5: "Smes'". Pp. 19–25.

________. 1886. *Istoricheskoe obozrenie Sibiri* [A Historical Review of Siberia]. Bks. 1–2. St. Petersburg: Tipografiya I. N. Skorokhodova.

Slyunin, N. V. 1895. "Sredi chukchei" [Among the Chukchi]. *Zemlevedenie* [Geography]. Bk. 4. Moscow. Pp. 1–46.

________. 1900. *Okhotsk-Kamchatskii krai* [The Okhotsk-Kamchatka Region]. Vols. I–II. St. Petersburg: Tipografiya A. S. Suvorina.

Smirnov, Yu. I. 1991. *Russkaya epicheskaya poeziya Sibiri i Dal'nego Vostoka* [Russian Epic Poetry of Siberia and the Far East]. Compiled by Yu. I. Smirnov. (*Pamyatniki fol'klora narodov Sibiri i Dal'nego Vostoka* [Monuments of Folklore of the Peoples of Siberia and the Far East]). Novosibirsk.

Smith, V. L. 1970. "Intercontinental Aboriginal Trade in the Beringian Straits Area." *Proceedings of the 8th Congress of Anthropological and Ethnological Science, 1968*. Tokyo and Kyoto. Ed. by B. Endo, H. Hoshi, and Sh. Masuda. Vol. 2. Tokyo: Science Council of Japan. Pp. 236–238.

Sokolov, A. P. 1851. "Pervyi pokhod russkikh k Amerike 1732 goda" [The First Expedition of Russians to America in 1732]. *ZGDMM*. Pt. 9. Pp. 78–107.

________. 1851a. *Severnaya ekspeditsiya 1733–1743* [The Northern Expedition of 1733–1743]. St. Petersburg: [B. i.].

________. 1852. "Ekspeditsiya na Medvezh'i ostrova serzhanta Andreeva, 1763–4" [The Expedition to Medvezhii Island of Sergeant Andreev, 1763–4] *ZGDMM*. Pt. 10. Pp. 104–118.

________. 1852a. "Reka Anadyr' po opisaniyam Lapteva i Plenisnera, 1742 i 1763" [The Anadyr' River Based on Descriptions of Laptev and Plenisner, 1742 and 1763]. *ZGDMM*. Pt. 10. Pp. 160–166.

Sredi chukchei. 1899. "Sredi chukchei" [Among the Chukchi]. *VO*. No. 38 (18 February). P. 3; No. 39 (19 February). P. 2.

Stebnitskii, S. N. 1934. "Nymylanskii (koryatskii) yazyk" [The Nymylanskii (Koryak) Language]. *Yazyki i pis'mennost' narodov Severa* [Languages and Writing of the Peoples of the North]. Pt. III. *Yazyki i pis'mennost' paleoaziatskikh narodov.* (*Nauchno-issledovatel'skaya assotsiatsiya In-ta narodov Severa TsIK SSSR. Trudy po lingvistike*. T. III) [Languages and Writing of the Paleoasiatic Peoples. (Science-Research Association of the Institute of Peoples of the North TsIK SSR. Works on Linguistics. Vol. III)]. Moscow; Leningrad: OGIZ. Pp. 47–84.

________. 1938. "Nymylany-alyutortsy: K voprosu o proiskhozhdenii olenevodstva u yuzhnykh koryakov" [The Nymylany-Alyutor People: On the Question of the Origin of Reindeer Herding among the Southern Koryak]. *SE*. Issue 1. Pp. 129–144.

________. 1994. "Koryatskii istoricheskii fol'klor i zarozhdayushchayasya koryatskaya literatura" [Koryak Historical Folklore and Emerging Koryak Literature]. In Stebnitskii, S. N. *Ocherki po yazyku i fol'kloru koryakov*

[Essays on the Language and Folklore of the Koryak]. St. Petersburg: MAE RAN. Pp. 18–101.

Steller, G. W. 1995. *Dnevnik plavaniya s Beringom k beregam Ameriki, 1741–1742* [Diary of a Voyage with Bering to the Shores of Amerika, 1741–1742]. Translated by E. L. Stanyukovich. Moscow: PAN.

Stepanov, N. N. 1957. "Malye narody Yakutii v 1861–1917 gg." [Ethnic Peoples of Yakutia in 1861–1917]. *Istoriya Yakutskoi ASSR* [The History of Yakutia ASSR]. Vol. II. Moscow: Izd-vo AN SSSR. Pp. 324–337.

Stolknovenie . . . 1896. "Stolknovenie mezhdu chukchami i russkimi" [Clashes between the Chukchi and the Russians]. *Tomskii listok* [Tomsk Leaf]. No. 93 (2 May). P. 2.

Strelov, E. D. 1925, "Materialy k istorii chukoch" [Materials on the History of the Chukchi]. *Sb. trudov issledovatel'skogo obshchestva "Saqa Keskile"* [Collection of Works of the "Saqa Keskile" Research Society]. Issue 1 (August). Yakutsk: [B. i.]. Pp. 13–19.

Sverdlov, L. M. 1992. "Russkoe poselenie na Alyaske v XVII v." [A Russian Settlement in Alaska in the 17th Century]. *Priroda* [Nature]. No. 4. Pp. 67–69.

Sverdrup, G. U. 1930. *Plavanie na sudne "Mod" v vodakh morya Laptevykh i Vostochno-Sibirskogo.* (*Materialy Komissii po izucheniyu Yakutskoi avtonomnoi sovetskoi sotsialisticheskoi respubliki.* Vyp. 30) [The Voyage of the Ship *Mod* in the Waters of the Laptev and East Siberian Seas. (Materials of the Commission on the Study of the Yakutsk Autonomous Soviet Socialist Republic. Issue 30)]. Leningrad: Izdatel'stvo AN SSSR.

Syroechkovskii, E. E. 1986. *Severnyi olen'* [Northern Reindeer]. Moscow: Agropromizdat.

Takakava, M. K. 1974. *Kto samyi sil'nyi na Zemle? Chukotskie skazki* [Who's the Strongest on Earth? Chukchi Stories]. Compiled by M. K. Takakava. Magadan: Magadanskoe knizhnoe izdatel'stvo.

Tan, N. A. 1898. "Iz zapisnoi knizhki" [From the Notebook]. *VO*. Year XVII. No. 114 (27 September). Pp. 2–3; No. 116 (2 October). P. 2; No. 118 (7 October). Pp. 2–3.

Tan-Bogoraz, V. G. 1930. "Chukotskii obshchestvennyi stroi po dannym fol'klora" [Chukchi Social Structure Based on Data from Folklore]. *Sovetskii Sever: Obshchestvenno-nauchnyi zhurnal* [The Soviet North: Social Science Journal]. Moscow. No. 6. Pp. 63–79.

________. 1935. "Drevneishie elementy v yazyke aziatskikh eskimosov" [The Earliest Elements in the Language of the Asiatic Eskimos]. *Akademiya nauk SSSR: XLV akademiku N. Ya. Marru* [Academy of Sciences of the USSR: 45th of Academician N. Ya. Marr]. Moscow; Leningrad: Izdatelstvo AN SSSR. Pp. 353–366.

Tein, T. S. 1983. *Taina Chertova Ovraga* [The Mystery of Devils Gorge]. Magadan: Magadanskoe knizhnoe izdatel'stvo.

________. 1991. Drevneeskimosskaya kul'tura Severnoi Chukotki, vklyuchaya ostrova Vrangelya i Ratmanova [The Old Eskimo Culture of Northern

Chukotka, Including Wrangel and Ratmanova Islands]. Abstract of dissertation for candidate in historical sciences. Leningrad: [B. i.].

________. 1992. *I bylo tak. . . . Ocherki traditsionnoi kul'tury aziatskikh eskimosov* [Thus It Was. . . . Essays on the Traditional Culture of the Asiatic Eskimos]. Magadan: Magadanskii oblastnoi tsentr narodnogo tvorchestva i dosuga.

Tekki Odulok. 1933. *Na krainem Severe* [In the Extreme North]. Leningrad: Molodaya gvardiya.

Tevlyankau, E. M. 2008. "Obrazovanie sela Lavrentiya" [Education of Lavrentiya Village]. *Aiverette: repertuarno-metodicheskii sb. Vyp. 12* [Aiverette: Repertory-Methodical Collection Issue 12]. Anadyr'. Pp. 16–17.

________. 2011. "Prazdnik 'Keretkun'" [The "Keretkun" Festival]. *Aiverette: repertuarno-metodicheskii sb. Vyp. 14* [Aiverette: Repertory-Methodical Collection Issue 14]. Anadyr'. Pp. 48–49.

Tikhmenev, P. A. 1861. *Istoricheskoe obozrenie obrazovaniya Rossiisko-Amerikanskoi kompanii i deistvii ee do nastoyashchego vremeni* [A Historical Review of the Formation of the Russian-American Company and Its Activities up to the Present Time]. Pt. I. St. Petersburg: Tipografiya E. Veimara.

Tokarev, S. A. 1948. "Neskol'ko dannykh XVII veka o yakutakh i ikh sosedyakh" [Some Data of the 17th Century on the Yakut and Their Neighbors]. *Sb. materialov po etnografii yakutov* [Collection of Materials Based on the Ethnography of the Yakut]. Yakutsk: Yakutskoe kn. izd-vo. Pp. 6–16.

Tolmachev, I. P. 1911. *Po Chukotskomu poberezh'yu Ledovitogo okeana* [Along the Chukotka Coast of the Arctic Ocean]. St. Petersburg: Ekonomicheskaya tipolitografiya.

Tret'yakov, P. I. 1869. "Turukhanskii krai" [The Turukhansk Territory]. *ZIRGO po obshchei geografii* [ZIRGO on General Geography]. Vol. 2. St. Petersburg. Pp. 215–530.

Trifonov, A. 1872. "Zametki o Nizhne-Kolymske" [Notes on Nizhne-Kolymsk]. *ISOIRGO*. Vol. III. No. 3. Pp. 160–166.

Tugolukov, V. A. 1979. Kto vy, yukagiry? [Who Are You, Yukaghir?]. Moscow: Nauka.

________. 1986. "Predaniya ob ukhode yukagirov na ostrova i 'za more'" [Legends about the Departure of the Yukaghir to the Islands and "beyond the Sea"]. *SE*. No. 6. Pp. 94–100.

Tunkina, I. V. 2013. "Ekspeditsiya D. G. Messershmidta 1719–1727 gg. v Sibir' v rossiiskoi istoriko-nauchnoi istoriografii XX–nachala XXI v." [The Expedition of D. G. Messerschmidt in 1719–1727 in Siberia in Russian Historical-Scientific Historiography of the 20th–Beginning 21st Century]. *Millerovskie chteniya: K 285-letiyu Arkhiva Rossiiskoi akademii nauk: sb. nauch. st. po materialam Mezhdunar. nauch. konf. 23–25 aprelya 2013 g., Sankt-Peterburg* [The Miller Reading: On the 285th Year of the Archive of the Russian Academy of Sciences: Collection of Science Articles Based on Materials of the International Science Conference of 23–25 April 2013, St.

Petersburg]. Edited by I. V. Tunkina. (Ad fonts: *Materialy i issledovaniya po istorii nauki.* Vyp. 4 [Ad fonts: Materials and Research on the History of Science. Issue 4]). St. Petersburg: Nestor-Istoriya. Pp. 238–250.

Tyneru, E. I. 2005. "Kak proizoshli nazvaniya gor, rek" [How the Names of Mountains and Rivers Originated]. *Aiverette: repertuarno-metodicheskii sb. Vyp. 11* [Aiverette: Repertory-Methodical Collection Issue 11]. Anadyr'. P. 59.

________. 2005a. "Potomki Kunlelyu" [Descendants of Kunlelyu]. *Aiverette: repertuarno-metodicheskii sb. Vyp. 11* [Aiverette: Repertory-Methodical Collection Issue 11]. Anadyr'. Pp. 49–50.

________. 2005b. "Khatkana" [Khatkana]. *Aiverette: repertuarno-metodicheskii sb. Vyp. 11* [Aiverette: Repertory-Methodical Collection Issue 11]. Anadyr'. Pp. 86–89.

________. 2008. "Legendy Chaunskogo raiona" [Legends of the Chaun Region]. *Aiverette: repertuarno-metodicheskii sb. Vyp. 12* [Aiverette: Repertory-Methodical Collection Issue 12]. Anadyr'. P. 9.

________. 2009. "V. I. Tynagirgin" [V. I. Tynagirgin]. *Aiverette: repertuarno-metodicheskii sb. Vyp. 13* [Aiverette: Repertory-Methodical Collection Issue 13]. Anadyr'. Pp. 50–51.

Ushakov, G. A. 1972. *Ostrov metelei* [Island of Snowstorms]. Leningrad: Gidrometeorologicheskoe izdatel'stvo.

Vakhtin, N. B. 2000. *Yazyk sirenikskikh eskimosov: teksty, grammaticheskie i slovarnye materialy* [The Language of the Sireniki Eskimos: Texts and Grammatical and Vocabulary Materials]. (Lincom Studies in Asian Linguistics. 33). München: Lincom Europa.

Vakhtin, N. B., E. V. Golovko, and P. Shvaittser [Schweitzer]. 2004. *Russkie starozhily Sibiri: Sotsial'nye i simvolicheskie aspekty samosoznaniya* [Russian Old Dwellers of Siberia: Social and Symbolic Aspects of Identity]. Moscow: Novoe izdatel'stvo.

Val'skaya, B. A. 1961. "Puteshestvie Andreya Argentova na severovostok Sibiri v 1851 godu" [The Journey of Andrei Argentov in Northeastern Siberia in 1851]. *Strany i narody Vostoka* [Countries and Peoples of the East]. Moscow: Vostochnaya litiratura. Issue II, pp. 172–187.

Vdovin, I. S. 1943. "Chertezhi Chukotki 1742 i 1746 gg." [Sketches of Chukotka 1742 and 1746]. *IVGO* 75(4): 52–54.

________. 1944. "Rasselenie narodnostei Severo-Vostoka Azii vo vtoroi polovine XVII i XVIII vv." [Resettlement of the Peoples of Northeast Asia in the Second Half of the 17th and 18th Century]. *IVGO* 76(5): 250–265.

________. 1950. "K istorii obshchestvennogo stroya chukchei" [On the History of the Chukchi Social System]. *Uchen. zap. LGU* [Scholarly Notes of LGU], no. 115: Department of Peoples of the North. Issue 1: *Istoriya i ekonomika Krainego Severa SSSR* [The History and Economy of the Extreme North of the USSR]. Leningrad. Pp. 73–100.

________. 1959. "Anadyrskii ostrog: istoricheskii ocherk" [The Anadyrsk Stockade: A Historical Outline]. KZOKM 2: 20–27.

________. 1960. “Anadyrskii ostrog: istoricheskii ocherk” [The Anadyrsk Stockade: A Historical Outline]. *KZOKM* 3: 31–48.

________. 1961. “Eskimosskie elementy v kul’ture chukchei i koryakov” [Eskimo Elements in the Culture of the Chukchi and Koryak]. *SES*, Issue III (TIE, vol. 64). Pp. 27–63. Moscow; Leningrad.

________. 1962. “Vaegskie chukchi” [The Vaegskie Chukchi]. *SES*, Issue IV (TIE, new series, vol. 28). Pp. 153–164.

________. 1964. “Iz istorii otnoshenii chukchei i eskimosov Alyaski” [From the History of Relations of the Chukchi and Eskimos of Alaska]. *VII Mezhdunar. kongress antropologicheskikh i etnograficheskikh nauk (Moskva, avgust 1964)* [7th International Congress of Anthropological and Ethnographic Sciences (Moscow, August 1964)]. Moscow: Nauka.

________. 1964a. “Torgovye svyazi naseleniya severo-vostoka Sibiri i Alyaski (do nachala XX v.)” [Trade Connections of the Population of Northeastern Siberia and Alaska (To the Beginning of the 20th Century)]. *Letopis’ Severa* [Chronicle of the North]. Vol. IV, pp. 117–127. Moscow: Mysl’.

________. 1965. *Ocherki istorii i ethnografii chukchei* [Essays on the History and Ethnography of the Chukchi]. Moscow; Leningrad: Nauka.

________. 1970. “O sootnoshenii fol’klora s istoriko-etnograficheskimi dannymi” [On the Correlation of Folklore with Historical-Ethnographic Data]. *Fol’klor i etnografiya* [Folklore and Ethnography]. Edited by B. N. Putilov. Pp. 16–24. Leningrad: Nauka.

________. 1971. “Eskimosskie elementy v kul’ture chukchei i koryakov” [Eskimo Elements in the Culture of the Chukchi and Koryak]. *SES*. Vol. III (TIE, vol. LXIV). Pp. 27–63. Leningrad; Moscow.

________. 1972. “Yukagiry v etnicheskoi istorii koryakov i chukchei” [The Yukaghir in the Ethnic History of the Koryak and Chukchi]. *Etnicheskaya istoriya narodov Azii* [The Ethnic History of the Peoples of Asia]. Edited by S. M. Abramzon and R. F. Its. Pp. 99–112. Moscow: Nauka.

________. 1973. *Ocherki etnicheskoi istorii koryakov* [Essays on the Ethnic History of the Koryak]. Leningrad: Nauka.

________. 1975. “Istoricheskie osobennosti formirovaniya obshchestvennogo razdeleniya truda narodov Severo-Vostoka Sibiri” [The Historical Features of the Formation of Social Division of Labor of Peoples of Northeastern Siberia]. *Sotsial’naya istoriya narodov Azii* [The Social History of Peoples of Asia]. Edited byA. M. Reshetov and Ch. M. Taksami. Moscow: Nauka. Pp. 143–157.

________. 1977. “Religioznye kul’ty chukchei” [Religious Cults of the Chukchi]. *Pamyatniki kul’tury narodov Sibiri i Severa (vtoraya polovina XIX–nachalo XX v.)* [Monuments of the Culture of Peoples of Siberia and the North (Second Half of the 19th–Beginning of 20th Century]. Edited by I. S. Vdovin. (SMAE. Issue 33). Leningrad: Nauka. Pp. 117–171.

________. 1987. “Istoriya i kul’tura chukchei v dooktyabr’skii period–s XVII v do 1917 g.” [The History and Culture of the Chukchi in the Pre-October Period—from the 17th Century to 1917]. *Istoriya i kul’tura chukchei:*

istoriko-etnograficheskie ocherki [The History and Culture of the Chukchi: Historical-Ethnographic Essays]. Edited by A. I. Krushanova. Leningrad: Nauka. Pp. 31–146.

________. 2010. "Chukchi. Etnogenez i etnicheskaya istoriya" [Chukchi. Ethnogenesis and Ethnic History]. *Narody Severo-Vostoka Sibiri* [Peoples of Northeastern Siberia]. Edited by E. P. Bat'yanova and V. I. Turaev. (*Narody i kul'tury* [Peoples and Cultures]). Moscow: Nauka. Pp. 510–516.

Vize, V. Yu. 1933. "Zemlya Andreeva" [Andreev Land]. *Arctica.* Bk. I. Leningrad: Izd-vo Arkticheskogo in-ta. Pp. 7–30.

________. 1948. *Russkie polyarnye morekhody iz promyshlennykh, torgovykh i sluzhilykh lyudei XVII–XIX vv.: Biograficheskii slovar'* [Russian Polar Mariners of the Hunting, Trading, and Service People of the 17th–19th Centuries: A Biographic Dictionary]. Moscow; Leningrad: Izdatel'stvo Glavsevmorputi.

Vladimirov, I. N., and M. I. Tsiporukha. 1992. *Chelovek stroit korabl': Ocherki po istorii sudostroeniya i morekhodstva (ot istokov do XVII v.)* [Man Builds Ship: Essays on the History of Ship Construction and Seafaring (From the Origins to the 17th Century)]. Moscow: Nauka.

Vol'vson, A. G. 1991. *Proiskhozhdenie chukotsko-koryakskoi kul'tury olenevodstva* [The Origin of the Chukchi-Koryak Culture of Reindeer Herding]. Vladivostok: DVO AN SSSR.

Voskoboinikov, M. G., and G. A. Menovshchikov. 1951. *Skazki narodov Severa* [Stories of the Peoples of the North]. Compiled by M. G. Voskoboinikov and G. A. Menovshchikov. Moscow; Leningrad: Khudozhestvennaya literatura.

________. 1959. *Skazki narodov Severa* [Stories of the Peoples of the North]. Edited by M. G. Voskoboinikov and G. A. Menovshchikov. Moscow; Leningrad: Khudozhestvennaya literatura.

Wahlen, A. 1844. *Moeurs, usages et costumes de tout les peuples du monde* [Manners, Customs, and Costumes of All the Peoples of the World]. Vol. III. Bruxelles: La Librairie historique artistique.

Wardwell, A. 1986. *Ancient Eskimo Ivories of the Bering Strait.* New York: Hudson Hills Press.

Weinstein, C. 2004. *The Chukchi Myths and Folklore Texts Collected by Vladimir Bogoraz.* Re-transcribed, checked, and translated into French by Ch. Weinstein. (Endangered Languages of the Pacific Rim, A2–046). Kyoto: Nakanishi Printing Co.

Whymper, F. 1869. *Travel and Adventure in the Territory of Alaska.* New York: Harper and Brothers.

Wrangell, F., von. 1835. "O torgovykh snosheniyakh norodov Severo-Zapadnoi Ameriki mezhdusoboi i s chukchami" [On Trade Relations of the Peoples of Northwestern America between Themselves and the Chukchi]. *Teleskop: Zhurnal sovremennogo prosveshcheniya* [Telescope: A Journal of Modern Education]. Pt. 26, pp. 604–613. Moscow.

________. 1839. *Statistische und ethnographische Nachrichten über die Russischen Besitzungen an der Nordwestküste von Amerika. (Beiträge zur*

Kentniss des Russischen Reiches und der angränzenden Länder Asiens. Bd. I) [Statistical and Ethnographical Information about the Russian Possessions on the Northwest Coast of America. (Contributions to the knowledge of the Russian Empire and the Adjoining Countries of Asia]. St. Petersburg: Buchdruckerei der Kaiserlichen Akademie der Wissenschaften.

________. 1846. "Okhota za olenyami" [Reindeer Hunting]. *Zhurnal dlya chteniya vospitannikam voenno-uchebnykh zavedenii* [Journal for Reading to Pupils of Military-Educational Institutions]. Vol. 63, no. 249, pp. 94–99. St. Petersburg.

________. 1848. *Puteshestvie po severnym beregam Sibiri i po Ledovitomu moryu, sovershennoe v 1820, 1821, 1822, 1823 i 1824 gg.* [A Voyage along the Northern Shores of Siberia and along the Arctic Ocean, Carried out in 1820, 1821, 1822, 1823, and 1824]. 2nd ed. Moscow: Izdatel'stvo Glavsevmorputi.

Yanikov, G. V. 1949. *Velikaya severnaya ekspeditsiya* [The Great Northern Expedition]. Moscow: Gos. izd-vo geograficheskoi literatury.

Yatgyrgyn, V. V. 1963. *Lymn'ylte eigysk'ykin.* [translation?]. Magadan: Magadanskoe knizhnoe izdatel'stvo.

Zadorin, V. I. 2009. "Triumf i padenie epokhi Antonina Kymytval' i ee chukotskogo naroda" [The Rise and Fall of the Epic of Antoninu Kumutval' and Her Chukchi People]. Severo-Vostochnyi nauchnyi zhurnal [The Northeastern Science Journal]. No. 1, pp. 11–20. Magadan.

________. 2009a. "'Esli olen' ot cheloveka ukhodit, vmeste s olenem ukhodit schast'e . . .'" ["If the Reindeer Leaves a Person, Together with the Reindeer Goes Good Fortune . . ."]. *Chukotka v proshlom i nastoyashchem* [Chukotka in the Past and Present]. Edited by A. M. Tarunov. (*Nasledie narodov Rossiiskoi Federatsii.* Vysh. 11 [The Heritage of Peoples of the Russian Federation. Issue 11). Moscow: Nauchno-informatsionnyi izdatel'skii tsentr. Pp. 118–127.

Zagoskin, L. A. 1956. *Puteshestviya i issledovaniya leitenanta Lavrentiya Zagoskina v Russkoi Amerike v 1842-1844 gg.* Moscow: Gos. izd-vo geograficheskoi literatury. [Published in English as *Lieutenant Zagoskin's Travels in Russian America, 1842–1844*. Arctic Institute of North America: University of Toronto Press, 1967.]

Zamechanie o chukchakh. 1835. "Zamechanie o chukchakh" [Comment on the Chukchi]. *Sankt-Peterburgskie vedomosti* [St. Petersburg Gazette]. No. 120 (31 May). Pp. 483–485; No. 121 (1 June). Pp. 487–489.

Zametki o chukachakh. 1877. "Zametki o chukachakh" [Notes on the Chukchi]. *Sibir': Ezhenedel'naya gazeta* [Weekly Gazette]. Irkutsk. No. 12 (20 March). Pp. 4–5.

Zapiska o chukotskom narode. 1858. "Zapiska o chukotskom narode, obitayushchem okolo beregov Ledovitogo morya, iz imeyushchikhsya v Senatskom sekretanom arkhive svedenii" [A Note on the Chukchi People Who Live near the Shores of the Arctic Ocean, from Information Available in the Secret Senate Archive]. *Chteniya v Imp. o-ve istorii i drevnostei rossiiskikh pri Moskovskom universitete* [Readings of the Imperial Society of History and Russian Antiquities of Moscow University]. Bk. 4, pp. 100–108 (fifth pagination). Moscow.

Zavadskii-Krasnopol'skii, A. K. 1873. *Russkoe tsarstvo: Koryako-chukotskii krai i Kamchatka* [Russian Tsarism: The Koryak-Chukchi Region and Kamchatka]. St. Petersburg: Knizhnyi magazin dlya inogorodnikh.

Zhikharev, N. A. 1992. *Povest' ob Afanasii D'yachkove* [The Story of Afanasii D'yachkov]. Magadan: Magadanskoe knizhnoe izdatelstvo. Pp. 5-159.

Zhukova, A. N. 1988. *Materialy i issledovaniya po koryakskomu yazyku* [Materials and Investigations on the Koryak Language]. Leningrad: Nauka.

Zolotarev, A. M. 1938. "Iz istorii etnicheskikh vzaimootnoshenii na severo-vostoke Azii" [From the History of Ethnic Relations in Northeast Asia]. *Izv. Voronezhskogo pedagogicheskogo in-ta.* T. IV: *Nauchnye raboty molodykh uchenykh* [News of the Voronezh Pedagogical Institute. Vol. IV. Scientific Works of Young Scholars]. Voronezh. Pp. 73–87.

Zonov, B. V. 1931. Opisanie basseina reki Omolon. (Materialy partii po issledovaniyu reki Kolymy. Vyp. IV) [A Description of the Omolon River Basin. (Materials of the Party on Investigation of the Kolyma River. Issue IV)]. Irkutsk: [B. i.].

Zubov, N. N., and K. S. Badigin. 1953. *Razgadka tainy zemli Andreeva* [Unraveling the Secrets of Andreev Land]. Moscow: Voenno-morskoe izdatel'stvo.

Zuev, A. S. 1997. "Russko-chukotskii mirnyi dogovor 1778 g." [Russian-Chukchi Peace Negotiations of 1778]. *Gumanitarnye issledovaniya: itogi poslednikh let: sb. tezisov nauch. konf., posvyashchennoi 35-letiyu gumanitarnogo fakul'teta NGU* [Humanities Research: The Results of Recent Years: Collection of Theses of the Scientific Conference Dedicated to the 35th Year of the Humanities Faculty of NGU]. Novosibirsk: Izdatel'stvo NII MIOO NGU. Pp. 76–79.

________. 1998. "'Nemirnykh chukchei iskorenit' vovse'" ["To Eradicate the Non-Peaceful Chukchi Altogether"]. *Rodina: Rossiiskii istoricheskii illyustrirovannyi zhurnal* [Homeland: An Illustrated Russian Historical Journal]. No. 1, pp. 38–41.

________. 2001. "Pokhod D. I. Pavlutskogo na Chukotku v 1731 g." [The Campaign of D. I. Pavlutskii in Chukotka in 1731]. *Aktual'nye problemy sotsial'no-politicheskoi istorii Sibiri (XVII–XX veka): Bakhrushinskie chteniya 1998 g.: mezhvuzovskii sb. nauch. trudov* [Topical Problems of the Sociopolitical History of Siberia (17th–20th Century): Bakhrushin Readings of 1998: Interuniversity Collection of Scholarly Works]. Edited by V. I. Shishkin. Novosibirsk: Novosibirskii gosudarstvennyi universitet. Pp. 3–38.

________. 2001a. "'Konkvistadory imperii': russkie zemleprokhodtsy na Severo-Vostoke Sibiri" ["Conquistadores of the Empire": Russian Explorers in Northeastern Siberia]. *Ab imperio. Kazan'*. No. 4, pp. 81–108.

________. 2002. *Russkie i aborigeny na krainem severo-vostoke Sibiri vo vtoroi polovine XVII–pervoi polovine XVIII v.* (*Trudy Gumanitarnogo fakul'teta*

NGU. Seriya I) [Russian and Aborigines in Extreme Northeastern Siberia in the Second Half of the 17th–First Half of the 18th Century. (Works of the Humanities Faculty at NGU. Series I)]. Novosibirsk: Novosibirskii gos. un-t.

________. 2002a. "Nachalo deyatel'nosti Anadyrskoi partii i russkokoryakskie otnosheniya v 1730-kh godakh" [The Beginning of Action of the Anadyrsk Party and Russian-Koryak Relations in the 1730s]. *Sibir' v XVII–XX vekakh: Problemy politicheskoi i sotsial'noi istorii: Bakhrushinskie chteniya 1999–2000 gg.: mezhvuzovskii sb. nauch. trudov* [Siberia in the 17th–20th Centuries: Problems of Political and Social History: Bakhrushin Readings of 1999–2000: Interuniversity Collection of Scholarly Works]. Edited by V. I. Shishkin. Novosibirsk: Novosibirskii gos. un-t. Pp. 53–82.

________. 2003. "Promemorii D. I. Pavlutskogo o deyatel'nosti Anadyrskoi partii v nachale 1730-kh gg." [D. I. Pavlutskii's Official Paper on the Activity of the Anadyrsk Party at the Beginning of the 1730s]. *Vestnik Novosibirskogo gos. un-ta*. T. 2. *Seriya: Istoriya, filologiya*. Vyp. 2: *Istoriya* [Bulletin of the Novosibirsk State University. Vol. 2. Series: History and Philology. Issue 2: History]. Vol. 2. Pp. 133–139.

________. 2003a. "T. I. Shmalev kak issledovatel' russko-aborigennykh otnoshenii na Severo-Vostoke Sibiri v XVIII v." [T. I. Shmalev as an Investigator of Russian-Native Relations in Northeastern Siberia in the 18th Century]. *Problemy istorii, istoriografii i istochnikovedeniya Rossii XII–XX vv.* [Problems of History, Historiography, and Source Studies of Russia in the 12th–20th Centuries]. Edited by A. N. Zheravin. Tomsk: Tomskii gos. un-t. Pp. 78–84.

________. 2004. "Iz istorii prisoedineniya Severo-Vostochnoi Sibiri: Russko-koryakskie otnosheniya v XVII–XVIII vekakh" [From the History of the Annexation of Northeastern Siberia: Russian-Koryak Relations in the 12th–18th Centuries]. *Yakutiya—forpost osvoeniya Severo-Vostoka Sibiri, Dal'nego Vostoka i Russkoi Ameriki (XVII–XX veka)* [Yakutia—Outpost in the Development of Northeastern Siberia, the Far East, and Russian America (17th–20th Century)]. Edited by V. N. Ivanov. Yakutsk: YaF izd-va SO RAN. Pp. 59–81.

________. 2004a. "Pokhody D. I. Pavlutskogo na Chukotku v 1744–1747 gg." [The Campaigns of D. I. Pavlutskii in Chukotka in 1744–1747]. *Voennoe delo narodov Sibiri i Tsentral'noi Azii. Vyp. 1. (Trudy Gumanitarnogo fakul'teta Novosibirskogo gos. un-ta. Seriya 2: Sborniki nauc. trudov)* [Warfare of the Peoples of Siberia and Central Asia. Issue 1. (Works of the Humanities Faculty of Novosibirsk State University. Series 2: Collections of Scientific Works)]. Novosibirsk: NGU. Pp. 207–230.

________. 2005. Prisoedinenie krainego Severo-Vostoka Sibiri k Rossii: voenno-politicheskii aspekt (vtoraya polovina XVII–XVIII vek) [Annexation of Extreme Northeastern Siberia to Russia: The Military-Political Aspect (Second Half of the 17th–18th Century)]. Abstract of dissertation for doctor of Historical Sciences]. Tomsk: [B. i.].

________. 2005a. "F. Kh. Plenisner kak issledovatel' Severo-Vostoka Sibiri" [F. Kh. Plenisner as Investigator of Northeastern Siberia]. *Aus Siberien-2005: Nauchno-informatsionnyi sbornik* [From Siberia-2005: Science-Information Collection]. Tyumen': Ekspress. Pp. 57–58.

________. 2006. "Otsenochnoe vospriyatie russkimi chukchei i koryakov (vtoraya polovina XVII–XVIII vek)" [Evaluative Perception by Russians of the Chukchi and Koryak (Second Half of the 17th–18th Century)]. *Problemy arkheologii, etnografii, antropologii Sibiri i sopredel'nykh territorii* [Problems of Archaeology, Ethnography, and Physical Anthropology of Siberia and Adjacent Territories]. Edited by A. P. Derevenko and V. I. Molodin. Vol. XII, pt. 2. Novosibirsk: Izdatel'stvo In-ta arkheologii i etnografii SO RAN. Pp. 96–101.

________. 2006a. "Russko-chukotskie peregovory 1778 goda i prinyatie chukchei v poddanstvo Rossii" [Russian-Chukchi Negotiations of 1778 and Acceptance of the Chukchi into Citizenship of Russia]. *Vestnik Novosibirskogo gosudarstvennogo universiteta. T. 5. Seriya: Istoriya, filologiya. Vyp. 3. (Prilozhenie 1)* [Bulletin of Novosibirsk State University. Vol. 5. Series: History, Philology. Issue 3. (Appendix 1)]. Pp. 26–37.

________. 2007. "'Amanatov dat' po ikh vere grekh": Otnoshenie chukchie k russkoi praktike zalozhnichestva (XVII–XVIII vv.) ["For Them to Give Amanaty Is a Sin": The Relation of the Chukchi to the Russian Practice of Hostage Taking (17th–18th Centuries)]. *Arkheologiya, etnografiya i antropologiya Evrazii* [Archaeology, Ethnography, and Physical Anthropology of Eurasia]. No. 2 (30). Novosibirsk. Pp. 155–159.

________. 2007a. "Russkaya taktika vedeniya boya v polevykh srazheniyakh s sibirskimi 'inozemtsami' (po materialam Severo-Vostoka Sibiri)" [Russian Tactics of Conducting Battle in Field Combat with the Siberian "Foreigners" (Based on Materials of Northeastern Siberia)]. *Russkie pervoprokhodtsy na Dal'nem Vostoke v XVII–XIX vv.: Istoriko-arkheologicheskie issledovaniya* [Russian Explorers in the Far East in the 17th–19th Centuries: Historical-Archaeological Investigations]. Edited by A. R. Artem'ev and N. G. Artem'eva. Vol. 5, pt. 1. Vladivostok: In-t istorii, arkheologii i etnografii narodov Dal'nego Vostoka DVO RAN. Pp. 284–295.

________. 2008. "O boevoi taktike i voennom mentalitete koryakov, chukchei i eskimosov" [On the Battle Tactics and the Fighting Mentality of the Koryak, Chukchi, and Eskimos]. *VNGU. Seriya: Istoriya, filologiya* [VNGU. Series: History, Philology]. Vol. 7, issue 1, pp. 35–43.

________. 2009. *Prisoedinenie Chukotki k Rossii (vtoraya polovina XVII–XVIII vek)* [The Annexation of Chukotka to Russia (Second Half of the 17th–18th Century)]. Novosibirsk: Izd-vo SO RAN.

________. 2009a. "Yasak, poddanstvo i dogovornoi tovaroobmen: chukotskii variant (XVII–XIX vv.)" [Tribute, Citizenship, and Contractual Exchange of Goods: The Chukotka Variant (17th–19th Centuries)]. *Ural'skii istoricheskii vestnik* [The Ural Historical Bulletin]. No. 2 (23), pp. 84–92.Ekaterinburg.

Zuev, A. S., and N. A. Minenko. 1992. *Sekretnye uzniki sibirskikh ostrogov (ocherki istorii politicheskoi ssylki v Sibir' vtoroi chetverti XVIII v.)* [Secret Prisoners of the Siberian Islands (Essays on the History of Political Exiles in Siberia in the Second Quarter of the 18th Century)]. Novosibirsk: Nauka.

LIST OF ABBREVIATIONS

AAN—Sankt-Peterburgskii filial Arkhiva RAN [St. Petersburg branch of the Archive of the RAN]. St. Petersburg.

AII—Nauchno-istoricheskii arkhiv Sankt-Peterburgskogo instituta istorii RAN [Science-History Archive of the St. Petersburg Institute of History of the RAN]. St. Petersburg.

AN—Akademiya nauk [Academy of Sciences].

ASSR—avtonomnaya sovetskaya sotsialisticheskaya respublika [Autonomous Soviet Socialist Republic].

AST—The abbreviation "AST" is the name of a publishing company; the name formed from the names of the founders of the company Andrey Gertsev, Sergey and Tatyana Derevyanko.

BashGU—Bashkirskii gosudarstvennyi universitet [Bashkir State University].

B. i.—Bez (ukazaniya) izdatel'stva [without (the guidance of) a publisher].

B. m.—Bez mesta [without a place].

DAI—*Dopolneniya k Aktam istoricheskim* [Supplements to the Historical Acts]. St. Petersburg: Tipografiya E. Pratsa.

DBK—Drevneberingomorskaya kul'tura [Old Bering Sea Culture].

DVO—Dal'nevostochnogo otdeleniya [Far Eastern Division].

DIB—*Dal'nevostochnaya istoricheskaya biblioteka* [Far Eastern Historical Library]. Magadan: Magadanskoe kn. izd-vo.

EM—*Etnograficheskie materialy Severo-Vostochnoi geograficheskoi ekspeditsii. 1785–1795 gg.* [Ethnographic Materials of the Northeastern Geographic Expedition. 1785–1795]. Ed. by I. S. Vdovin (DIB). Magadan: Magadanskoe knizhnoe izdatel'stvo, 1978.

EO—*Etnograficheskoe obozrenie* [Ethnographic Review]. Moscow: Nauka.

GME—Gosudarstvennyi muzei etnografii [State Museum of Ethnography].

GMINV—Gosudarstvennyi muzei iskusstva narodov Vostoka [State Museum of Art of Peoples of the East]. Moscow.

IIRGO—*Izvestiya Imperatorskogo russkogo geograficheskogo obshchestva* [News of the Imperial Russian Geographic Society]. St. Petersburg: Tipografiya V. Bezobrazova i Komp.

IMLI—Institut mirovoi literatury imeni A. M. Gor'kogo Rossiiskoi akademii nauk [A. M. Gorky Institute of World Literature of the Russian Academy of Sciences].

IPTs—Izdatel'sko-proizvodstvennyi tsentr [Publishing and Production Center].

ISOIRGO—*Izvestiya Sibirskogo otdela Imperatorskogo russkogo geograficheskogo obshchestva* [News of the Siberian Division of the Imperial Russian Geographic Society]. Irkutsk: Tipografiya N. N. Sinitsina.

IVGO—*Izvestiya Vsesoyuznogo geograficheskogo obshchestva* [News of the All-Union Geographic Society]. Leningrad: Izd-vo AN SSSR.

IVSOIRGO—*Izvestiya Vostochno-Sibirskogo otdela Imperatorskogo russkogo geograficheskogo obshchestva* [News of the East-Siberian Division of the Imperial Russian Geographic Society]. Irkutsk: Tipolitografiya I. I. Makushina.

KPMGYa—*Kolonial'naya politika moskovskogo gosudarstva v Yakutii v XVII v.: Sbornik dokumentov* [Colonial Politics of the Moscow Government in Yakutia in the 17th Century: Collection of Documents]. Ed. by Ya. P. Al'kor and B. D. Grekov. Leningrad: Izd-vo In-ta narodov Severa TsIK SSSR im. P. G. Smidovicha, 1936.

KPTs—*Kolonial'naya politika tsarizma na Kamchatke i Chukotke v XVIII veke: sbornik domumentov* [Colonial Politics of Tsarism in Kamchatka and Chukotka in the 18th Century: Collection of Documents]. Ed. by Ya. P. Al'kor and A. K. Drezen. Leningrad: Izd-vo In-ta narodov Severa TsIK SSSR, 1935.

KSIE—*Kratkie soobshcheniya Instituta istorii etnografii AN SSSR* [Brief Reports of the Institute of History of Ethnography of the AN SSSR]. Moscow-Leningrad.

KSIIMK—*Kratkie soobshcheniya Instituta istorii material'noi kul'tury AN SSSR* [Brief Reports of the Institute of History of Material Culture of the AN SSSR]. Moscow-Leningrad.

KZOKM—*Kraevedcheskie zapiski Oblastnogo kraevedcheskogo muzeya* [Regional Notes of the Regional Museum]. Magadan: Magadanskoe kn. izd-vo.

LGU—Leningradskii gosudarstvennyi universitet [Leningrad State University].

MAMNH JE—*Memoir of the American Museum of Natural History. Publications of the Jesup North Pacific Expedition*. Ed. by F. Boas. Leiden: E. J. Brill; New York: G. E. Stechert.

MAZ—*Mir arkticheskikh zveroboev: Shagi v nepoznannoe. Katalog vystavki* [The World of Arctic Sea Mammal Hunting: Steps into the Unknown. Catalog of the Exhibition]. M. M. Bronshtein, K. A. Dneprovskii, and E. S. Sukhorukova. Moscow-Anadyr': Departament kul'tury, sporta, turizma i informatsionnoi politiki Chukotskogo avtonomnogo okruga; GMIV, 2007.

MIOO—Matematiko-informatsionnykh osnov obucheniya [Mathematical-informational foundations of learning].

MIYa—*Materialy po istorii Yakutii XVII veka. (Dokumenty yasachnogo sbora)* [Materials on the History of Yakutia in the 17th Century. (Documents on Tribute Collection)]. Pts. 1–3. Moscow: Nauka, 1970.

MVD—Ministerstvo vnutrennikh del [Ministry of Internal Affairs].

NGU—Novosibirskii gosudarstvennyi universitet [Novosibirsk State University].

NII—Nauchno-issledovatel'skii institut [Science-Research Institute].

NKVD—Narodnyi komissariat vnutrennikh del [The People's Commissariat of Internal Affairs].

OGIZ—Ob`edinenie gosudarstvennykh knizhno-zhurnal'nykh izdatet'stv [Association of State Book and Magazine Publishers].

OS—Old Style or Julian calendar.

PSI—*Pamyatniki sibirskoi istorii XVIII veka* [Monuments of Siberian History of the 18th Century]. Bks. 1–2. St. Petersburg: Tipografiya MVD, 1882–1885.

RAN—Rossiiskoi akademii nauk [Russian Academy of Science].

REM—Rossiiskii etnograficheskii muzei [Russian Ethnographic Museum].

RGADA—Rossiiskii gosudarstvennyi arkhiv drevnikh aktov [Russian State Archive of Ancient Acts]. Moscow.

SE—*Sovetskaya etnografiya* [Soviet Ethnography]. Moscow: Izd-vo AN SSSR/Nauka.

SES—*Sibirskii etnograficheskii sbornik* [Siberian Ethnographic Collection]. Moscow: Izd-vo AN SSSR/Nauka.

SMAE—*Sbornik Muzeya antropologii i etnografii* [Collection of the Museum of Anthropology and Ethnography]. Moscow-Leningrad: Nauka.

SMNV—*Skazki i mify narodov Vostoka* [Stories and Myths of the Peoples of the East]. Moscow: Nauka.

SPb—Sankt-Peterburg [St. Petersburg].

SSSR—Soyuz Sovetskikh Sotsialisticheskikh Respublik [Union of Soviet Socialist Republics].

SVKNII—Severo-Vostochnyi kompleksnyi nauchno-issledovatel'skii institut [Northeastern Interdisciplinary Science-Research Institute].

TB—*Tropoyu Bogoraza: Nauchnye i literaturnye materialy* [On the Trail of Bogoras: Scientific and Literary Materials]. Ed. by L. S. Bogoslovskaya, V. S. Krivoshchёkov, and I. I. Krupnik. Moscow: Institut Naslediya-Geos, 2008.

TIE—*Trudy instituta etnografii im. N. N. Miklukho-Maklaya* [Works of the Institute of Ethnography Named for N. N. Miklukho-Maklaya]. Moscow: Nauka.

TsIK—Tsentral'nyi Ispolnitel'nyi Komitet [Central Executive Committee].

UZLGPI—*Uchenye zapiski Leningradskogo gosudarstvennogo pedagogicheskogo instituta im. A. I. Gertsena* [Scholarly Notes of the Leningrad State Pedagogical Institute Named for A. I. Gertsen]. Leningrad: [B. i.].

VIRGO—*Vestnik Imperatorskogo russkogo geograficheskogo obshchestva* [Bulletin of the Imperial Russian Geographic Society]. St. Petersburg: Tipografiya E. Pratsa.

VNGU—*Vestnik Novosibirskogo gosudarstvennogo universiteta* [Bulletin of Novosibirsk State University]. Novosibirsk: NGU.

VO—*Vostochnoe obozrenie* [Eastern Review]. Irkutsk.

ZChKM—*Zapiski Chukotskogo kraevedcheskogo muzeya* [Notes of the Chukotka Regional Museum]. Magadan.

ZGDMM—*Zapiski Gidrograficheskogo departamenta Morskogo ministerstva* [Notes of the Hydrographic Department of the Maritime Ministry]. St. Petersburg: Morskaya tipographiya.

ZhMNP—*Zhurnal Ministerstva narodnogo prosveshcheniya* [Journal of the Ministry of National Education] St. Petersburg: Senatorskaya Tipografiya.

ZhMVD—*Zhurnal Ministerstva vnutrennikh del* [Journal of the Ministry of Internal Affairs]. St. Petersburg: Tipografiya K. Kraiya.

ZIRGO—*Zapiski Imperatorskogo russkogo geograficheskogo obshchestva* [Notes of the Imperial Russian Geographic Society]. St. Petersburg: Tipografiya V. Bezobrazova i Komp.

ZPOIRGO—*Zapiski Priamurskogo otdela Imperatorskogo russkogo geograficheskogo obshchestva* [Notes of the Priamur'e Division of the Imperial Russian Geographic Society]. Khabarovsk: Tipolitografiya pri kantselyarii priamurskogo general-gubernatora; St. Petersburg: Tipografiya Imperatorskoi akademii nauk.

ZSOIRGO—*Zapiski Sibirskogo otdela Imperatorskogo russkogo geograficheskogo obshchestva* [Notes of the Siberian Division of the Imperial Russian Geographic Society]. St. Petersburg.

ZVSOIRGO—*Zapiski Vostochno-Sibirskogo otdela Imperatorskogo russkogo geograficheskogo obshchestva* [Notes of the East-Siberian Division of the Imperial Russian Geographic Society]. Irkutsk: Tipografiya shtaba Irkutskogo voennogo okruga.

GLOSSARY

(-i/-y are common Russian plural markers)

amanat—a hostage, usually offered as security.
artel'—a Russian cooperative association.
baidara—a skin boat, like the umiaq of the Eskimos.
desyatnik—leader of ten men, usually translated as foreman.
erem—a man of high status, strongman, leader among the Chukchi.
ezov'—a weir, woven dam.
Foot Chukchi—that group of Chukchi who did not own reindeer.
kamus, kamos—winter footwear made of reindeer shank hide.
karbas—a type of sailing-rowing boat.
kargina—a reindeer intended for slaughter.
koch—a type of sailing-rowing boat
korga—shoals or rocks.
korol'ki—small pieces of coral.
kukhlyanka—a kind of shirt of fur.
letov'e—a summer pasture.
narodovolets—a Russian revolutionary in the 19th century, member of a secret populist organization.
nosovyi—"nose"—referring to the residents of Chukotskii Nos, that is, the Chukchi Peninsula.
polog—a canopy or bed curtain, in essence, a compartment. Fur tent for one family in yaranga.
promyshlennik—a hunter/trapper/trader.
pud—1 pud equals 36.11 lbs.
pyatidesyatnik—a military rank among the Cossacks indicating a commander of 50 men.
raznochinets—a person who does not belong to any of the established Russian classes.
red fish—a name commonly used for any salmon.
Reindeer Chukchi—those Chukchi who used reindeer to ride and to pull sleds. The nomadic Chukchi.
rotastye—"mouthed." "Rot'" = "mouth," "lip." Referring to the labrets worn by the Bering Strait Eskimos.
sazhen'—7 feet or 2.13 m.
Settled Chukchi—that group of Chukchi who did not own reindeer and whose lifestyle was much like the Asian Eskimos.

shitik—in the 18th century a small sailing-rowing single-mast vessel, the planking of which was fastened ("sewn," from which comes the name [shit'—Russian to sew]) with baleen, thongs, or willow withes.
sivodushek—a cross between a black-brown and red fox or silver-black and red fox.
sotnik—a military rank among the Cossacks indicating a commander of 100 men.
toyon, toion—chief.
urasa—a cone-shaped type of Yukaghir summer dwelling.
Wagenburg—from German, "wagon fort." In this case, a sled fort.
versta—a distance equal to 1.067 km or 0.6629 miles.
vetka—a very light Siberian boat made of several layers of birch bark sewn together with cords twisted from thin branches of larch, the seams being covered with a layer of resin.
yaranga—the traditional mobile dwelling of some northern nomadic peoples.
yukola—dried fish.
zaimka—open lands for hunting or fishing.
zalozhnik—a hostage forcibly taken.
zolotnik—one zolotnik equals 4.25 grams or 0.150 oz.

INDEX

L

M

N

O

R

S

T

Y

Z

www.ingramcontent.com/pod-product-compliance
Lightning Source LLC
LaVergne TN
LVHW010718210926
848485LV00008B/1436/J